Plasma Polymer Films

Plasma Polymer Films

Editor

Hynek Biederman

Charles University, Czech Republic

ICP

Imperial College Press

Published by

Imperial College Press
57 Shelton Street
Covent Garden
London WC2H 9HE

Distributed by

World Scientific Publishing Co. Pte. Ltd.
5 Toh Tuck Link, Singapore 596224
USA office: 27 Warren Street, Suite 401-402, Hackensack, NJ 07601
UK office: 57 Shelton Street, Covent Garden, London WC2H 9HE

British Library Cataloguing-in-Publication Data
A catalogue record for this book is available from the British Library.

PLASMA POLYMER FILMS

ISBN-13 978-1-86094-467-3
ISBN-10 1-86094-467-1

CONTENTS

PREFACE

The area of plasma polymerization is now well recognized as an important part of material science. Plasma polymerization spans over several fields of science, namely chemistry, physics and extends in recent years to biology. This book is intended not for beginners but for advanced students, PhD students and researchers from both academia and the industry who are already more or less in the field. The authors of the book try to give the readers the present state of the art and future trends of plasma polymerization.

However, in the first chapter **Introduction,** the accidental reader, who never came across this field before, may obtain very basic information and is directed to more detailed and basic books and review papers. The definition of the term plasma polymer is given, deposition systems, process parameters and models of plasma polymerization are described.

The second chapter **Plasma deposition of fluoropolymer films in different glow discharge regimes** focuses on continuous, afterglow and modulated (pulsed) discharges and discusses active species in fluorocarbon plasmas. Diagnostics techniques, namely OES (Optical emission spectroscopy) and AOES (Actinometric optical emission spectroscopy) are described. Active species in fluorocarbon plasmas are presented and process parameters mentioned. The activated growth model is introduced and the applications of fluoropolymers briefly reviewed.

In the third chapter **Deposition of silicon containing films and FTIR diagnostics** (Deposition from organosilicon/oxygen mixtures), most of the attention is devoted to the $SiO_xH_yC_z$ (z,y are almost zero) film deposition processes: the nature of the most frequently used precursors and their physical characteristics; type of plasma generation and the description of certain reactors. However, a summary of the literature on $SiN_xH_yC_z$ is presented. The second part deals with the characterization of organosilicon — oxygenated gas mixture created plasmas. The principal species detected by the optical emission spectrometry, the mass spectrometry and the infrared absorption spectrometry (FTIR) are introduced. The application of FTIR in situ diagnostics is explained in more detail using the above examples.

In the fourth chapter **Corpuscular diagnostics of plasma polymerization processes**, experimental equipment for corpuscular diagnostics is described. Detection of neutral species and ions during the process are discussed in great detail. Examples are presented in terms of the discussion of particle fluxes to the growing film and the influence on the film morphology.

In the fifth chapter **Electrical and optical properties of plasma polymers**, at first, the electronic structure of plasma polymers, investigated by various methods, such as electronic spectroscopies, photoinjection, optical absorption, etc, is described. The idea of two types of plasma polymers, namely amorphous semiconductors and amorphous insulators is discussed. For general view both inorganic films (as a-Si:H) and organic films (deposited from organic precursors) are presented. The electric conductivity, its mechanisms, as well as carrier generation and transport processes are further discussed. Attention is also paid to practical applications of the films in electronics such as field-effect transistors and amorphous "superlattice" structures. Basic dielectric parameters of the films and their dependence on deposition conditions, post-treatment of deposited films, aging processes, and environment influences (e.g. humidity) are also mentioned. At the end optical properties of plasma polymers are briefly summarized.

The sixth chapter **Pulsed plasma polymerizations** describes fundamental aspects of modulated plasmas, discusses potential advantages and limitations of this approach, and provides a thorough review of work to date in this area.

One of the most important features of the pulsed plasma approach is the convenient and high level of film chemistry controllability it provides during polymer formation. The inherent compositional controllability afforded by this approach is illustrated in this chapter with data from several laboratories involving a variety of monomers. This chapter also identifies and discusses differences in several important plasma variables when contrasting CW and pulsed plasma polymerizations and the potential influences on these variables on the respective film formation processes. The chapter concludes with a brief review of current and future applications which take particular advantage of the molecular surface tailoring possibilities provided by the pulsed plasma surface modification technique.

In the seventh chapter **Stability of plasma polymers and plasma polymer coatings**, thermal stability of plasma polymers (also versus the stability of conventional polymers) is discussed. Chemical (including photochemical) stability of plasma polymers is presented. Mechanism of physical and chemical processes during storage and use (long term stability, environmental stability) is discussed in detail. Stability of coating systems (layer adhesion, interphase adhesion) is described in detail.

In the eighth chapter **Application of atmospheric discharge for plasma polymer processes**, the basics of the atmospheric pressure glow plasma (APG) are presented. The studies of the deposition process of mainly organic (plasma polymer) films using emission spectrometry, IR absorption spectra and molecular beam (MB) mass spectrometry are described. The deposition system with parallel plate electrodes covered by glass insulator is described. Along with helium as carrier gas C_2H_4 and C_2F_4, TEOS and some other monomers were used. Deposition processes for the first two monomers are discussed in detail.

The ninth chapter **Hard plasma polymers, composites and plasma polymer films prepared by rf sputtering of conventional polymers** pays attention at first to hydrocarbon plasma polymers and to the effect of energetic positive ion bombardment on the film hardness. Fluorocarbon plasma polymers are briefly mentioned. The basics of composite metal/plasma polymer films will be concisely presented: modes of their preparation, structure and morphology, anomalous electrical and optical properties. Composites: metal/hard plasma polymer and metal/hard carbon are described and their temperature and time stability compared to usual metal/plasma polymer composites.

Sputtering of polytetrafluoroethylene (PTFE) in argon and various gases as well as in self-sputtering mode is described. Composition of species in plasma volume and the growth of the fluorocarbon film in dependence on deposition parameters is discussed. The morphology (SEM), structure (FTIR, ESCA) and the electrical properties of the fluorocarbon plasma polymer films are described in detail. The main results in rf sputtering of polyethylene and some other polymers are briefly described.

The tenth chapter **Biomedical applications of plasma-deposited thin films** discusses the applications. The requirements for biocompatibility and how plasma deposited films meet those requirements are introduced. Following applications are described: Applications for cell culture, for controlled release of drugs, for biosensors, for implanted biomaterials and for contact lenses. Biorecognition and plasma-deposited films as well as patterned plasma deposited films are considered. In conclusion the perspectives are discussed.

The editor wishes to thank all his colleagues – and the co-authors of the book for their effort. Actually, the idea and need for such a book have been stimulated to a great extent by the Action COST 527: Plasma Polymers and Related Materials. The editor is further indebted to Dr. Pavel Kudrna for formatting the manuscript and to Eric Fuocco and Andrei Choukourov for reading and correcting it. The editor also thanks Dr. Christian Oehr for valuable discussions and comments related to the Chapter 1 – Introduction.

CHARLES UNIVERSITY
Prague, 31st July, 2002 Hynek Biederman

CHAPTER 1

INTRODUCTION

HYNEK BIEDERMAN
Department of Macromolecular Physics
Faculty of Mathematics and Physics, Charles University,
Prague, Czech Republic

1.1. History

The first reports of the deposits known today as plasma polymers appeared in the second half of the 19th century [1, 2]. Several papers treating these films and related to plasma polymerization processes were published in the 1930s [3-6] and more attention was paid to the subject in the 1950s [9-10]. König and Helwig [7] deliberately deposited films that would be described nowadays as hydrocarbon plasma polymers. Basically, they used a glass bell-jar with parallel plate electrodes (made of Al) that were powered by singly rectified 50 Hz voltage up to 2 kV. Their "pulsing" DC glow discharge was operated in benzene vapours and films were deposited on various substrates resting on the anode. Films up to 1 μm thick on glass were yellowish in transmitted light and their density was about 1.4 g/cm^3 and electrical conductivity was 10^{-9} Ω^{-1} cm^{-1}. After annealing these films started to change at 100 °C and at 400 °C turned fully into carbonaceous films. The use of these films for carbon foils was proposed. Brokes and König [8] deposited films not only from benzene but also from methane and presented their absorption IR spectra. The absorption IR spectra, and the absorption in the visible region of light of plasma polymerized benzene films deposited at the increased voltage 500 V, 800 V and 3000 V, showed the transition from a hydrocarbon plasma polymer to a hard plasma polymer (now mostly denoted as C:H) as a consequence of the increased energetic ion bombardment. A more detailed study of the deposition process was published by Heisen [9]. Energetic particles from the discharge (positive ions and electrons) bombard the adsorbed benzene molecules on the substrate surface. Originated fragments react with the other molecules and a plasma polymer film grows. The above described early period of the plasma polymer field has been nicely reviewed by Pagnia [11].

In the 1960s, especially during the early development of the microelectronics research and industry, rapid growth of plasma polymerization started that was inspired by the hope to use a plasma polymer as a dielectric film [12]. Bradley and Hammes [13], who continued previous work [12], plasma polymerized up to 40 monomers in a parallel plate electrode bell jar arrangement using AC voltage at 10 – 50 kHz frequency. These monomers included acetylene, styrene, naphthalene, etc. and certain metalorganics, such as ferrocene, diphenyl mercury etc. The authors measured the electrical properties of the prepared films that were sandwiched between Al electrodes. The intensive attempts to develop the new dielectric film were continued by Stuart [14,15], Bashara and Doty [16], Wiliams and Hayes [17], Gregor [18] and Hirai and Nakada [19]. The two reviews [20, 21] appeared that discussed glow-discharge (plasma) polymerization, however, they also considered polymer films deposition techniques such as electron bombardment of organic vapors or their exposure to UV light. Another proposed technique was simple vacuum evaporation [22-23]. When heated to the evaporation temperature, most polymers fragment. The gaseous fragments then condense on the substrate without creating a polymer network. Without additional e.g. UV light irradiation this deposition technique is applicable only in special cases (low molecular weight polymers). This subject was reviewed in detail by Hogarth and Igbal [24]. In recent years the ionization-assisted method (IAD) was developed [25]. Species evaporated from the evaporation source are ionized by the electron shower (across the beam of evaporated particles) and are accelerated towards the substrate. The ion concentration is typically from 0.1 to 1 % of the depositing material and ion energy around several hundred of eV.

Once the plasma polymerization process was recognized, it started to be investigated because of many potential applications. These included, not only electrical but also, optical field, various surface modifications especially for biomedical applications, etc. The number of published studies has increased, and several reviews [26-36] and monographs appeared [37-41].

1.2. What is a Plasma Polymer?

The term plasma polymer denotes a material that is created as a result of a passage of an organic gas or vapor through the glow discharge - more generally one should say an electric discharge in an organic gas. Usually, it is deposited in the form of a thin film. In spite of the use of the word polymer, the plasma polymer means a new class of material that has only a little in common with the conventional polymer that has regularly repeating units (please see Fig.1.1). In the case of the plasma polymer, the chains are short and in addition they are randomly branched and terminated with a high degree of crosslinking (please see Fig.1.2). In most cases, a great number of free radicals are trapped within the network. These radicals cannot recombine rapidly and therefore when the plasma polymer is exposed to the open atmosphere they react with the oxygen and water vapor. These free radicals are the cause of often observed aging processes in plasma polymers. Plasma polymers possess a rather disordered structure and this depends on the intensity and energy of the species bombarding the growing film.

The plasma polymerization process involves reactions between plasma species, between plasma and surface species, and between surface species [31]. Usually a free radical mechanism is considered and the two cases are distinguished: 1) plasma-state polymerization and 2) plasma induced polymerization.

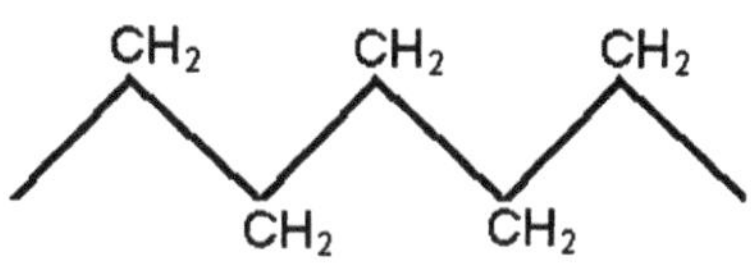

Figure 1.1 HDPE

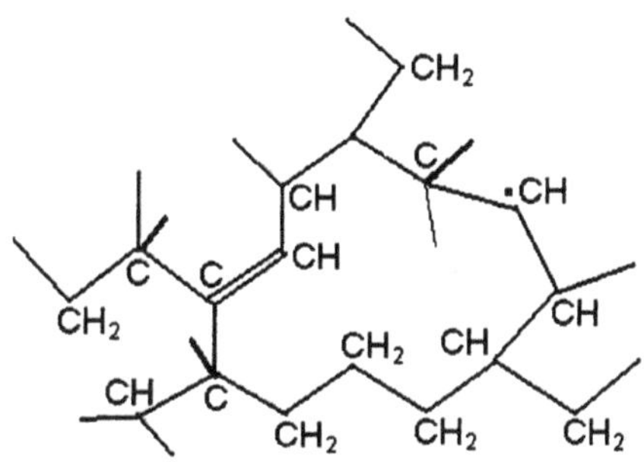

Figure 1.2 Hypothetical structure of a hydrocarbon plasma polymer

Plasma-state polymerization occurs only in a plasma of an electric discharge in the presence of organic molecules. Energetic particles but mainly energetic electrons produce, in the collisions with organic molecules, a number of species (fragments) i.e. also free radicals. These undergo chain reactions and free radical termination reactions. As a result, any organic compound in the gaseous state may be plasma polymerized.

Plasma induced polymerization is essentially free radical induced polymerization of molecules including unsaturated carbon-carbon bonds. In most cases, the term of plasma polymer is used in the case of plasma-state polymerization.

The terms plasma polymerization and plasma polymer originated hystorically and are commonly accepted. However, some researchers suggest that it is more accurate to use the terms PECVD (plasma enhanced chemical vapor deposition) of organic compounds and plasma deposited coatings, respectively. If one has in mind the definitions of these terms any of them are correct to apply. Throughout this book mostly the former ones are used.

1.3. Deposition Systems for Plasma Polymerization

The plasma polymerization process is carried on usually in a low pressure, low temperature plasma. An atmospheric plasma is becoming more and more popular for deposition including plasma polymers. This topic will be specially treated in Chapter 8.

A great number of deposition systems using low pressure plasma was described in the literature and summarized in Refs 26-41. Generally, the three classes of plasma polymerization deposition systems (reactors) can be distinguished, as pointed out by Shi [33]: (a) internal electrode reactors, (b) external electrode reactors, and (c) electrodeless microvawe or high frequency reactors.

The leading configuration in the group (a) is a bell-jar-type reactor with internal parallel plate electrode arrangement (Fig.1.3). The discharge can be excited either by DC or AC and/or RF voltage. DC is used in special cases or in the case when a planar magnetron is used (please see Chapter 9). The magnetron electrode serves as a cathode and is at high negative potential (several hundred volts) while the counterelectrode serves as an anode and is usually connected to the stainless steel-bell-jar and earthed (asymmetrical arrangement). The "anode" (substrate holder) may be also isolated and in this case biased.

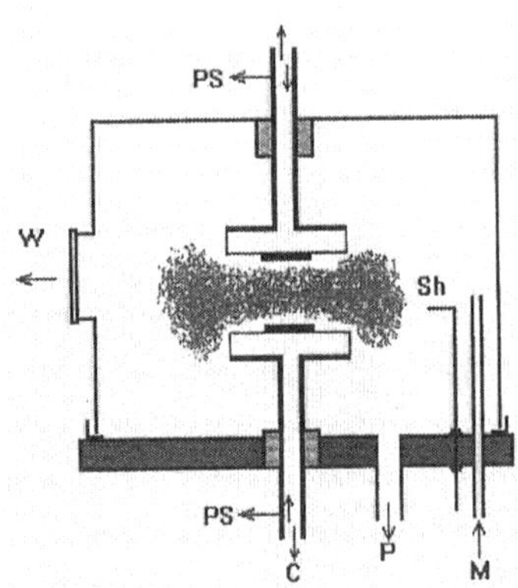

Figure 1.3. Deposition systems: (a) parallel plate (internal) electrode reactor (PS -to power supply, C - cooling, W - window, P - to pumps, M - monomer, Sh - shutter, S - substrate

For a plasma polymerization process, more typical is the discharge excitation by

AC and RF voltages in so called symmetrical arrangement. If the frequency of the excitation voltage is above about 50 kHz (100 kHz), the power must be delivered through a matching unit and determined by a power meter. In the case of the asymmetrical arrangement where the excitation electrode is connected to the power supply through a blocking capacitor and matching unit, a DC negative bias is developed on this electrode. In this case, the surface (target) of the excitation electrode is sputtered by positive ions (RF sputtering). In some processes this may be the reason for concern as there is a danger of impurities sputtered off from the excitation electrode. In other cases this may be welcome e.g. simultaneous plasma polymerization and co-sputtering of the metallic target for composite metal/plasma polymer film deposition (please see Chapter 9).

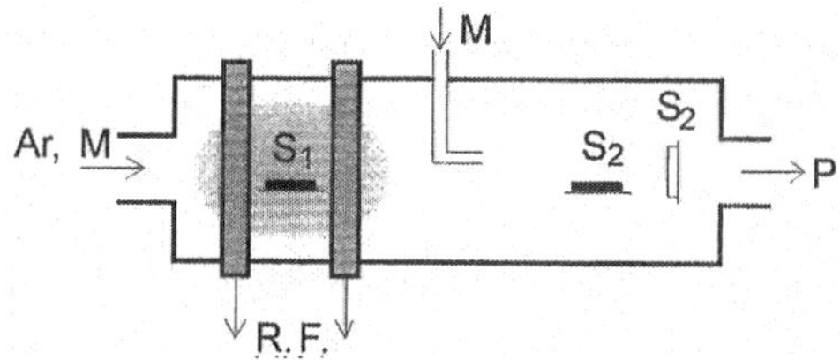

Figure 1.4. Tubular glass or silica deposition system Ar-argon, M - monomer, S_1,S_2 – substrates, P - pumps)

External electrode reactors use ring electrodes (Fig.1.4) or an external coil around a tube from glass or silica. Rf power may be coupled into the plasma inductively but in most cases the coupling is of a capacitive nature [42]. Therefore when the external coil is used instead of the term inductive coupling one could use inductive excitation to point out the use of the coil but to avoid the confusion as the coupling is likely capacitive. This reactor then cannot be considered as electrodeless as the DC negative bias appears on the inner reactor wall under the coil turn or under the ring electrode (please see the discussion in Ref. 41) The substrate is positioned inside the plasma or downstream of the plasma.

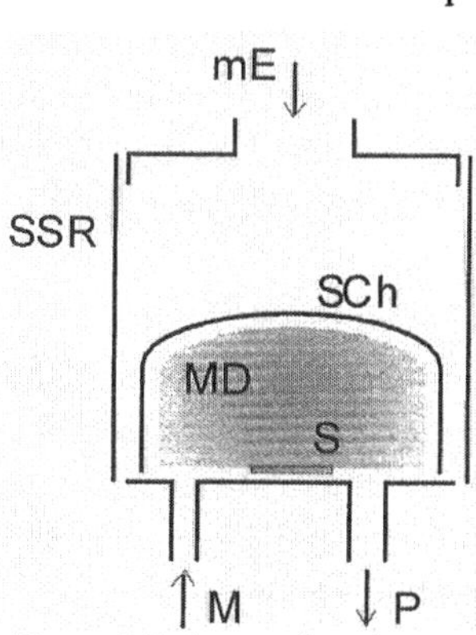

Figure 1.5. A type of a microwave reactor (mE - microwave energy, MD - microwave discharge, SSR - stainless steel resonator, SCh - silica chamber, S - substrate, M – monomer, P - to pumps)

Electrodeless microwave reactors are usually composed from a silica tube that passes through the resonant cavity coupled to a microwave power supply (typically 2.45 GHz).

Plasma is generated in the tube within the cavity and substrates are positioned downstream of the plasma. The avity actually acts instead of the RF excitation coil or ring electrodes in Fig.1.4. In Fig.1.5 another type of microwave reactor is shown that resembles a "microwave oven" of more sophisticated resonator type shape (not shown in schematic). Some other types of microwave reactors were designed – please see Refs. 41 and 43.

1.4. Deposition (Plasma) Process Parameters

The structure and composition, physical and chemical properties of a plasma polymer and its deposition rate depend on many parameters for a given monomer (feed gas) or gas mixtures:

(a) The type of the reactor and its geometrical configuration
(b) Frequency of the discharge excitation voltage (for RF continuous wave mode or pulsing mode)
(c) Power delivered to the discharge (0.001 – 2.5 W/cm^3)
(d) Flow rate of the monomer (1 – 100 cm^3 (STP)/min)
(e) Working gas pressure (1 – 100 Pa)
(f) Substrate temperature
(g) Substrate size and its position

If the type of reactor and frequency of the excitation voltage are chosen, substrate position and size are fixed, then one should optimize the deposition rate in dependence on power, flow rate and working gas pressure. Usually, one of these parameters is changing while the other two are kept constant. In this case we find that the deposition rate increases with power until all available monomer at a concentration given by working pressure and supplied at the constant flow rate is consumed. Further increase of power has no effect and the deposition rate is constant. Resulting deposition rate is given by the growth rate and competing ablation rate for some fluoro(chloro)carbons and hydrocarbons containing oxygen, the deposition rate may decrease for higher powers as the ablation start to dominate.

Similar deposition rate curve is obtained when pressure is the variable. When pressure and power are kept constant and the flow rate is increased – the deposition rate also increases. When the flow becomes too high at a given power level not all supplied monomer can be excited. Some of the excited species are even drawn away from the substrate and cannot take part in the deposition. The deposition dependence on the flow rate has a maximum.

As may be seen, for proper optimization of the deposition rate, very large amount of measurements are needed. This can be speeded up by a "process analysis". Also homogenous thickness deposition should be optimized in respect to the substrate size.

The substrate temperature is important. With decreasing temperature the deposition rate of a plasma polymer increases. This is because plasma polymer deposition carried out at decreased pressure is governed by the adsorption processes. However, not only the deposition rate but the structure of the plasma polymer changes. For certain monomers used at decreased substrate temperature, the plasma polymer retains more of the molecular structure of the monomers. But physical structure may also change giving rise to so called columnar structure – please see Chapter 9.

Our last remark is to the frequency of the excitation voltage. As mentioned before, DC voltage is rarely used. If AC voltage is applied with a frequency less than 100 Hz, it is basically of DC type with changing polarity, but it switches off between half cycles. If the frequency is further increased up to the point and above when the time needed for positive ions to move, from the momentary anode to cathode becomes equal to 1/2 of the period of the voltage, not all positive ions originating at anode can transit to the cathode before the voltage is reversed. This causes partial retention of the positive space charge from one half cycle to the next and helps reinitiation of the discharge. Because of the low mobility, the positive ions are unable to move during each half cycle while the electrons are swept though the interelectrode space by the voltage on each half cycle. As a result, a pulsating negative sheath voltage on the walls (electrodes covered by a dielectric film) appears. We are in the range of RF discharge. DC negative bias will appear on the surface (electrode) covered by a dielectric or connected to RF power supply through the blocking capacitor. Generation of positive ion – electron pairs takes place in the plasma volume and does not depend on the processes on the electrodes. In contrast, DC type discharges are dependent strongly on the processes at the cathode where the electrons are emitted by bombarding positive ions. After the acceleration through the dark cathode space, the electrons, in ionization collisions with the neutral gas, create pairs electron-positive ion that is necessary for sustaining the DC discharge. In a microwave discharge, the transfer of power to the plasma particles and especially electrons is much more efficient than in RF discharge and as a result, more electrons with higher energies are present. Deposition rate is increased as a result of increased efficiency of the plasma polymerization process in this case. For more details, please, see Refs. 27, 28, 33, 41.

In order to be able to transfer the deposition process to another apparatus or to perform the scale-up, the detailed description of plasma-chemical processes is needed. This is done by means of in situ diagnostics such as OES (optical emission spectroscopy), IR (infrared) absorption spectroscopy of the plasma,

MS (mass spectroscopy) or corpuscular diagnostics and LP (Langmuir probes). Because of their importance, the first three methods will be discussed in detail in the Chapters 2, 3 and 4, respectively.

1.5. Models of Plasma Polymerization

Williams and Hayes [17] excited an AC (10 kHz) discharge in styrene at a pressure 0.1 – 3 torr using a bell-jar-type reactor with a parallel plate electrode system. They found, that at constant substrate temperature and constant flow of styrene, the polymerization rate at first increased and then reached saturation with either increased pressure at constant discharge current or with increased current at a constant pressure. They also found that the polymerization rate increases with decreasing substrate temperature. Plasma polymer formation proceeds as follows. Positive ions are formed by collisions between the energetic electrons and monomer molecules in the negative glow. These ions are attracted to the electrodes and bombard in the first stage of growth adsorbed monomer molecules. These are activated (fragmented) and react with the unactivated monomer molecules. When the first layers of plasma polymer are created not only adsorbed monomer molecules are activated but also polymer surface molecules. These active species propagate the polymerization reaction by addition of unactivated molecules.

Denaro et al. [46-48] proposed the explanation of plasma polymerization of styrene (at 1 torr) and some other monomers in a 2 MHz discharge. The initiation of the polymerization reaction was done by energetic electrons bombarding the polymer film that was covering the electrodes and resulted in the creation of surface free radicals that react with the adsorbed monomer molecules. They obtained a good agreement with the experiment in case of allylalcohol [48]. A similar model was proposed by Westwood [49] who used a parallel-plate electrode system in RF discharge and relied in his model on activation by bombarding positive ions.

Poll et al. [50,51] developed a detailed model based mainly on the processes on the surfaces adjacent to the discharge region, e.g. adsorption, desorption and polymerization. He also considered processes leading to the formation of new compounds by the transformation of the monomer under the influence of the discharge (for more details see also Ref. 41). Comparison to the experiment is done for C_2F_4.

Lam et al. [52] developed a model of plasma polymerization of styrene in a RF (800 kHz) discharge using a capacitively coupled parallel plate electrode system. Three steps were considered: (1) Initiations of monomers by electron impact was followed by (2) propagation and (3) termination as in conventional polymerization. Free radical polymerization was assumed. Four mechanism models were examined considering the above mentioned three steps taking place either in the gas phase or on the surface. Comparing the theory and experiment, the following model seemed to be the most probable: Monomers are activated in the gas phase via collisions with energetic electrons. They diffuse to the surface where they react in propagation reactions with adsorbed monomers and subsequently terminate.

A similar model as above for plasma polymerization of unsaturated hydrocarbons in a RF discharge was proposed by Tibbitt et al. [52].

The overall scheme of plasma polymerization based on Poll's et al. [51] work was proposed by Yasuda and Hsu [52] that involves the competitive ablation and polymerization mechanism (CAP). A monomer passing through the plasma zone is turned into the complex mixture of untact monomer, its excited or ionized fragments and gaseous products that do not participate in plasma polymerization. Yasuda suggests that plasma polymerization can be explained by a bicyclic step-growth mechanism. The first cycle relies on monofunctional species while the second one on bifunctional species. For more details, please, see Refs. 29 and 41.

1.6. Types of Plasma Polymers and their Characterization

According to the type of monomer used several types of plasma polymers are distinguished [30, 33]: hydrocarbon, fluoro(chloro)carbon nitrogen containing plasma polymers and composites.

A number of characterization techniques are used that include: FTIR, XPS, AES, SIMS, RHEED, NMR, ESR, ERDA, RBS, NRA, DSC, TGA, XRD(SAXS), SEM, AFM (STM). Most frequently used are FTIR, XPS, RBS, ERDA, SEM, AFM.

Electrical properties are studied by conventional DC techniques, AC properties by dielectric spectroscopy.

Optical properties in the visible region use transmission and absorption and ellipsometric measurements.

Surface properties such as wettability (contact angle) [41], adhesion (Scotch tape test or various scratch testers), and mechanical properties such as internal stress, hardness, abrasion and friction are measured by routine thin film methods (for more details see [53]).

1.7. Conclusions

A number of applications were proposed, some of them mainly as protective and optical films and layers for packaging and in the biomedical field. The applications are summarized in several reviews and monographs e.g. [29,31,33, 34,35,39,41]. However, some features of plasma polymers should be improved such as suppression of ageing etc. for their wider applications. Therefore, further investigations are needed in order to employ all potentially attractive properties of plasma polymers. In the following chapters, the reader will find the state of the art of the field and its current developments.

References

1. v. Wilde M.P., *Berichte Deutsche Chem. Ges.* **7** (1874), 352.
2. Thenard A., Hebd C.R., *Seances Acad. Sci.*, **78** (1874), 219.
3. Eisenhut O., Conrad R., *Z. Elektrochem.* **36** (1930), 654.
4. Harkins W.B., Jackson J. M., *J. Chem. Phys.* **1** (1933), 37.
5. Harkins W.B., *Trans. Faraday Soc.* **30** (1934), 221.
6. Conrad R., *Trans. Faraday Soc.* **30** (1934), 215.
7. König H., Helwig G., *Z. Physik* **129** (1951), 491.
8. Brockes A., König H., *Z. Physik* **152** (1958), 75.
9. Heisen A., *Analen der Physik* **2** (1958), 23.
10. Schmellenmeier H., *Exp. Techn. Phys.* **1** (1953), 49.
11. Pagnia H., *Progress in Colloid & Polym. Sci.* **78** (1988),143.
12. Goodman J., *J. Polymer. Sci.* **44** (1960), 551.
13. Bradley A.P., Hammes J. P., *J. Electrochem. Soc.* **110** (1963), 15.
14. Stuart M., *Nature,* **199** (1963), 59.
15. Stuart M., *Proc. Inst. Electr. Eng.* **112** (1965), 1614.
16. Bashara N.M. and Doty C.T., *J. Appl. Phys.* **35** (1964), 3498.
17. Williams T. and Hayes M.W., *Nature* **209** (1966), 769.
18. Gregor L.V., *Thin Sol. Films* **2** (1968), 235.

19. Hirai T. and Nakada O., *Jpn. J. Appl. Phys.* **7** (1968), 112.
20. Mearns A.M., *Thin Sol. Films* **3** (1969), 201.
21. Gregor L.V., *IBM Journal,* March (1968), 140.
22. White M., *Vacuum* **15** (1965), 449.
23. Luff P.P., White M., *Vacuum* **18** (1968), 437.
24. Hogarth C. and Igbal ., *Phys. Stat. Sol.* **A65** (1981), 12.
25. Usui H., *Thin Sol. Films* **365** (2000), 22.
26. Yasuda H., in *Thin Film Processes*, eds Vossen J. V., Kern W. (Academic Press, New York, 1978), 361.
27. Shen M., Bell T., *ACS Symp.* **108** (1979), 1.
28. Bell A.T., in *Current Chemistry* , *Vol. 94, Plasma Chemistry*, eds. Veprek S. and Venugopalan M. (Springer Verl., Berlin, Heidelberg, New York, 1980), 43.
29. Yasuda H., Gazicky M., *Plasma Chem. and Plasma Process*, **3** (1983), 279.
30. Biederman H., *Vacuum*, **37** (1987), 367.
31. Osada Y., Biederman H., *Plasma Chemistry of Polymers, Advances Polymer Science* **95** (1990), 57.
32. Morosoff N., in *Plasma deposition, treatment and etching of polymers*, ed. a'Agostino R.,. (Academic Press , Boston, 1990) 1-93 and other chapters.
33. Shi F.F, *J.M.S.-Rev. Macromol. Chem. Phys.*, **C36**(4) (1996),795.
34. Favia P., d'Agostino R., *Surface and Coatings Technology*, **98** (1998),1102.
35. Biederman H., Slavinska D., *Surf. and Coat. Technol.* **125** (2000), 371.
36. Tkaczuk B.V., Kolotyrkin M. V., *Polutzenije tonkich polimernych plenok iz gazovoj fazy*, (Chimija, Moskva, 1977, in Russian).
37. Yasuda H., *Plasma Polymerization*, (Academic Press, Orlando, 1985).
38. Osada Y., *Plasma Polymerization*, (Tokyo, 1986, in Japanese).
39. Tyczkowski J., *Cienkie warstwy polimerow plazmovych*, (Wydawnictwa Naukowo-Techniczne, Warsaw 1990, in Polish).
40. Inagaki N., *Plasma Surface Modification and Plasma Polymerization*, (Technomic Publishing, Lancaster, Basel 1996).
41. Biederman H.and Osada Y., *Plasma Polymerization Processes*, (ELSEVIER, Amsterdam, 1992).
42. Vossen J., *J. Electrochem. Soc.* **126** (1979) 319.
43. Wrobel A.M. and Wertheimer M.R., in *Plasma deposition, treatment and etching of polymers*, ed. a'Agostino R., (Academic Press , Boston, 1990),1.
44. Wilson J., Bonnar M.P., Burnside B.M., in *Proc. of Plasma Polymerization Sem. IEE* (Savoy Place, London, 1999), 5/1.

45. Yin G.Z., Jilie D.W., *Solid State Technol.*, **30** (1987), 127.
46. Denaro A.R., Owens P.A. and Crawshaw A., *Europ. Polymer J.,* **4** (1968), 93.
47. Denaro A.R., Owens P.A. and Crawshaw A., *Europ. Polymer J.,* **5** (1969), 471.
48. Denaro A.R., Owens P.A. and Crawshaw A., *Europ. Polymer J.,* **6** (1970), 487.
49. Westwood A.R., *Europ. Polymer J.*, 7 (1971), 363.
50. Poll H.U., *Z. angew. Physik*, **29** (1970), 260.
51. Poll H.U., Artz M. and Wickleder K.H., *Europ. Polymer J.*, **12** (1976), 505.
52. Yasuda H, Hsu T., *Surface Sci.,* **76** (1978), 232.
53. Ohring M., *The Material Science of thin Films,* (Academic Press, Inc., Boston, 1992).

CHAPTER 2

PLASMA DEPOSITION OF FLUOROPOLYMER FILMS IN DIFFERENT GLOW DISCHARGE REGIMES

PIETRO FAVIA
Department of Chemistry, Istituto di Metodologie Inorganiche e dei Plasmi, CNR
University of Bari, Bari, Italy

2.1. Introduction

Fluorocarbon gases or vapors can be utilized as feed in glow discharges aimed to deposit thin films of CF_x composition, or to perform the etching and/or the fluorination of inorganic materials and organic polymers. PE-CVD (Plasma Enhanced Chemical Vapor Deposition) and plasma etching processes have been developed for about thirty years; continuous (CW) plasmas have been utilized first for both processes, while afterglow (AG) discharges became popular in the late eighties. More recently, modulated (pulsed) plasmas have been investigated, mostly for PE-CVD, and revealed to be a source of very interesting surfaces.

Thin films deposited in glow discharges fed with fluorocarbons, alone or mixed with other gases, are of interest for modifying the surfaces of materials in different discharge regimes due to their surface energy, wettability, friction coefficient, chemical inertness, dielectric (low-k) constant and interactions with biological systems; all these properties have generated a number of applications in different fields of surface and material science, from microelectronics to biomaterials, from textiles to corrosion protection.

Plasma-deposited fluorocarbon coatings are known also as fluoropolymers; other definitions are teflon-like coatings and PPFM (Plasma Polymerized Fluorinated Monomers). Their composition is characterized by a CF_x $(0 < x \leq 2)$ stoichiometry; other atoms, e.g., oxygen and hydrogen, may be present, depending on the particular deposition conditions. Structure (cross-linking, retention of the monomer structure, grafting and orientation of chains) and surface morphology (roughness, nano/micro structures) of the coatings are also highly dependent on the deposition conditions.

The literature of plasma-deposited fluorocarbons is very rich; the first investigations date back to the 70s, when CF_4 and other feeds started to be utilized in microelectronics for dry etching processes of silicon, SiO_2 and other materials. In certain conditions, the deposition of fluorocarbon coatings occurred, which was investigated separately. Many points related to plasma and surface diagnostics, deposition kinetics and applications of fluorocarbon coatings have been reviewed [1-6]; the matter is still in evolution, though, and new aspects often emerge.

The aim of this contribution is to review, from the experience of the author, old and new aspects in PE-CVD processes of fluoropolymer films from continuous, afterglow and modulated glow discharges. In particular, it will be emphasized how it is possible to tune, in the different deposition regimes, the retention of the monomer structure in the coatings, and the related surface properties.

2.2. On the Jargon of PE-CVD Processes

The words *plasma polymer*, *monomer* and *plasma polymerization* are imported from polymer science, and utilized also in the jargon of plasma science, albeit with different meanings; words like plasma deposited coating, gas feed, and plasma deposition should be utilized, more correctly, in their place. Plasma coatings, in fact, almost never show a regular repeating structural unit, as conventional polymers do rather, they can be described as networks of homologous chemical groups. Fluoropolymers, plasma-deposited in most of the experimental conditions, for example, consist of $–CF_3$, $-CF_2-$, $>CF-$, and $>C<$ moieties assembled in a random network rather than in ordered $-(CF_2-CF_2)_n$ polytetrafluoroethylene (PTFE) chains, as shown in Figure 2.1.

The word monomer indicates a compound whose stoichiometry is regularly repeated in a conventional polymer after a conventional polymerization reaction. Tetrafluoroethylene ($CF_2=CF_2$, TFE), for example, is the monomer from which PTFE is synthesized. In PE-CVD, instead, the word monomer indicates the compound fed and fragmented in the plasma, from which different neutral (atoms, radicals, molecules) and charged (electrons, ions) species are generated. All these *active species*, rather than the monomer alone, concur through different routes to the overall heterogeneous deposition kinetics, which are almost completely different from conventional homogeneous polymerization reactions. In spite of all these differences, and even though some confusion still is generated in scientists approaching PE-CVD for the first time, the words *plasma polymer*, *monomer* and plasma polymerization are accepted and widely utilized in the plasma literature.

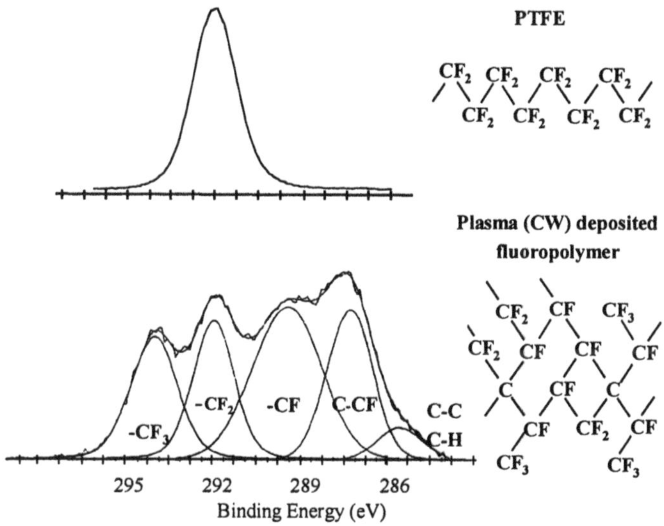

Figure 2.1. Chemical structure and ESCA (Electron Spectroscopy for Chemical Analysis) C1s spectra for PTFE and a fluorocarbon coatings plasma-deposited in CW conditions.

2.3. Continuous, Afterglow and Modulated PE-CVD

Continuous, afterglow and modulated plasma regimes can be easily activated in glow discharges for PE-CVD; each regime can originate coatings with very different characteristics. Radiofrequency (RF) at 13.56 MHz is the most utilized field frequency to generate homogeneous plasma in many possible different reactor geometries; lower (KHz) and higher (GHz) frequencies are also utilized. In CW processes, the discharge is kept continuously switched on; the substrate is directly exposed to the glow, thus it is subjected to direct interaction with neutral (chemical reactions) and ionic (positive ion bombardment) species generated in the plasma. Both chemical reactions and ion bombardment play an important role in CW deposition kinetics, and lead to coatings with variable composition and cross-linking, where the composition and structure of the starting monomer is almost completely lost. Parallel plate reactors are among the most used systems for CW PE-CVD.

Substrates positioned downstream with respect to the plasma zone, in the so called afterglow region, experience the chemical effect of long-lived atom and radical species, but not the ion-bombardment [7-10]. For this reason, AG PE-CVD processes result in films with a different chemistry, deposition rate, structure and properties compared to CW conditions. Generally, the deposition rate is lower, and the degree of structure retention of the monomer in the coating can be very high, due to a deposition mechanism which very likely has many aspects in common with a conventional polymerization. Since in these processes the flow dynamics of the reactor can play a major role, AG PE-CVD processes are investigated mostly in tubular reactors, with many different plasma sources. The distance of the substrates from the plasma zone in the flow direction has a key role in AG deposition processes.

In modulated glow discharges, the power input is delivered periodically to the plasma for micro-milliseconds (t_{ON}), and switched off, or reduced to a certain fraction, for 10-100 times longer time (t_{OFF}) [11-16]. The substrates are positioned in the plasma zone, thus experiencing ion bombardment and interact with unstable species from the plasma during t_{ON}, and with unreacted monomer molecules and long-lived radicals/atoms during t_{OFF}. Modulated PE-CVD processes are investigated as a function of the period (t_{ON} + t_{OFF}) and of the Duty Cycle (DC% = t_{ON} x 100 / t_{ON} + t_{OFF}) of the discharge, which tune the power delivered to the system, thus the fragmentation of the monomer, and finally the retention of the monomer structure in the coating. Most of the experiments are investigated keeping the power at zero value during t_{OFF}.

Plasma pulses in modulated PE-CVD will likely result in a mixed deposition process, which can probably be approximated by a combination of CW deposition kinetics with conventional polymerization reactions. Coatings deposited in modulated PE-CVD can exhibit a high degree of monomer structure retention, as we will see in the next section for fluoropolymers, and a micro-nano surface morphology. As for CW, modulated PE-CVD processes are prevalently investigated in parallel plate reactors.

2.3.1. Fluorocarbon Monomers

The choice of the monomer influences the process to be performed, since it is the primary source of reactive fragments and film precursors in the plasma. Volatile fluoroalkanes (C_nF_{2n+2}), -alkenes, -alkynes, cyclic and aromatic fluorocarbon compounds can be utilized as monomers for plasma deposited fluorocarbon coatings. All compounds are, in principle, suitable for CW deposition processes, while unsaturated ones such as TFE and hexafluoropropylene oxide (HFPO) are preferred for AG and modulated discharges, where reactions very similar to conventional polymerization are very likely important. Compounds able to produce a high density of radicals, rather than atoms, should be preferred in CW PE-CVD, since F-atoms and CF_x radicals have been demonstrated to be the main etching and deposition precursor species, respectively.

Fluorocarbons can be ranked as a function of their tendency to etch silicon and other materials; this inclination can be approximated by the F/C formula ratio [2,17] of the fluorocarbon compound. For PE-CVD it can be said that the lower the F/C ratio of the monomer, the lower the F/C overall ratio of the fragments in the plasma, the lower the tendency to deposit a coating. CF_4, for example, should not be considered as a monomer for PE-CVD, while it is a good choice for etching or for fluorination processes, since it ensures a remarkable production of F atoms in the plasma [2,6,17-19]. Of course, this criterion is applicable for homologous monomers in the same experimental conditions.

Concerns for the environment can also drive the choice of the gas feed in plasma processing techniques. Recently, in fact, great research efforts have been spent to optimize new effective etching and PE-CVD processes in microelectronics, with monomers allowing a reduced emission of greenhouse gases in the exhaust [20-22]. This is due to the evidence that most fluorocarbons show a Global Warming Potential similar or higher than that of CO_2.

2.4. Diagnostic Techniques for PE-CVD Processes of Fluoropolymers

Plasma deposition, treatment and etching processes can be thoroughly understood and optimized in the lab, then duplicated in industrial processes, only when meaningful correlations can be found between the "internal" features of the discharge, and the final surface property requested in each application. A complete investigation requires extensive use of plasma and surface diagnostics, as well as valid material tests.

Due to multivariate relationships among the "internal" parameters of the discharge, that are the key parameters to tune for driving a process (density of species, ionization degree, intensity of ion bombardment, etc.), and the "external" variables (pressure, power, configuration, etc.) that can be adjusted by the operator, reproducing the same set of external variables does not guarantee to replicate exactly the process in question [1,2,6,17-19,12-14,23,24]; the key internal parameters should be reproduced, instead, under accurate control by plasma/surface diagnostics, to achieve the transfer of the same process from one reactor to another, and from the lab to the industrial scale. This approach to *process control* is quite often under considered in plasma processing of materials for applications. Only monitoring *in situ* the internal parameters with a proper set of diagnostics, and reproducing them by tuning the external ones, can guarantee the reproduction of the process and the maximization of the desired surface property. The same approach allows, obviously, to investigate in detail the fundamental chemistry of the modifications.

In this section, examples are given on diagnostic techniques utilized to investigate PE-CVD of fluoropolymers, with emphasis given to ESCA and actinometry.

2.4.1. Surface and Bulk Analysis

Due to the large *chemical shift* induced by F atoms on C1s photoelectrons, ESCA [25-27] is by far the most applied technique to probe the surface chemistry of plasma-deposited fluorocarbon coatings, since it allows to clearly distinguish all the CF_x moieties ($0 < x \leq 3$) present in the C1s photoemission spectrum of the material. Table 2.1. lists the Binding Energy (BE) values of the CF_x peak components generally utilized to best-fit the C1s spectrum of plasma-deposited fluoropolymers.

An example of C1s best-fitting has been shown in Figure 2.1. The $-CF_2$ peak component is normally used as BE reference; a –C–(C,H) component is usually added when the coating is deposited from hydrogen-containing feed (e.g., from C_2F_6/H_2), or it is thinner than the ESCA sampling depth (≈ 100 Å) and it is deposited on some hydrogen-containing substrate. In fluorocarbon coatings, the C(C)H component may be found at values slightly higher than the usual 285.0 eV, due to the chemical shift induced by the fluorine moieties.

The area ratio between the F1s (BE ≈ 689 eV) and C1s spectra, properly corrected with their atomic sensitivity factors [25], allows to calculate the F/C stoichiometry ratio; the same parameter can be calculated also from the best-fitting of the C1s spectrum using equation (1), where CF_3%, CF_2% and CF% are the contributions of the CF_x ($x = 1, 2, 3$) components to the total C1s area. The F/C ratio allows to probe the structure retention of the monomer in a plasma-deposited fluoropolymer; beside this, it is also a measure of the degree of cross-linking (the lower F/C, the higher the degree of cross-linking).

$$F / C = (3\ CF_3\ \% + 2\ CF_2\ \% + CF\ \%) / 100 \tag{2.1}$$

Table 2.1. Binding Energy of the CF_x peak components in C1s spectra of plasma-deposited fluoropolymers.

Peak component	Binding Energy (eV)
$-\underline{C}F_3$	294.0–294.8
$-\underline{C}F_2$	292.0 (reference)
$-\underline{C}F$	288.6–290.0
$-\underline{C}$(H,C)-CF,	286.5–287.8
$-\underline{C}$(C,H)	285.0–285.8

In addition to ESCA, SIMS (Secondary Ion Mass Spectrometry, [8,10,27,28]) in the static mode, CA (Contact Angle [29,30]), AFM (Atomic Force Microscopy [31]), and other surface analysis techniques are extensively utilized to probe chemical composition, morphology and properties of plasma-deposited fluoropolymers. NEXAFS (Near Edge X-rays Absorption Fine Structure) has been utilized with success by Castner *et al.* [8,32,33], combined with other techniques, to investigate the oriented-chain surface structure of AG plasma-deposited fluoropolymers. Cross-linking and chemical structure of plasma-deposited fluoropolymer bulk can be finely probed by solid state NMR (Nuclear Magnetic Resonance) [34,35], FT-IR (Infrared) and Electron Spin Resonance (ESR) spectroscopy.

The deposition rate should also be measured to fully characterize PE-CVD processes of fluoropolymers. Accurate weight measurements of the substrate before and after the deposition, and simple masking techniques followed by step measurements allow to measure *ex situ* thickness and deposition rate of PE-CVD coatings; Laser Interferometry (LI) and Quartz Crystal Microbalance (QCM) allow more accurate in situ measurements, provided that the refractive index and the density of the coatings are known, respectively.

2.4.2. Plasma Diagnostic Techniques

Compared to techniques such as Langmuir Electrostatic Probes (LEP) and Mass Spectrometry (MS), optical techniques such as Optical Emission Spectroscopy

(OES), UV Absorption Spectroscopy and Laser Induced Fluorescence (LIF) allow to probe *in situ* the species produced in the homogeneous phase of a plasma process in a non intrusive way. Time Resolved Optical Emission Spectroscopy (TR-OES) can be utilized to investigate modulated plasmas.

Among optical techniques, Actinometric Optical Emission Spectroscopy (AOES, actinometry [1,6,23,24,36,37]), allows to deduce semi-quantitative density trends of light emitting species as a function of the experimental variables, as well as those of electrons of different energies. Provided that electron impact is the main excitation channel, and that the energy excitation thresholds, E_{th}, match for the emitting species X under scrutiny and for the actinometer (*i.e.*, the two species are excited by electrons of close energy), the density of the species X in the ground state can be considered proportional to the emission intensity of X itself ratioed by that of an actinometer, an inert gas (noble gases, N_2) added in small, constant amount to the discharge. Table 2.2. lists the spectral features of the emitting species detected in discharges fed with fluorocarbons, alone or with O_2 or H_2, and with He, Ar and N_2, the most utilized actinometers.

Table 2.2. Emission lines and bands detected in fluorocarbon based RF glow discharges. From [1,38].

Species	System	Spectral feature (Å)	Threshold Energy (eV)
CF	$B^2\Delta$-$X^2\Pi$	1970-2200	6.0
CF_2	A^1B_1-X^1A_1	2200-2800	4.5
F	$3s^2P_{3/2}$-$3p^2P_{3/2}$	7037	14.8
CH	$A^2\Delta$-$X^2\Pi$	4314	2.9
H	$2p^2P^0$-$3d^2D$	6563	12.1
O	$3s^3S^0$-$3p^3P$	8446	11.0
He	$2s^3S$–$3p^3P$	3889	23.1
Ar	4s'[1/2]-4p'[1/2]	7503	13.5
N_2	$C^3\Pi_u$-$B^3\Pi_g$	3371	11.1

The actinometric approach is described in equation (2), where k includes the density of the actinometer, I_X and I_{act} are the emission intensities of X and of the actinometer, respectively, and n_X is the density of the species X in their ground state. The minimum flow rate of the actinometers, and the limit beyond which it is perturbative, can be easily set with OES and surface analysis.

$$I_X / I_{act} = k \; n_X \qquad (2.2)$$

Adding different actinometers allows to probe electrons of different energy, thus species X with different E_{th}. When the relative intensity trends of the actinometers match as a function of the external parameter under investigation (*i.e.*, their intensity ratio is constant), it can be assumed that the energy electron distribution is not affected by the parameter in question, thus AOES can be applied to all emitting atoms and radicals, where the E_{th} values of the latter are usually lower. If the trends of the actinometers do not coincide, equation (2) can be applied only to species whose E_{th} matches with a single actinometer, a situation that does not permit to probe the density of radicals with sufficient accuracy.

AOES is a semiquantitative method; specially when used for *in situ* control, it permits to obtain relative density trends of unstable precursors in a simple way. Combined with a proper reactor design and with the use of surface diagnostics, for example, actinometry can be used for fundamental studies, for optimizing experimental parameters and for reproducing, control and transfer plasma-processes. Actinometry has been fully exploited by d'Agostino [1, 6, 37-40] for investigating CW fluorocarbon plasmas, and for proposing the Activated Growth Model (AGM, see section 2.7) to explain PE-CVD kinetics of fluoropolymers.

2.5. Active Species in Fluorocarbon Plasmas

As evident from plasma diagnostics [1,2,17-19,23,36-41], a fluorocarbon plasma is populated by CF_x ($3 \geq x \geq 1$) radicals, F and C atoms, ions and electrons produced by the fragmentation of the monomer, as well as by heavier species originated by recombination reactions among different fragments and monomer molecules. The monomer itself may participate in heterogeneous reactions, and it is likely among the film precursor species in AG and modulated PE-CVD processes.

All *active species* are present in different excitation states, and their distribution depends highly on the experimental parameters of the discharge. The distribution of species in the plasma and in its vicinity drives the interactions with the substrate along with its nature, temperature, ion bombardment and position in the reactor, and controls structure and composition of the coating or of the etched/fluorinated layer.

Since F atoms and CF_x radicals have been demonstrated to be the main etching and deposition precursor species in CW fluorocarbon plasmas, the fluorine to radical F/CF_x density ratio in the plasma has been proposed as one of the most relevant internal simplifying parameters to describe the deposition of fluoropolymers [1,6,17]. The same parameter turns out to be very important when a selective etching process, *e.g.* SiO_2 in the presence of Si, is requested [17-19]. F atoms, in fact, etch both materials, albeit at different rates, while CF_x radicals can etch only SiO_2.

The F/CF_x density ratio is, obviously, an internal parameter of the discharge; for this reason it has to be optimised per each single process and reactor by tuning the external parameters. The relative density of F and CF_x species can be influenced by many experimental parameters, but it can be tuned under spectroscopic control, because it can be monitored by OES and related plasma diagnostic techniques. In this way, it is possible to correlate the density of species in the glow with the chemistry of the processed surface, an approach that allows to easily study the deposition process, and to control its optimization and transfer.

2.5.1. Additives to Drive the F/CF_x Density Ratio of a Fluorocarbon Discharge

The attitude of the monomer to etch a material (production of F atoms) or to deposit fluorocarbon coatings (production of CF_x radicals) can be further driven in both directions with the addition of reactive additives to the gas feed, such as O_2 or H_2, or by performing the discharge in the presence of certain reactive materials. The distribution of active species can be easily driven in this way. Adding inert buffer gases such as Ar and He also influences the fragmentation patterns of the monomer, resulting in a changed F/CF_x ratio in the plasma, and tuned surface composition.

Adding O_2 to a fluorocarbon plasma increases the density of F atoms, due to the combustion of CF_x radicals with oxygen atoms and excited oxygen molecules according to reaction (3); as a consequence, the F/CF_x density ratio of the gas, thus its etching capability for silicon, resists and other materials, increases [1,2,6,17-19,38].

$$CF_x + O\,(O_2) \rightarrow CO\,(CO_2) + x\,F \qquad (2.3)$$

H_2 can be added, instead, to lower the F/CF_x density ratio and favour the deposition of coatings with variable cross-linking degree and chemical

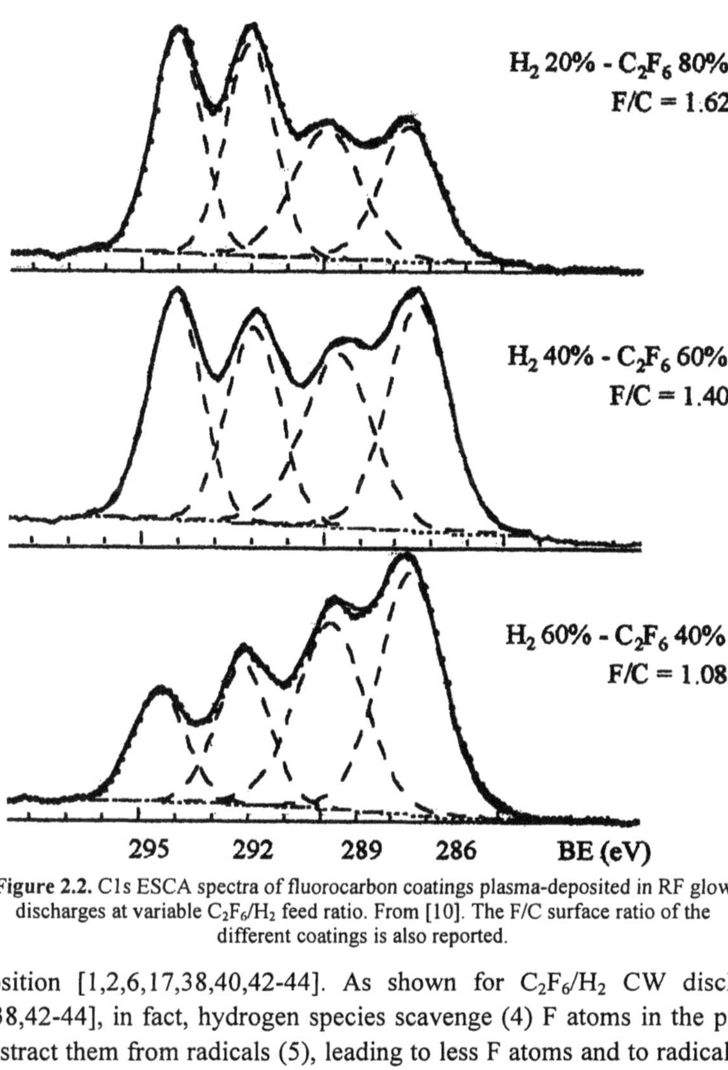

Figure 2.2. C1s ESCA spectra of fluorocarbon coatings plasma-deposited in RF glow discharges at variable C_2F_6/H_2 feed ratio. From [10]. The F/C surface ratio of the different coatings is also reported.

composition [1,2,6,17,38,40,42-44]. As shown for C_2F_6/H_2 CW discharges [1,10,38,42-44], in fact, hydrogen species scavenge (4) F atoms in the plasma, and abstract them from radicals (5), leading to less F atoms and to radicals with a lower fluorination degree. HF formed in dry conditions do not contribute to surface reactions. Scavenging of F atoms can also be achieved when silicon, polymers, or certain metals are present in the reactor, due to the etching reaction (6), and others similar. With a terminology imported from etching systems, it is said that the PE-CVD process is performed in loading conditions for F atoms.

$$H\ (H_2) + F \rightarrow HF + (H) \tag{2.4}$$

$$H\,(H_2) + CF_x \rightarrow CF_{x-1} + HF + (H) \tag{2.5}$$

$$Si + x\,F \rightarrow SiF_x \tag{2.6}$$

C_2F_6/H_2 feeds have been utilized, at variable ratio, to switch CW discharges from etching ($80 \leq C_2F_6\% \leq 100\%$) [38] through the deposition of fluoropolymers ($20 \leq C_2F_6\% \leq 80$) [10,38,40,42] to the deposition of fluorinated carbon films ($0 < C_2F_6\% \leq 20$) [43,44]. Figure 2.2 shows how the addition of H_2 to a C_2F_6 feed changes the composition (distribution of groups, lower F/C surface ratio) of coatings deposited in CW RF glow discharges, as a

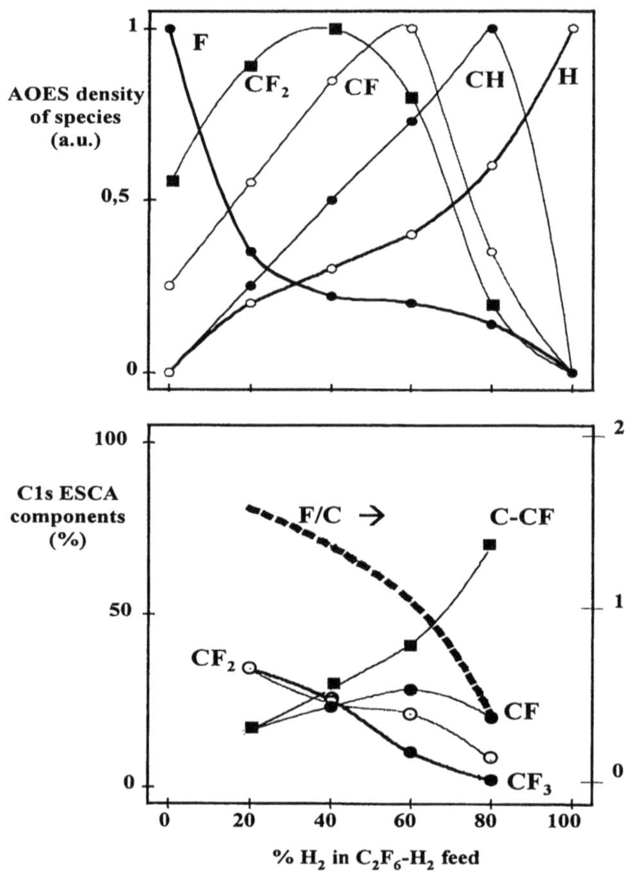

Figure 2.3. AOES density trends of emitting species detected in C_2F_6/H_2 RF glow discharges as a function of the H_2 content of the feed, along with the corresponding trends of the C1s ESCA components and of the F/C surface ratio of the plasma-deposited coatings. Adapted from [38].

function of the C_2F_6/H_2 feed ratio. As a result of the lower F/C ratio, the advancing Water Contact Angle of the coatings is lowered.

It is relevant to note now that the same feed in the same reactor can generate a different distribution of active species in the glow (CW processes) with respect to its afterglow; as a consequence, different processes takes place in two different zones of the same reactor. A RF glow discharge feed with pure C_2F_6, for example, as reported in Ref. [10], etches polymer substrates on the ground electrode of a parallel plate tubular reactor, with no deposition, but generates oriented teflon-like coatings in the AG region of the same discharge (see section 2.6.3).

Figure 2.3. show how the AOES distribution of species depends on the H_2 content in C_2F_6/H_2 RF glow discharges performed in a parallel plate reactor [38], and how this distribution is reflected in that of the C1s best fitting ESCA components and in the F/C ratio of the films.

When highly fluorinated CF_x radicals dominate the distribution (low H_2 feed content), fluoropolymers with high F/C ratio and low cross-linking degree are deposited, while less fluorinated radical distributions (high H_2 feed content) produce coatings with lower F/C ratio and higher cross-linking degree. We will see, in the next sections, how the hydrophobic character of teflon-like coatings depends on the F/C surface ratio; the distribution of radicals in the plasma can thus be adjusted to obtain coatings with pre-determined F/C ratio and surface properties.

2.6. Relevant Parameters in PE-CVD Processes of Fluoropolymers

The effect of internal parameters such as the choice of the monomer and additives to the feed has been described in the previous sections, with particular reference to CW PE-CVD processes. Some other parameters are described in this section, which may act, at different extents, in CW, AG and modulated PE-CVD.

2.6.1. Ion Bombardment and Substrate Temperature

Albeit the ionization degree of low pressure plasmas almost never exceeds 10^{-5} ions/neutrals, the contribution of positive ions must always be considered in plasma deposition/etching mechanisms, since they bombard all surfaces exposed to the plasma. Due to the negative *self bias* potential that develops between the surfaces exposed and the plasma itself, a positive-ion bombardment is induced on any surface exposed to a low pressure plasma, with intensity and energy depending on the geometry of the reactor and on the experimental parameters. The electrical conditions and the position of the substrate with respect to the electrodes can determine, among other parameters, the extent of the ion bombardment. The substrate is not exposed to ion bombardment in AG processes, while it is drastically reduced during t_{ON}, and during t_{OFF}, in modulated discharges [45-48]. Ion bombardment of the growing coating can thus be considered a peculiarity of CW PE-CVD processes; in the other regimes it can be considered negligible.

As a rule of thumb it can be said that the ion bombardment increases with the power delivered to the plasma, and when the pressure of the system is lowered. The geometry of the reactor also influences ion bombardment; in a capacitively coupled parallel plate RF reactor, for example, the most intense bombardment develops at the surface of the smallest electrode according to a general law expressed by Koenig and Maisel [47,49]. The role of ion bombardment in PE-CVD processes is very important for describing plasma-surface interactions, and should be investigated separately from any other effect. When the substrates are positioned on the third electrode of a triode reactor, the ion bombardment can be varied independently from the other internal plasma parameters, such as the density of the neutral species, by simply tuning the power delivered to the third electrode, thus the bias potential that develops there.

CW PE-CVD experiments performed in a triode reactor can be evidence of the effect of the ion bombardment on the deposition/etching rates of plasma processes, provided that the distribution of active species in the plasma (*e.g.*, the F/CF_x ratio) is kept constant [6,50,51]. Increasing the positive-ion bombardment in the low energy (bias) range results, generally, in an increased (activated) deposition rate when the F/CF_x ratio is low (*e.g.*, for feeds as C_2F_4), and in an increased (activated) etching rate when the F/CF_x ratio is high (*e.g.*, for feeds as CF_4/O_2).

Increasing the bias may depress the deposition rate of a fluoropolymer, due also to the competition of ion-activated etching and sputtering processes. Beyond a certain bias threshold, the etching of the plasma deposited layer can occur, as schematically shown in Figure 2.4. The trend shown is rather general, and can be explained in detail, as it will be shown, by using the AGM kinetics [38]. The etching rate of plasma deposited fluoropolymers in O_2 discharges, where O atoms are the etchants, is also ion-activated.

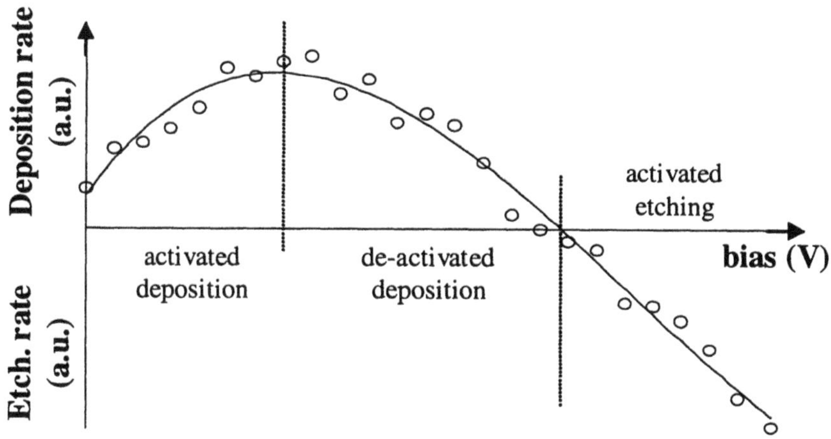

Figure 2.4. Example of the effect of the bias-induced positive-ion bombardment on the deposition/etching rate of fluoropolymers plasma-deposited in CW RF glow discharges. Adapted from [51].

The temperature of the substrate influences the adsorption-desorption equilibrium of the film precursors, characterized by a negative enthalpy ($\Delta H < 0$, exothermic), as well as the reaction kinetics of the precursors with the coating during its growth. Both effects are rather general, and characterize all PE-CVD processes. The overall effect of increasing the temperature of the substrate is that of decreasing the deposition rate of the coating. A very high temperature would also influence its composition and structure, due to the reduced adsorption of precursors species and to decomposition reactions.

2.6.2. Power, Flow Rate, Pressure and Combined Parameters

The flow rate F and the pressure P of the gas feed, as well as the power W delivered, all external parameters, influence almost all the internal parameters of each particular plasma system; their effects depend highly on the configuration of the reactor. The power input, for example, heavily influences the Electron Energy Distribution Function (EEDF), the density of all the active species, as well as the relevant electric features (plasma potential, bias potential, etc.) of the discharge. Increasing the power results always in an increased fragmentation of the monomer in the plasma, thus in an increased production of both atoms and radicals. The flow rate of the feed influences CW and modulated processes, and

AG processes particularly, since it triggers the rate of arrival of active precursor species on the substrate positioned far from the plasma. The arrival rate of species on the substrates is also determined by their lifetime.

The volume density W/V_{disch} of power delivered is also utilized, sometimes, to compare PE-CVD processes, provided the volume V_{disch} of the discharge zone is determined with good accuracy. The internal features are influenced also by F and P which, together with V_{disch}, determine the residence time $\tau = P\ V_{disch}/F$ of the species in the discharge. Any change of these parameters should be investigated per each single deposition/etching system with proper diagnostics, to predict their effects.

The W/FM ratio has been utilized very often for many years, where M is the molecular weight of the monomer. This composite parameter, proposed by Yasuda [3,52-54] to describe CW PE-CVD processes from different homologous monomers (e.g. CH_4, C_2H_6, C_3H_8, etc.), including fluorocarbons, describes the energy delivered per mass unit of monomer. Plotting features like deposition rate, or others, as a function of the W/FM ratio allows, in effect, to compare the behaviour of different homologous monomers in different power conditions, provided that the reactor configuration is the same, the pressure remains constant and the ion bombardment at the surface of the substrates does not change too much. Other composite parameters based on F and W have also been proposed to rationalize CW PE-CVD processes [55].

2.6.3. Distance of the Substrate from the Glow

In AG PE-CVD the substrates are positioned at some distance (centimeters) from the glow, downstream, in the direction of the feed flow, where no electrons are present, the density of ions is very low, and the ion bombardment of the substrate is negligible. In these conditions the interaction with the substrates of the monomer molecules, and of the unstable species (CF_x and other radicals, atoms) generated in the plasma, depends on the life-time of the species and on their residence time in the plasma reactor. Species with a long life-time have the highest possibility to interact with the substrate and, eventually to deposit a coating. The optimum plasma-substrate distance has to be found, per each single case, to deposit coatings with the requested composition, cross-linking and monomer structure retention degree, at an acceptable deposition rate. A low distance from the glow may result in coatings very similar to those deposited in the glow (G-coatings), or in substrate etching, depending on the density of species in that particular position of the reactor. Due to the absence of the ion-activation step (see next section), in AG PE-CVD the rate of deposition is 1-3 orders of magnitude lower than in CW and modulated PE-CVD. In certain conditions (high residence time of the species in the reactor), the deposition of coatings can occur also upstream from the glow, in the so called pre-glow positions [1].

The deposition of very thin AG teflon-like coatings with high retention of the monomer structure and F/C ratio very close to 2 has been performed with C_2F_4 [7-9,32], C_2F_6 [10,32] and C_2F_6/H_2 (80/20, [10,32]) feeds. The distance from the glow in each condition was the key parameter to select between cross-linked films very similar to G-coatings, and AG teflon-like coatings. The structure of AG-coatings has been characterized by ESCA, SIMS and NEXAFS, and it has been found that AG-coatings are characterized by highly oriented -CF_2-$(CF_2)_n$-CF_3 chains grafted to the surface of the substrate (silicon, polymers), as shown in Figure 2.5. Depending on the deposition conditions, the orientation of the chains and the density of the (few) cross-linking sites can change. The deposition rate is very low, in the orders of a few Å per minute; this can lead to a partial coverage of the substrate with the coating. As a further note, it can be reported that, due probably to the low surface energy properties associated with terminal -CF_3 groups of the chains, this AG deposition process has been found self-limited: after a certain deposition time, in fact, the coating does not grow

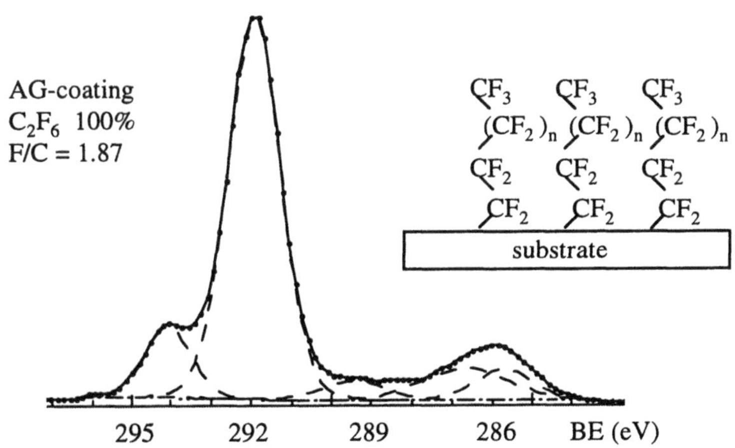

Figure 2.5. C1s spectrum and structure of a very thin (< 100 Å) AG teflon-like coating deposited in the afterglow of a RF glow discharge fed with from C_2F_6 [10]. The features at 286-289 eV of BE attest the presence of some cross-linking sites. The feature at lower BE is due to the substrate underneath the very thin coating.

anymore, albeit the discharge is still on.

AG chain-oriented teflon-like coatings display an hydrophobic character very close to that of PTFE (WCA ≈ 115°); due to their peculiar structure, they show interesting adsorption-retention properties for proteins, that has stimulated a certain interest in the field of biomaterials. With reference to AG-coatings generated from C_2F_6 RF glow discharges, it is to remark here that this kind of discharge produces, in the glow region, an high density of F atoms (see Figure 2.3), which induces the etching of any Si or polymer substrate rather than the deposition of G-coatings. In AG positions, evidently, the distribution density of the active species becomes completely different from the glow, due to their different life-time, or to different consumption rates with the walls of the reactor, and the deposition of very thin, highly fluorinated chain-oriented coatings occurs.

2.6.4. Duty Cycle and Period in Modulated Plasmas

In Section 2.3. Duty Cycle and period have been defined for modulated PE-CVD. Albeit modulated plasma have been investigated since the 70s [56,57], many papers have been produced in the last years on this subject [11-16,21,33,34,41,48,58-66], most of them dealing with plasma deposited fluoropolymers from unsaturated monomers.

With all the other parameters constant, the power delivered to the discharge, and the Duty Cycle, determine the degree of fragmentation of the monomer in the gas phase and, consequently, the degree of retention of the monomer structure in the coating. The "equivalent", or "average" power $W_{eq} = W [t_{ON} / (t_{ON} + t_{OFF})]$ is often utilized to describe modulated plasma processes, where W is the power value delivered during the pulse by the power supply. In this contest, the deposition rate of the coatings is frequently expressed in terms of thickness per unit input energy or per pulse. Many experiments performed by different authors at constant t_{ON} and increasing t_{OFF} values have shown that the coating also grows during t_{OFF}, often with a higher deposition rate with respect to t_{ON}. This is due to heterogeneous reactions of the radicals formed during t_{ON} with the substrate and also, very likely, of the intact monomer molecules with the substrate activated during t_{ON}, in a sort of grafting-polymerization reaction. This mixed deposition mechanism, that can be approximated to a combination of AGM growth (see next section) during t_{ON}, and of a conventional polymerization during t_{OFF}, is still not completely understood, but can lead sometimes to very high deposition rates, up to many hundreds Å/min.

Within certain limits, working at low W_{eq}, DC% and t_{ON} values generally allows to deposit coatings with the highest fluorination degree, where the structure of the monomer is highly retained, as attested by ESCA, NMR, FT-IR, SIMS and other surface/material diagnostic tools. Such conditions may result in different topographic features of nanometric-micrometric size at the surface of the coatings. Labelle et al. [64] describe "cauliflower-like" nodular structures at the surface of coatings plasma-deposited in modulated conditions from HFPO, from 1,1,2,2-tetrafluoroethane, and from difluoromethane. We have observed "ribbon-like" interpenetrating structures, as those shown in Figure 2.6, that characterize fluoropolymer coatings deposited from TFE at low W_{eq} and DC% $\leq$ 7-10, depending on the plasma reactor and the other conditions [16,67]; R. B. Timmons [68] reports of similar structured surfaces originated from modulated PE-CVD of fluorocarbons. Most probably the growth of the structures, which is indeed characterized by a low rate, is superimposed to the deposition of a conventional "background" fluorocarbon smooth coating, as preliminary SIMS

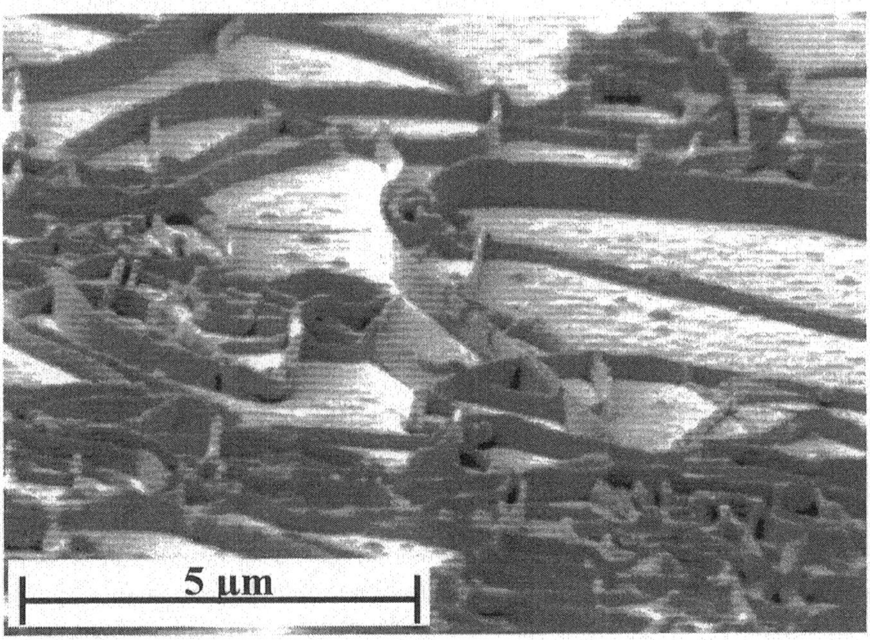

Figure 2.6. Tilted SEM picture of a teflon-like coating plasma-deposited from TFE on a silicon substrate in a modulated RF glow discharge (DC $\leq$ 7%). The coating is characterized by microns long, hundreds of nanometers high ribbon-like interconnected topographic features [16,67].

chemical maps seems to show in our case.

The particular features described for modulated PE-CVD is quite original with respect to CW regimes, where smooth coatings are usually deposited, and with respect to AG regimes, where surface non-homogeneity is usually generated by the incomplete coverage of the substrate. The mobility of the film precursor species on the substrate may play a role in the generation of surface topography; probably this effect is triggered only below a certain density threshold of the precursors in the gas phase. The formation of large molecules in the homogeneous phase, and/or eventually of dust particles, should also be considered. Accurate time resolved plasma diagnostics may help in understanding this structuring effect, which influences the hydrophobic properties of the coating. Super imposing surface roughness beyond a certain limit to a fluorinated hydrophobic material, in fact, results in a marked increase of its hydrophobic character, as we will see in section 2.8.

Our coatings have been investigated also with X-Ray Diffraction (XRD) [16,67], and it has been found that the presence of the structures on the coatings are associated with that of a definite feature at 18° in the 2θ XRD spectra, which indicates the occurence of hexagonal crystalline PTFE in the ribbon-like structures. Similar result are reported also in Ref. [13], where it is not clearly associated to the presence of surface microstructures, though. The presence of crystalline PTFE in fluoropolymers plasma-deposited at low DC% is quite surprising, in effect, and opens many questions on the nature of the deposition kinetics of such coatings.

Plasma diagnostic techniques have clearly shown that CFx radicals are the main precursor species of fluorocarbon coatings [1,2,17,18,36-42,69,70] deposited in CW conditions. Time resolved diagnostic techniques, such as TR UV Adsorption Spectroscopy [41] and TR-OES [16,67] have highlighted a major role of CF_2 radicals in modulated PE-CVD processes from TFE and HFPO monomers. CF_2 radicals revealed to be formed also during pyrolytic plasma-less decomposition routes of HFPO [71], that lead to coatings with many structural similarities compared to those obtained in modulated PE-CVD [13].

We have performed TR-OES experiments on emitting CF_2 radicals (A^1B_1 - X^1A_1 system, 2200-2800 Å) during modulated PE-CVD processes from TFE [16, 67]. CF radicals and F atoms were not detected at all in the experimental conditions utilized. In particular, it has been found, at constant period (100 ms), that the emission profile of CF_2 species in the very first milliseconds of t_{ON} are drastically different when the Duty Cycle of the discharge is switched from low values, when the deposition of structured ribbon-like, crystalline, highly

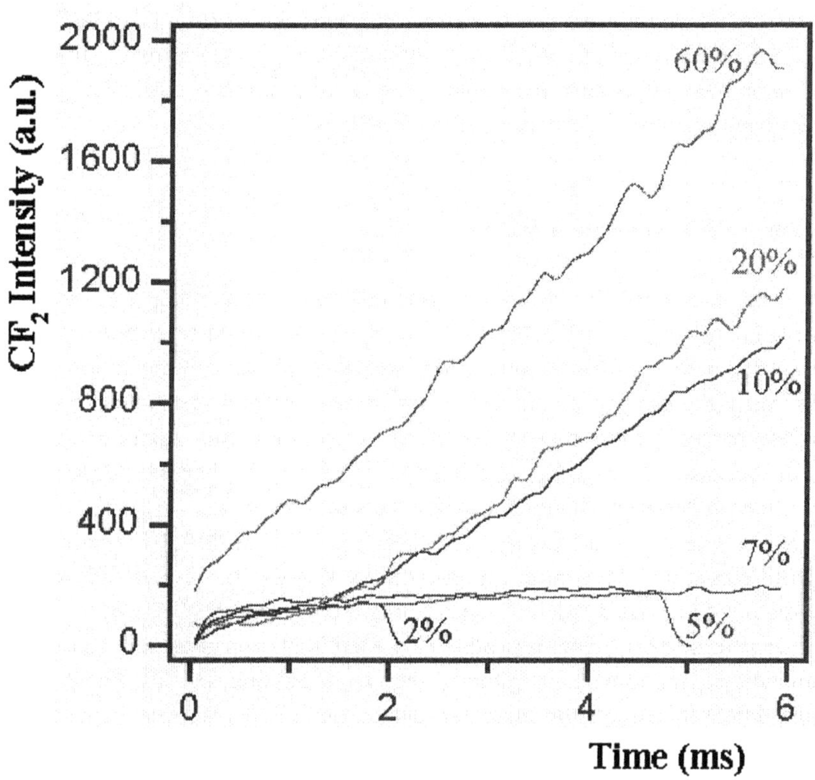

Figure 2.7. TR-OES trends of CF_2 radicals in TFE modulated RF glow discharges (30 Watt, 300 mTorr, 6 sccm TFE) performed at constant period values (100 ms) and variable Duty Cycle in a parallel plate reactor [67]. DC conditions ≤ 7% lead to ribbon-like, crystalline structured fluorocarbon coatings, like that shown in Fig. 2.6.

fluorinated coatings occurs, to high values, when the deposited coatings are smooth and resemble those deposited in CW conditions. The CF_2 emission trends are shown in Fig. 2.7.

The deposition kinetics of structured coatings is still under scrutiny while this chapter is being written, but some hypotheses may be expressed. At low DC values the emission intensity of the radicals is very low and constant, while it becomes much higher at increased DC. The low, constant emission intensity is most probably due to the low density of the CF_2 radicals, as originated in low fragmentation (and substrate ion bombardment) plasma conditions. This situation allows, very likely, enough surface mobility to the few CF_2 radicals present in the system, so they can assemble on the substrate in ordered PTFE chains starting from the few active sites of the surface. The deposition rate of our structured coating is very low, and a "conventional" coating is probably also formed, as seen by means of AFM at different stages of the growth. Unreacted TFE monomer molecules may also play a role in this particular growth mechanism, by means of radical grafting reactions.

2.7. The Activated Growth Model

The Activated Growth Model is probably the most complete proposed deposition mechanism for CW PE-CVD processes of fluorocarbon coatings, since it explains the role of precursor species from the fragmentation of the monomer in the gas phase, and that of ion bombardment and temperature of the substrate during the growth of the coating. Very likely, this model mechanism can be adapted to various experimental situations to explain also plasma deposition processes in AG and modulated conditions.

Homogeneous and heterogeneous reactions in CW PE-CVD processes of teflon-like coatings have been rationalized by d'Agostino [1,6,38,40], and the ion-Activated Growth Model of deposition has been proposed. AGM, which has also been utilized to rationalize other CW PE-CVD processes, like those from organosilicon monomers, combines the contributions of low-energy ion-bombardment to activate the substrate surface, and of CF_x radicals formed in the glow discharge as *building blocks* of the coating. AGM took form after many investigations performed under actinometric control of the plasma composition, with bias-activated ion-bombardment of temperature-controlled substrates in triode reactors. It must be said that AGM has been tested by changing many external parameters of the discharges, such as feed chemistry, pressure, power and substrate bias. A simplified AGM scheme is reported in the following reactions:

fragmentation of the monomer in the plasma

$$\text{monomer} \rightarrow n\,CF_x\ (1 \le x \le 3) \tag{2.7}$$

ion-activation of the coating (substrate)

$$I^+\ (\text{low energy}) + (\text{coating})_n \rightarrow (\text{coating})_n^* \tag{2.8}$$

growth of the coating

$$CF_x \leftrightarrow CF_{x\,(\text{adsorbed})} \tag{2.9a}$$

$$CF_{x\,(adsorbed)} \;+\; (coating)_n^{*} \;\rightarrow\; (coating)_{n+1} \tag{2.9b}$$

Reaction (2.7) describes schematically the production of the CF_x precursors from the monomer. A scenario similar to that described in Figure 2.3. can be imagined, depending on the experimental conditions. If many F atoms are produced in this step (this can be easily controlled by plasma diagnostics), an etching route can be also established, which would compete with the deposition. This possibility has been described by Yasuda in his Competitive Ablation-Polymerization (CAP, [3]) model. Another possibility that can occur is an intense production of radicals in the plasma, as it can happen in high power-density and high pressure (> 1 Torr) conditions; in this case, the radicals can polymerize also in the plasma, not only on the growing coating, thus leading to oligomers and, eventually, to heavy particles. These conditions lead to inhomogeneous coatings of scarce quality.

The ion-activation step (2.8) depends highly on the energy of the ions bombarding the growing film (or the substrate, in the early deposition stages), thus on the related external parameters (pressure, power, bias, geometry). At low energy (see also Figure 2.4), the ions create surface defective sites (*e.g.*, dangling bonds) that act as preferential adsorption sites for the precursor CF_x radicals from the plasma. If the energy of the ions overcomes a certain threshold, though, the desorption of the precursors is induced. When the ion energy becomes very high the sputtering of the coating may also be activated. Step (2.8) greatly influences the deposition rate, the composition and the cross-linking of the coating. The ion bombardment is one of the plasma parameters that most influences the retention of the monomer structure in the coating; it is evident that it is extremely important in CW processes, while it is practically absent in AG processes, and drastically reduced in modulated ones, depending on the Duty Cycle applied.

Steps (2.9a) and (2.9b) describe, respectively, the adsorption-desorption equilibrium of the CF_x radicals on the substrate (coating), and the following reaction of polymerization with the surface active sites originated by the ion-bombardment. The temperature influences heavily both reactions, but in a different way. PE-CVD of fluoropolymers are generally performed with substrates kept at room temperature; due to its $\Delta H < 0$ the equilibrium (9a) is moved to the desorption side by the increase of temperature, while the polymerization kinetics (2.9b) is favoured. As an overall effect of the increased substrate temperature in the 25-100°C range, a reduced deposition rate is usually recorded [10,38] and a reduced F/C ratio (increased cross-linking) in the deposited coating.

2.8. Applications of Plasma-Deposited Fluoropolymers

Thin fluoropolymer coatings plasma-deposited in different discharge regimes have been proposed, through the last thirty years, for many purposes in materials technology. A complete list of tentative and effective applications proposed for fluoropolymers in the last years would be extremely difficult to be completed, due to the very high number of references and patents produced, and to the many different fields where the properties of fluorocarbon coatings are or could be exploited. In the remaining part of this section some applications of fluorocarbon coatings will be described, as they come out from the literature, in the field of textiles and paper, and for biomedical applications, where these materials have stimulated particular interest.

Other uses of fluoropolymers have also been proposed, for example as low dielectric constant (low-k) coatings for integrated circuits [34,60,72,73], which are not described in this review.

2.8.1. Wettability of Fluorocarbon Coatings and Applications for Textiles and Paper

As reported previously, the F/C composition of fluoropolymers can be easily tuned with the plasma conditions, and this results also in tuned hydrophobic character and surface energy of the coating, a feature that is of obvious interest in industry, particularly when it is requested to impart hydrophobicity and/or oleophobicity and stain resistance to paper and textiles, or to other substrates [14-16,74-77]. When the coating is deposited in certain modulated conditions (see Section 2.6.4.), and/or rough substrates are utilized, the effect of surface roughness is added to the effect of the fluorocarbon surface chemistry, resulting in very high advancing WCA values, in the 140-170° range, often reported as "super hydrophobic" character.

A direct correlation has been found between the hydrophobic character of fluoropolymers deposited in different glow discharges and their F/C surface ratio, which has been reported in Refs. [10,23]. WCA values comparable and even higher than PTFE have been reported for smooth coatings deposited in low fragmentation regimes. Figure 2.8 reports the data, including those relative to modulated processes.

A clear linear trend results for all coatings, as a function of the F/C fluorination degree, that includes conventional fluorinated polymers, except for those deposited in low DC modulated conditions. Due to the presence of ribbon-like structures at their surfaces, such highly fluorinated structured coatings also display a remarkable surface roughness, that induces the "super hydrophobic" character.

Plasma-deposited fluorocarbon coatings appear thus suitable for applications that rely on their hydrophobic character. Long term durability of their properties has to be assessed, however, in that some chemical changes of their chemical composition have been reported during the aging, at least for coatings deposited under CW conditions [78]. This is true, evidently, for all

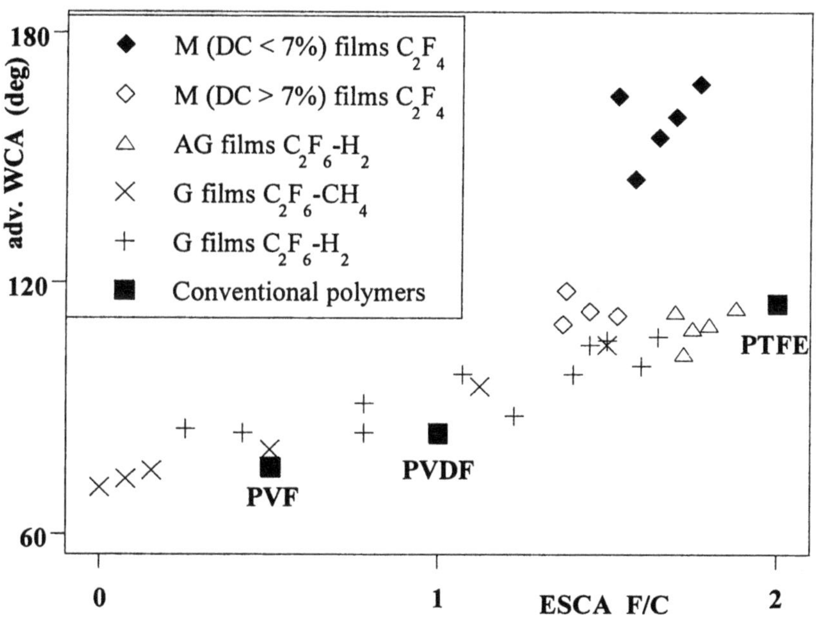

Figure 2.8. Advancing WCA values of fluoropolymers deposited in different plasma conditions and regimes, as a function of their F/C surface ratio. Adapted and updated after [10,16,23,67]. PVF = polyvinylfluoride; PVDF = polyvinyldifluoride.

application of fluoropolymers and other plasma-deposited coatings.

2.8.2. Biomedical Applications Proposed for Plasma-Deposited Fluoropolymers

Plasma-deposited coatings and plasma treated surfaces have been tested for many different uses in biomedical applications, where it is necessary to adapt the response of material surfaces to different biological media, *in vitro* and *in vivo* [79]. Among many other PE-CVD processes, the deposition of fluoropolymers has been applied often for biomedical applications.

Fluorocarbon coatings have been plasma-deposited from CW plasmas at the inside walls of Dacron vascular grafts and at the surface of other conventional polymers [7,9,10,79-81], and positive results have been published due to their non-thrombogenic character, resistance to platelet adhesion and activation, blood compatibility and adsorption-retention properties toward blood proteins such as fibrinogen and albumin. Similar coatings gave interesting results when deposited on Intra Ocular Lenses (IOLs), since they induced a reduced loss of endothelial cells in the cornea of rabbits [79], this a reduced corneal damage during the surgery.

Very recently, Gleason *et al.* [65,72] provided the surface of very small radius (< 100 μm) metal wires with an extremely flexible, highly dielectric, very thick (10 μm) fluorocarbon coating plasma-deposited in very high monomer structure retention conditions. HFPO modulated RF glow discharges have been utilized for this study. This finding opens the possibility of coating and insulating with fluoropolymers very small electrical parts such as electrodes, probes, wires, Micro Electro Mechanic Systems (MEMS) and other, to be utilized in biomedical practices.

Plasma-deposited fluoropolymers may also play a role in micro patterning techniques to be utilized in future biomedical applications, particularly in tissue engineering, sensors, diagnostic assays and bioseparation techniques, according to what is reported by Ratner et al. [82-84]. Smooth CW PE-CVD fluoropolymers, in fact, can be easily provided with geometric topographic patterns of micrometric size by means of photolithographic techniques or by direct patterning with AFM [82]. Further, CW plasma-deposited conformal fluoropolymer coatings have been utilized in a very elegant sequence of chemical processes to develop smooth surfaces imprinted with protein-recognition nanocavities [83,84]. The production sequence of such molecularly imprinted surfaces includes the following steps: protein adsorption on a mica sheet, coating with a polysaccharide solution, drying, CW PE-CVD of conformal a fluoropolymer layer, removal of the mica sheeth and wash-off of the template protein. A polysaccharide surface with the imprints of the protein is left, that can selectively recognize the template proteins including albumin, immunoglobulin G, lysozyme, ribonuclease and streptavidin, by means of hydrogen bonds.

2.9. Conclusive Remarks

Old and new aspects on process parameters, diagnostics, deposition mechanisms and applications of fluoropolymers deposited in continuous, afterglow and modulated plasma regimes have been reviewed from the literature and from the experience of the author. Particular emphasis has been given in describing how it is possible to adjust in a controlled way, in the different deposition regimes, the retention of the monomer structure in the coatings, and the related surface properties.

Investigating the deposition processes of fluorocarbon coatings has played a determining role in the development of Plasma Chemistry Science in the last thirty years, and the author has no doubt that these coatings will continue to raise interest and applications in the future, as attested by the continuously growing number of studies on their properties and on their deposition mechanisms in the different regimes.

Acknowledgements

Many funds have contributed to support the group of the author in investigating CW, afterglow and modulated PE-CVD processes; Procter & Gamble, in particular, is acknowledged for funding the research on the deposition of super-hydrophobic fluoropolymers from modulated discharges. Dr. Gianfranco Palumbo is acknowledged for the active participation in this research.

Co-working with Pasqua Rossini and Antonella Milella on this subject in the last years have been particularly productive and appreciated. Dr. Grazia Cicala and Dr. Fabio Palumbo (IMIP Bari, CNR) are gratefully acknowledged for a fruitful collaboration and for stimulating discussions. Many thanks are due also to Prof. Riccardo d'Agostino (University of Bari) for suggestions, discussions and motivations through many years of fertile collaboration.

References

1. d'Agostino R., Cramarossa F., Fracassi F., Illuzzi F., in *Plasma Deposition, Treatment and Etching of Polymers*, d'Agostino R. ed. (Acad. Press, 1990) 95.
2. Kay E., Coburn J. W., Dilks A., in *Topics in Current Chemistry* **94**, Veprek S., M. Venugopalan M. eds. (Springer Verlag, 1980) 1.
3. Yasuda H., *Plasma Polymerization* (Acad. Press, 1985).
4. Inagaki N., *Plasma Surface Modification and Plasma Polymerization* (Technomic, 1996).
5. Biedermann H., Osada Y., *Plasma Polymerization Processes* (Elsevier, 1992).
6. d'Agostino R., in *Plasma Processing of Polymers,* d'Agostino R., Favia P., Fracassi F. eds. (Kluwer Acad. Publ., NATO ASI Series, E: Appl. Sci., Vol. 346 (1997) 1.
7. Kiaei D., Hoffman A. S., Ratner B. D., Horbett T. A., *J. Appl. Polym. Sci.: Polym. Symp.* **42** (1988) 269.
8. Castner, D. G., Lewis, Jr. K. B., Fischer, D. A., Ratner, B. D., Gland, J. L., *Langmuir* **9** (1993) 537.
9. Kiaei D., Hoffman A.S., Horbett T.A., *J. Biomat. Sci. Pol. Ed.* **4** (1992) 35.
10. Favia P., Perez-Luna V. H., Boland T., Castner D. G., Ratner B. D., *Plasmas and Polym.* **1** (1996) 299.
11. Cicala G., Losurdo M., Capezzuto P., Bruno G., *Plasma Sources Sci. Tech.* **1** (1992) 156.
12. Panchalingam V., Poon B., Huo H. H., Savage C. R., Timmons R. B., Eberhart R. C., *J. Biomat. Sci. Polym. Ed.* **5** (1993) 131.
13. Limb S. J., Lau K. K. S., Edell D. J., Gleason E. F., Gleason K. K., *Plasmas and Polym.* **4** (1999) 21.
14. Panchalingam V., Chen X., Savage C. R., Timmons R. B., Eberhart R. C., *J. Appl. Polym. Sci.: Appl. Polym Symp.* **54** (1994) 123.
15. Coulson S. R., Woodward I. S., Badyal J. P. S., Brewer S. A., Willis C., *Langmuir* **16** (2000) 6287.
16. Favia P., Cicala G., Milella A., Palumbo F., Rossini P., d'Agostino R. in *Vol II Proc. 15th Int. Symp. on Plasma Chemistry, ISPC-15*. Orleans, FRA, Jul. 2001, 587.
17. Coburn J. W., Winters H. F., *J. Vac. Sci. Tech.* **16** (1979) 391.
18. Flamm D. L., Donnelly V. M., *Plasma Chem. Plasma Proc.* **1** (1981) 317.

19. Manos D. M., Flamm D.L. eds., *Plasma Etching: an Introduction* (Plasma-Materials Interaction Series, Acad. Press, 1989).
20. Fracassi F., d'Agostino R., *Plasmas and Polym.* **4** (1999) 147.
21. Labelle C. B., Karecki S. M., Reif R., Gleason K. K., *J. Vac. Sci. Tech. A* **17** (1999) 3419.
22. Shirafuji T., Kamisawa A., Shimasaki T., Hayashi Y., Nishino S., *Thin Solid Films* **374** (2000) 256.
23. Favia P., in *Plasma Processing of Polymers,* d'Agostino R., Favia P., Fracassi F. eds. (Kluwer Acad. Publ., NATO ASI Series, E: Appl. Sci., Vol. 346, 1997) 487.
24. Favia P., Creatore M, Palumbo F., Colaprico V, d'Agostino R., *Surf. Coat. Tech.* **142-144** (2001) 1.
25. ESCA Briggs D., Seah M.P. eds., *Practical Surface Analysis Vol 1, Auger and X-ray Photoelectron Spectroscopy* (second ed., Wiley, Salle+Sauerlander, 1990).
26. Ratner B. D., Castner D. G., *Coll. Surf. B: Bioint.* **2** (1994) 333.
27. Ratner B. D., Johnston E. E., *J. Elect. Spectrosc. Rel. Phen.* **81** (1996) 303.
28. Davies M. C., Lynn R. A. P., *Crit. Rev. in Biocomp.* **5** (1990) 297.
29. Mittal K. L. ed., *Contact Angle, Wettability and Adhesion* (VSP, Zeist, 1993).
30. Zisman W. A., in *Contact Angle, Wettability and Adhesion*, Fowkes F. M. ed. (ACS Adv. Chem. Ser. 43, Am. Chem. Soc., 1964) 1.
31. Klemberg-Sapieha J. E. in *Plasma Processing of Polymers,* d'Agostino R., Favia P., Fracassi F. eds. (Kluwer Acad. Publ., NATO ASI Series, E: Appl. Sci., Vol. 346, 1997) 233.
32. Castner D.G., Favia P., Ratner B. D., in *Surface Modifications of Polymeric Biomaterials*, Ratner B. D., Castner D.G. eds. (Plenum Press, 1996) 45.
33. Butoi C. I., Mackie N. M., Gamble L. J., Castner D.G., Barnd J., Miller A. M., Fisher E. R., *Chem. Mater.* **12** (2000) 2014.
34. Lau K. K. S., Gleason K. K., *Mat. Res. Soc. Symp. Proc.* **544** (1999) 209.
35. Lau K. K. S., Gleason K. K., *J. Electrochem. Soc.* **146** (1999) 2652.
36. Coburn J .W., Chen M., *J. App. Phys.* **51** (1980) 3134.
37. d'Agostino R., Cramarossa F., De Benedictis S., Ferraro G., *J. Appl. Phys.* **52** (1981) 1259.
38. d'Agostino R., Cramarossa F., Illuzzi F., *J. Appl. Phys.* **61** (1987) 2754.
39. d'Agostino R., Cramarossa F., De Benedictis S., *Plasma Chem. Plasma Proc.* **2** (1982) 213.

40. d'Agostino R., Favia P., Fracassi F., *J. Polym. Sci. A: Polym. Chem.* **28** (1990) 3387.
41. Cruden B. A., Gleason K. K., Sawin H. H., *J. Appl. Phys.* **89** (2001) 915.
42. Truesdale E. A., Smolinsky G., *J. Appl. Phys.* **50** (1979) 6594.
43. Lamendola R., Favia P., d'Agostino R., *Plasma Sour. Sci. Tech.* **1** (1992) 256.
44. d'Agostino R., Lamendola R., Favia P., Gicquel A., *J. Vac. Sci. Tech. A* **12** (1994) 308.
45. Chapman B., *Glow Discharge Processes* (Wiley, 1980) 160.
46. Kohler K., Coburn J. W., Horne D. E., *J. Appl. Phys.* **57** (1984) 59.
47. Fracassi F., in *Plasma Processing of Polymers,* d'Agostino R., Favia P., Fracassi F. eds. (Kluwer Acad. Publ., NATO ASI Series, E: Appl. Sci., Vol. 346, 1997) 47.
48. Zabeida O., Klemberg-Sapieha J. E., Martinu L., Morton D., *Mat. Res. Soc. Symp. Proc.* **544** (1999) 233.
49. Koenig H. R., Maisel L. I., IBM J. Res. Dev. **14** (1970) 168.
50. Fracassi F., Coburn J. W., *J. Appl. Phys.* **63** (1988) 1758.
51. Fracassi F., Occhiello E., Coburn J. W., *J. Appl. Phys.* **62** (1988) 3980.
52. Yasuda H., Hirotsu T., *J. Polym. Sci., Polym. Chem. Ed.* **16** (1978) 743.
53. Morosoff N., in *Plasma Deposition, Treatment and Etching of Polymers*, d'Agostino R. ed. (Acad. Press, 1990) 1.
54. Chen R., Gorelik V., Silverstein M. S., *J. Appl. Polym. Sci.* **56** (1995) 615.
55. Silverstein M. S., Chen R., Kesler O., *Polym. Eng. Sci.* **36** (1996) 2542.
56. Yasuda H., Hsu T., *J. Polym. Sci., Polym. Chem. Ed.* **15** (1977) 81.
57. Vinzant J. M., Shen M., Bell A. T. in *Plasma Polymerization* (ACS Symp. Series 108, 1979) 79.
58. Limb S. J., Edell D. J., Gleason E. F., Gleason K. K., *J. Appl. Polym. Sci* **67** (1998) 1489.
59. Labelle C. B., Limb S. J., Gleason K. K., *J. Appl. Phys.* **82** (1997) 1784.
60. Han C. M., Timmons R. B., Lee W.W., *J. Vac. Sci. Tech. B* **18** (2000) 799.
61. Mackie N. M., Castner D.G., Fisher E. R., *Langmuir* **14** (1998) 1227.
62. Coulson S.R., Woodward I., Badyal J. P. S., Brewer S.A., Willis C., *J. Phys. Chem. B* **104** (2000) 8836.
63. Labelle C. B., Gleason K. K., *J. Vac. Sci. Tech. A* **17** (1999) 445.
64. Labelle C. B., Gleason K. K., *J. Appl. Polym. Sci.* **74** (1999) 2439.
65. Limb S. J., Gleason K. K., Edell D. J., Gleason E. F., *J. Vac. Sci. Tech. A* **15** (1999) 1814.

66. Panchalingam V., Chen X., H. H. Huo, Savage C. R., Timmons R. B., Eberhart R. C., *ASAIO J.* **39** (1993) M305.
67. Cicala G., Favia P., Palumbo F., Milella A., d'Agostino R., unpublished results.
68. Timmons R. B., private communications.
69. Haverlag M., Stoffels W. W., Stoffels E., Kroesen G. M. W., de Hoog F. J., *J. Vac. Sci. Tech. A* **14** (1996) 384.
70. Booth J. P., Cunge G., Chabert P., Sadeghi N., *J. Appl. Phys.* **85** (1999) 3097.
71. Lau K. K. S., Gleason K. K., Trout B. L., *J. Chem. Phys.* **110** (2000) 4103.
72. Gleason K. K., Limb S. J. H., Gleason E. F., Sawin H. H., Edell D. J., *US patent* 5.888.591.
73. Lee W. W., Tyndall G., Zehringer R., Crowder, *Mat. Res. Soc. Symp. Proc.* **544** (1999) 161.
74. Chen W., Fadeev A. Y., Hsieh M. C., Oner D., Youngblood J., Mc Carthy T. J., *Langmuir* **15** (1999) 3395.
75. Coulson S. R., Woodward I. S., Badyal J. P. S., Brewer S. A., Willis C. R., *Chem. Mat.* **12** (2000) 2031.
76. Badyal J. P. S., Coulson S. R., Willis C. R., Brewer S. A., *WO patent* 98/58117.
77. d'Agostino R., Corzani I., Favia P., Lamendola R., Palumbo G., *WO patent* 14.297.
78. Gegenbach T. R., Griesser H. J., *Surf. Interf. Anal.* **26** (1998) 498.
79. Ratner B D., Chilkoti A., Lopez G. P., in *Plasma Deposition, Treatment and Etching of Polymers*, d'Agostino R. ed. (Acad. Press, 1990) 463.
80. Clarotti G., Schue F., Sledz J., Aoumar A. A. B., Geckler K. E., Orsetti A., Paleirac G., *Biomaterials* **13** (1992) 832.
81. Bohnert J. L., Fowler B. C., Horbett T. A., Hoffman A. S., *J. Biomat. Sci. Polym. Ed.* **1** (1990) 279.
82. Garrison M. D., Luginbuhl R., Overney R. M., Ratner B. D., *Thin Solid Films* **352** (1999) 13.
83. Shi H., Tsai W., Garrison M. D., Ferrari S., Ratner B. D., *Nature* **398** (1999) 593.
84. Ratner B. D., Shi H., *Current Opinions in Solid State and Material Science* **4** (1999) 395.

CHAPTER 3

DEPOSITION OF SILICON CONTAINING FILMS AND FTIR DIAGNOSTICS

YVAN SEGUI AND PATRICE RAYNAUD
Laboratoire de Génie Electrique,
CNRS/Université P. Sabatier
Toulouse, France

3.1. Introduction

Silicon dioxide films deposition is a key step in very large scale integration (VLSI) technology. Years ago, thermal oxidation of silicon met all the requirements; then, the process temperatures needed to be much lower (typically < 400 °C) and also dielectric deposition on surfaces other than silicon was needed (eg. intermetallic insulation). These new requirements were a driving force for PECVD (Plasma Enhanced Chemical Vapor Deposition) processes. In a first step, mixtures of silane (SiH_4) and an oxygenated gas (O_2, O_3, N_2O) were successfully used. The limits of the silane technology happened with the submicronic sizes and high aspect ratios (height / spacing) features. In these conditions, silane offers a poor step coverage. Organosilicon precursors in general and especially TEOS (Tetraethoxysilane) provided a new and promising way. Moreover, safety, easy handling and chemical stability provide additional interest for its use.

Besides microelectronic industry, the remarkable properties of plasma deposited silicon oxide films made this material very interesting for many large scale applications. Low gas and moisture permeability offer the opportunity to grow very thin (thinner than 20 nm), efficient gas barriers. This property can be applied to food or pharmaceutical packaging [1-3] or anti-corrosion treatments [4-5]. Silicon oxide like films can also be used as hard coatings on soft materials eg. PMMA or polycarbonate [6]. Ophthalmic lenses can be protected by such anti-scratch layers. The main interest is due to the combination of hardness and optical transparency in the visible range. More recently, very promising results were published in the field of bioengineering in particular for protein adsorption

[7]. Briefly speaking, an improvement of surface properties of different materials like glasses, polymers or metals can be obtained [8-10].

These emerging applications present some new requirements: needs to treat large areas or on line processes and also the substrate is no longer a silicon based material and consequently interface problems (Adhesion, diffusion, compatibility of thermal expansion coefficients) arise. Organosilicon precursors instead of SiH_4 provide some opportunities to solve these problems. This is due to the organic part of the precursor. As will be indicated below, it is always possible to preserve some organic groups by adjusting the ratio O_2/ Organosilicon, then a *soft* transition between a polymeric surface and a pure silicon oxide is possible. To do that, it is necessary to have good control and, if possible, a good understanding of the chemistry taking place in the gas phase and at the interface between plasma and substrate. It should be pointed out, this chemistry is very complex (more than 120 different species have been identified by Tachibana et al. in a CH_4 plasma [11]) and changes as a function of external parameters like pressure, flow, power or gas composition.

An excellent and exhaustive review of plasma processes in organosilicon and organo metallic vapors by A. Wrobel and M. R. Wertheimer was published in 1990 [12]. In this chapter we consider the situation in which an oxygen carrier gas is mixed with organosilicon monomer.

After a brief presentation of the most often used organosilicon molecules, the nature of species present in the gas phase and their evolution as a function of the external parameters will be discussed. Special attention will be paid to Infra-red absorption spectroscopy of the gas phase which is a relatively unusual but powerful tool compared to the more conventional mass or optical emission spectroscopy. Finally, a short discussion on the remaining open questions will be presented.

3.2. Precursors

Organosilicon molecules is a generic name for any molecule consisting of at least one atom of silicon and at least one organic group, like a methyl group CH_3. This family is obviously quite large, however a rapid overview of the literature shows that only few members of this family have been used for PECVD processes, leading to silicon oxide or silicon nitride deposition. The reasons to so drastically reduce the number of organosilicon monomers candidate to act as precursors for PECVD processes are based on rather simple criteria:

– For the sake of simplicity, the precursor should be a gas or a liquid at room temperature and atmospheric pressure. Solid precursors are not strictly excluded but the consequence is a very complex process involving in a first step the evaporation of the starting material.
– For liquid precursors, which is in fact the case of molecules with two or more silicon atoms, this liquid should have a vapor pressure high enough to proceed at room temperature or close to room temperature. In that case it is relatively easy to introduce the vapors of the monomer in the reactor by simply pumping down the chamber. This condition reduces our choice to molecules with no more than two silicon atoms, eventually four, for cyclic molecules.
– Precursors should be commercially available at a reasonable cost.
– Relatively often the objective is a deposition of SiO_2 or Si_3N_4 like films. For this reasons it is preferable that the environment of the silicon atom in the precursor gives the best chances to built a tetrahedral structure for the deposited film.

For all these reasons a very large majority of the published results (maybe more than 90 %) are related to two starting molecules: tetraethoxysilane also called tetraethylorthosilicate (TEOS) and hexamethyldisiloxane (HMDSO). For silicon carbide deposition, which is out of the scope of this chapter, tetramethylsilane (TMS) is largely used.

The list below gives of the main precursors involved in PECVD processes.

1. Hexamethyldisiloxane — HMDSO — $(CH_3)_3Si\text{-}O\text{-}Si(CH_3)_3$
2. Tetraethoxysilane — TEOS — $(C_2H_5\text{-}O)_4Si$
3. Tetramethyldisiloxane — TMDSO — $H\text{-}Si(CH_3)_2\text{-}O\text{-}(CH_3)_2Si\text{-}H$
4. Divinyltetramethyldisiloxane — DVTMDSO — 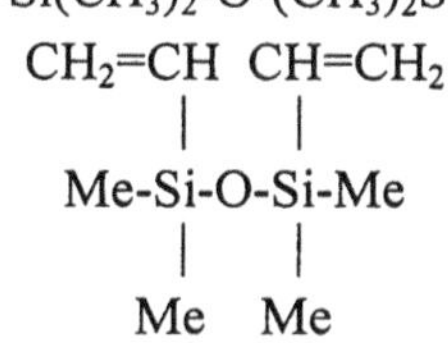
5. Methyltrimethoxysilane — MTMOS — 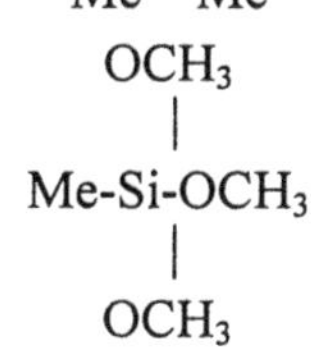
6. Octamethylcyclotetrasiloxane — OMCATS
7. Bis(Trimethylsilyl)methane — BTMSM — $(CH_3)_3Si\text{-}CH_2\text{-}Si(CH_3)_3$

8. Hexamethyldisilane	HMDS	$(CH_3)_3Si\text{-}Si(CH_3)_3$
9. Tetramethylsilane	TMS	$Si(CH_3)_4$
10. Hexamethyldisilazane	HMDSN	$(CH_3)_3Si\text{-}NH\text{-}Si(CH_3)_3$
11. Tris(dimethylaminosilane	TDAS	$[(CH_3)_2N]_3Si\text{-}H$
12. Tetrakis(dimethylamino)silane	TDMAS	$(N(CH_3)_2)_4Si$
13. Hexamethylcyclotrisilazane	HMCTSN	

From 1 to 6, the precursors present at least one Si-O bond in their structure; the last one (number 6) is cyclic. Precursors 1 and 4 are very similar; HMDSO is the most utilized, however TMDSO presents the advantage of having a weak Si-H bond easily dissociable in the plasma and, consequently, an expected higher deposition rate (Si-C: 104 kcal/mol; Si-H: 72 kcal/mol).

From 7 to 9, the monomer does not contain any oxygen or nitrogen atom thus addition of oxygenated or nitrogenated gas is necessary to form Si-O or Si-N bonds in the deposited layer.

From 10 to 13, all the precursors contain at least one Si-N bond in their chemical structure; the last one (number 13) is cyclic.

All the precursors listed here have been used under plasma conditions. The Table I below gives at least one of the plasma conditions employed for the deposition.

Let us now focus our interest on the two molecules most often used: TEOS and HMDSO.

Their chemical structure is shown below. TEOS exhibits the tetraedric silicon environment similar to that met in silicon dioxide. It can thus be expected to give favourable conditions to give good oxide films, provided that the structure of the monomer was preserved. HMDSO exhibits a Si-O-Si bridge which is particularly stable and, again, a preservation of this unit is expected.

Table 3.1. Plasma conditions for the organosilicon monomers listed above.

Precursors	Plasma System	Pressure/ Flow Rate	Power	Ref
1	RF, μm, LF	10^{-3} to 1 mbar	3W-100W	[13-17]
2	RF, μW	10^{-3} to 1 mbar	3W-100W	[18-22]
3	13.56 MHz, Inductive Coupling	1.3 Pa	25 W	[23]
4	13.56 MHz,	10^{-2} - 10^{-1} mbar	14-200 W	[24]

	Capacitve coupling Remote plasma			
5	13.56 MHz, Parallel Plate	0.110 Torr	300 W	[25]
6 (id 4)				
7	13.56 MHz, Inductive Coupling	1.3 Pa	50-150 W	[1]
8 (id 7)				
9 (id 3)				
10	13.56 MHz Extern. Electr.	0.25 Torr	7 W	[26]
11	ECR (2.45 GHz)		1000 W	[27]
12	CVD + Ozonator	300 cm^3/mn Pa	Ts = 40 °C	[28]
13	20 KHz Parallel Plate	0.2 Torr	2mA/cm^2	[29]

Figure 3.1 gives the vapor pressure of those two molecules as a function of temperature. It can be noted that HMDSO is a more volatile product than TEOS. An ambient temperature process is possible with the former whereas a heating unit will be necessary with TEOS.

Both are liquid at room temperature and atmospheric pressure, commercially available and relatively cheap. As stated in the introduction an important advantage is their chemical stability allowing easy handling and storage.

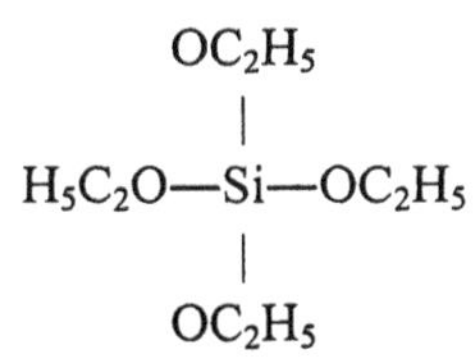

TEOS

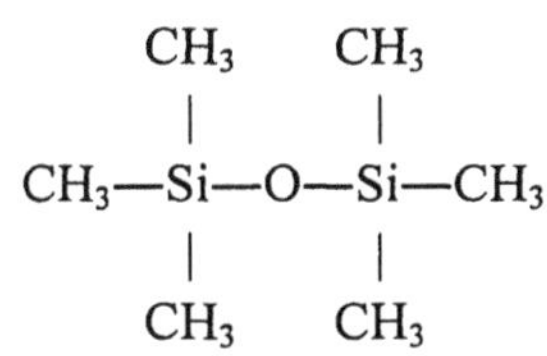

HMDSO

Oxygen or nitrogen carrier gases are added to organosilicon molecules. Most often, O_2 or N_20 are used. In some rare cases, due to some thermal CVD experiments, ozone O_3 can be introduced [30].

3.3. Species Detected in Organosilicon Plasmas

The aim of this section is to give the nature of the main species detected by Optical Emission Spectroscopy, Mass Spectroscopy and FTIR analysis of the gas phase. The two first diagnostic techniques are widely used in plasma processes and no comments will be added on the diagnostic itself. The third one is not so usual and some comments on advantages and limitation of this technique will be presented.

In all cases an attempt to point out the general trends for the variation of the concentration of the main species as a function of external parameters, power, pressure, flows, etc. will be made.

3.3.1. Optical Emission Spectroscopy (OES)

For several reasons, OES in the visible and near UV range is probably the most popular diagnostic technique: basic equipment is not very expensive, experiments are relatively easy, and a rather large body of results are available in the literature as well as handbooks for emission lines identifications.

Applied to organosilicon/oxygen mixtures, most of the results identify the same species. Differences can be shown when considering the variations of the emission lines as a function of parameters like gas composition, power, pressure.

Table 3.2 gives the species most often identified in optical emission

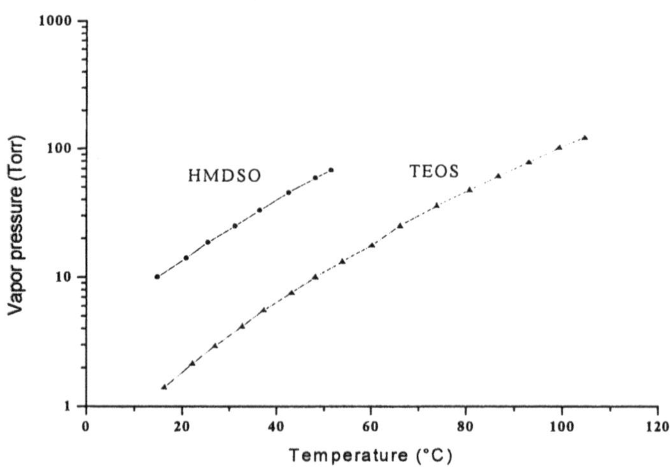

Figure 3.1. Vapor pressure of TEOS and HMDSO as a function of temperature.

spectra and the corresponding wavelength. Strictly speaking, each single specie gives rise to many emission peaks, details can be obtained in handbooks. The wavelengths reported in the table below are those falling in the wavelength range covered by the conventional experimental apparatus.

Table 3.2. Main species identified in a Organosilicon/oxygen plasma.

Species		Wavelength (nm)
H_α	Balmer system	656.2
H_β		486.1
H_γ		434
H_δ		410.1
C		193.1, 247.8
C_2		516.5
CH		431.4
CO		266
O		844
OH		282; 306
SiO		252; 216; 292
H_2		752

The wavelength of respective species can be found in handbooks. However the values reported here were taken from papers by Benissad et al. [31]; Aumaille et al. [32] and from the author's own work.

These species can be generated either by direct electronic impact on gas phase molecules or by recombination at the surface or in the gas. It is not an easy task to distinguish between these two pathways. A comparison of the results from two experimental conditions can be a way; for example Tochitani et al. [33] have shown that C_2, CO, CO_2 and CO^+ are not visible if N_2 is used instead of O_2 and in that case carbon is found in the film. Their conclusion was that carbonated species were not generated by direct electronic impact and resulted more likely from surface reactions although there is no evidence that the electron energy distribution function was the same in the both cases. The other question was the origin of the oxygen detected in combination or not with carbon - organosilicon molecule vs. molecular oxygen. OES cannot give a definitive answer on this point but interesting results have been reported by Flamm et al. [34] who used ^{18}O and mass spectroscopy. A short comment will be given in the next section on these results.

Most of the papers report the variations of Si, SiO, H, OH, C as a function of the ratio oxygen (or other gas acting as oxygen carrier) / organosilicon. Results can be different from one author to the other due to different plasma generation systems or pressure or power range. However some general trends can be pointed out.

- CH intensity is decaying as the oxygen content increases.
- H, OH, C and CO intensities increase first as oxygen content increases and then decrease beyond a maximum occurring between 40 and 70 % of O_2.

SiO emission was also found increasing as O_2 increased. The formation of this species through a heterogeneous reaction at the surface can be ruled out because no detection was reported if an O_2 plasma is run in a reactor contaminated by plasma organosilicon deposited on the walls [35].

It is difficult to draw conclusions on the chemistry in the gas phase from OES as some authors did. Firstly, OES is a measurement of excited species which are generally light and not necessarily key species for the deposition process; actinometry was not always validated for the detected species (compared to fluorinated plasmas, almost no LIF validation was reported for Si or C containing species). Secondly, changing the composition of the gas phase induced some changes in the electronic density and EEDF. Thirdly, results can be drastically changed due to the contamination of the reactor walls.

Besides the analysis of excited species OES, is a very efficient tool for monitoring plasma processes. Pai et al. [36] showed first that the stoichiometry of plasma deposited organosilicon where related to some emission lines. More recently Lamendola et al. [35] related for the first time an emission ratio, namely I_{CH}/I_{AR}, not only to the carbon content in the plasma film but also to the oxygen barrier property.

3.3.2. Mass Spectroscopy

Mass spectroscopy is not, strictly speaking, an in-situ technique. It is a sampling method; a small hole connects the mass analyzer to the plasma chamber. This small hole is necessary to maintain the difference between the working pressure in the plasma reactor ranging between 10 mBar to 10^{-3} mBar and the pressure in the mass analyzer is around 10^{-7} mBar. Generally, an electron beam ionized the gas sample and the nature of the ions can be identified through their weight. If the electron beam is turned off, the neutral species created in the plasma are not detected. Even this simple (too simple?) description raises several questions which can be considered to avoid erroneous conclusions.

- Some chemical reactions may occur during the transport from the plasma reactor to the mass analyzer. In particular, short lived species can recombine. In some experiments, the detection is made through an orifice located at the substrate level.
- The mass dependent transmission factor of the orifice is generally unknown and needs a calibration procedure before measurements.
- The electron beam itself cannot only ionize the existing species but also create new species by breaking some bonds. Two ways are used to control these artifacts: some equipment allows a change of the energy of the electron beam, then several mass spectra are recorded as a function of the energy. In all cases, it is highly recommended to compare the mass spectra with and without plasma to have insight on the species created by the plasma itself. For more details about this method see Chapter5.

Relatively few papers on mass spectroscopy of organosilicon/O_2 plasmas have been published so far. Basic experiments in very well controlled conditions are reported by the "Greiswald Group" [37- 40]. To the best of our knowledge, it is the unique reports on ionization cross sections of HMDSO and TEOS, pure or mixed with oxygen. An extended work, unfortunately not yet entirely published, was made by K. Aumaille [41 thèse Karine].

Table 3.3 is a summary of the main species, neutral and ions, reported in the literature. These results come from Fracassi et al. [42], Latreche et al. [43,44], Horii et al. [45], Nothe et al. [46] and Aumaille et al. [47]. They are related to TEOS/O_2 or HMDSO /O_2 plasmas.

Table 3.3. Summary of the main species (neutrals and ions) from Organosilicon/O_2 plasmas [42-47].

m/e	Ion	m/e	Ion	m/e	ion
2	H_2^+	*43*	$C_2H_3O^+$	107	$(OH)_2Si(OC_2H_5)^+$
12	C^+	44	CO_2^+, SiO^+, $C_2H_4O^+$	111	$(OH)_3SiOCH_3^+$
13	CH^+	45	$C_2H_5O^+$, $SiOH^+$	**117**	$(CH_3)_3SiOSi^+$
14	CH_2^+	46	$C_2H_6O^+$, $HsiOH^+$	*119*	$(C_2H_5O)_2SiH^+$
15	CH_3^+	47	$SiOH_3^+$	125	$SiO_2CH_5^+$
16	O^+, CH_4^+	**58**	$Si(CH_3)_2^+$	**131**	$(CH_3)_3SiOSiCH_2^+$
17	OH^+	61	$SiOCH_5^+$, SiO_2H^+	*135*	$(C_2H_5O)_2Si(OH)^+$

18	H_20^+	*63*	$Si(OH)_2H^+$	**147**	$(CH_3)_3SiOSi(CH_3)_2^+$
19	H_3O^+	**66**	$(CH3)_2Si_2O^{++}$	*149*	$(C_2H_5O)_2Si(OCH_3)^+$
26	$C_2H_2^+$	**73**	$(CH_3)_3Si^+$,$SiOSiH^+$	153	$SiO_5C_3H_9^+$
27	$C_2H_3^+$	*79*	$SiOH_3^+$	*163*	$Si(OC_2H_5)_3^+$
28	CO, $C_2H_4^+$, Si^+	**87**	CH_3SiOSi^+	*179*	$(C_2H_5O)_3SiO^+$
29	$C_2H_5^+$, CHO^+	*89*	$OSi(C_2H_5)^+$	*181*	$SiO_5C_5H_{13}^+$
30	$C_2H_6^+$, H_2CO	91	$OHSi(C_2H_5)^+$	*193*	$(C_2H_5O)_3Si(OCH_2)^+$
31	CH_3O^+, SiH_3^+	97	O_3SiOH^+	*207*	$(C_2H_5O)_3Si(C_2H_4)^+$
32	O_2^+	**102**	$(CH_3)_2SiOSi^+$	*208*	$(C_2H_5)_4Si^+$

In this table, bold characters are used for peaks only detected for HMDSO Plasmas, italic characters for peaks only detected for TEOS plasmas, normal characters for peaks detected for both precursors.

Figure 3.2 shows a mass spectra in HMDSO/O_2 plasmas without discharge (Fig. 3.2-a) and at the applied microwave power of 100 W (Fig 3.2-b). The most important feature is the strong increase of low molecular weight mass products

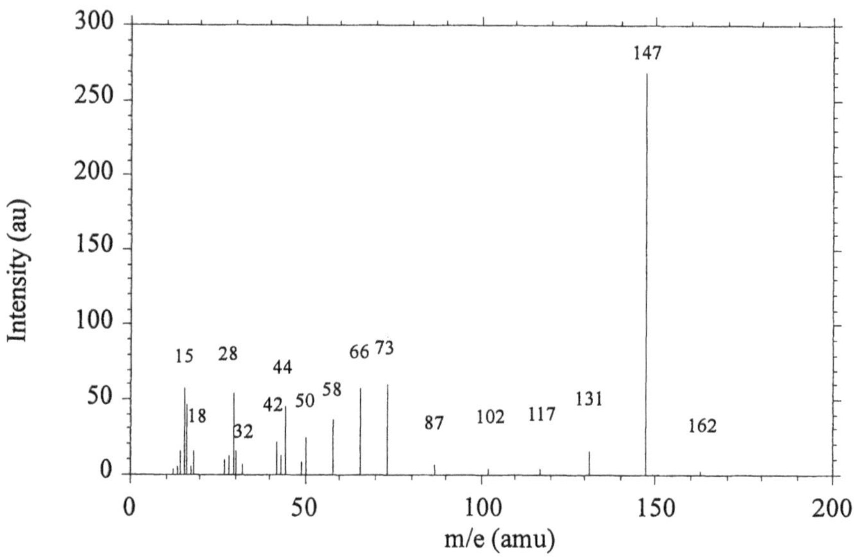

Figure 3.2-a. Mass spectra of a O/HMDSO mixture, discharge off from [44].

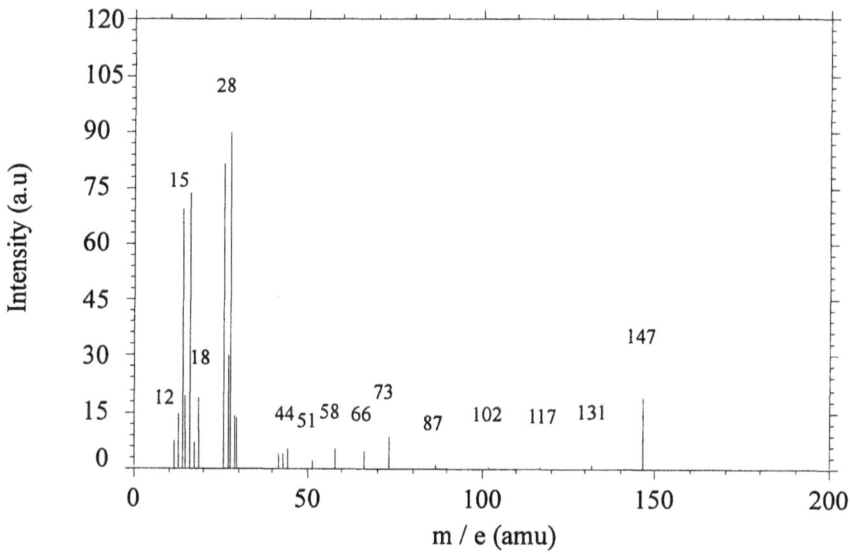

Figure 3.2-b. Mass spectra of a O_2/HMDSO plasma from [44].

when the discharge is on.

In some cases, peaks can be attributed to several species. The ambiguity can be ruled out if the equipment makes possible a variation of the energy of the electron beam. In that case, it is possible to identify the energy threshold of the ionic current which is specific to a given species. The database proposed by NIST [48] allows, in most cases, a proper identification (e.g. 28 amu can be attributed to CO or C_2H_2. However the threshold for CO is 14 eV whereas C_2H_2 appears at 10.5 eV). It was observed that increasing the power, the contribution of CO to the peak at 28 amu increases [41].

The parent peaks of TEOS and HMDSO appears at 208 and 162 amu respectively. A typical mass spectra can be described considering three zones.

The high molecular weights: parent ions of TEOS or HMDSO or ions related to these molecules dissociated by electron impact with the release of CH_3 (15 amu), C_2H_5 (29 amu) or OC_2H_5 (45 amu) and creation of TEOS or HMDSO fragments at 193, 179 and 163 for TEOS or 147, 131 and 117 for HMDSO.

These high weight species almost disappears from the spectra by increasing the power (which increases the electron density) or increasing the molecular oxygen contents in the gas mixture (which indicates that atomic oxygen also plays also an important role for the dissociation). The variation of

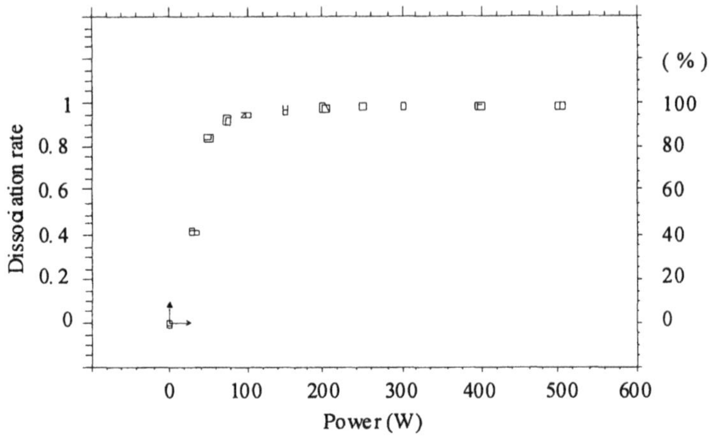

Figure 3.3. Dissociation rate of HMDSO in a microwave O_2/HMDSO plasma as a function of the microwave power from [43].

the amplitude of parent peaks gives the overall dissociation rate of TEOS or HMDSO in the plasma. Figure 3.3 below shows how the dissociation factor changes as a function of the power for a microwave DECR plasma.

The intermediate species, 30 to 45 amu, are attributed to some hydrocarbons partially or totally oxidized. Increasing the power, these species also tend to disappear.

Light species are directly created by the electronic impact of TEOS or HMDSO or also through a second step of dissociation, from the intermediate species. The intensity of all these peaks strongly increases when the plasma was on and when the power increases. Very few silicon containing light products have been detected. This fact is in good agreement with OES data.

One of the advantages of mass spectroscopy is the possibility to separate the isotopes of the same element. Then some isotopic labeling studies have been reported and their main goal was to gain insight on the role of oxygen in the deposition process, to know whether these atoms originates from the organosilicon or molecular oxygen. Then some authors [34, 45] using $^{18}O_2$ and mass spectroscopy were able to distinguish ^{16}O from ^{18}O in the oxygenated sampled species e.g. CO, CO_2, H_2O. Their conclusions were as follows.

Most of the intermediate mass number species can be explained by oxidative removal of hydrocarbon groups from TEOS: detection of $C^{18}O$, $C^{18}O_2$, $H_2{}^{18}O$ and etc. and $C^{16}O^{18}O$ was detected suggesting Si-O bond rupture. This is also in agreement with the detection of both ^{18}O and ^{16}O in the gas.

Oxygen atoms involved in Si-O bonds in the deposited films originate from the precursor.

3.3.3. Infrared Absorbtion Spectroscopy of the Plasma

The infrared (IR) range of light looks particularly promising because most of the species, heavy or light, have strong signatures in this wave range. Monomer dissociation, new species formation or powder formation can be studied [49-55]. Although emission or absorption spectra can be obtained, however, this presentation will focus on absorption spectroscopy. The IR absorption spectroscopy can be performed differently according to the nature of the IR source used, namely IR diode laser or the broadband sources of Fourier Transform Infrared spectrometers (FTIR).

IR diode laser spectroscopy presents the same advantages as LIF regarding time and space resolution. High wavelength resolution (10^{-3} cm^{-1}) makes possible the separation of rotational levels. Although this presentation focuses on FTIR absorption spectroscopy, it should be noted that excellent results where obtained by Magane et al. [56] probing SiH_4, CF_4 and CHF_3 plasmas by IR diode laser. They obtained quantitative evaluation of CF, CF_2, CF_3 and SiH_3 radical-concentrations. Moreover, examination of transients gave an estimation of the diffusion coefficients of these radicals. IR diode laser spectroscopy also has some drawbacks. Only one specie can be monitored and the diode has to be changed if another wavelength is to be monitored. These drawbacks are overcome if FTIR spectrometry is used. The counterparts are relatively low photon flux and some wavelength resolution problems occurring if a low quality spectrometer is used. Moreover, from a single spectrum, quantitative data belonging to one or more species can be extracted. FTIR complements the other optical diagnostics. Species not detected by FTIR absorption spectroscopy (short lived species, atoms...) are those most easily detected by OES or LIF.

3.3.3.1. Principles of FTIR Diagnostic of the Plasma Phase

In order to perturb as little as possible the measurement in a deposition plasma process, fast analyses are necessary. Infrared in situ analyses are achieved through windows which can be coated during the deposition. Most of the IR spectrometers commonly used are Fourier Transform Infrared spectrometers

(FTIR) allowing to acquire spectrum in a wide range of wave-number (i.e. 400 – 4000 cm^{-1}) and short time (< few seconds). Fourier Transform infrared spectroscopy is based on Michelson interferometer and presents the following main advantages:

If it is needed to record a spectrum for T seconds in a region (n_{max}-n_{min}), with a resolution d_n, the number of spectral elements will be M=(n_{max}-n_{min})/D_n.

With a mono-channel grating system, the signal and noise detected for each element are S and N respectively. With an interferometer, the signal and noise detected are S' and N' respectively, the signal to noise ratio is improved: S'/N' = $M^{1/2}$ (S/N). As a consequence, a spectrum of the same quality is obtained more rapidly with an interferometer than with a grating device. The larger the value of M, the more this advantage is emphasized (high resolution or extensive spectral range). This is the "Multiplexing" or Fellgett's advantage.

The second one is the Luminous flux or Jacquinot advantage. It allows to work with a smaller flux emitted by the source, which is very important for emission spectra where the source is replaced by the emitting body to be studied or for transmission spectrometry at high resolution.

3.3.3.2. The Absorption Spectra

Basic features of a light beam absorption are explained below. More details can be found in various books [57,58].

A molecule with energy states E_i and E_f is characterized by the wave functions Ψ_i and Ψ_f respectively. When the molecule receives an electromagnetic radiation, a molecular excitation is not necessarily produced. There is a probability of transition from the state i to the state f, $[A]_{if}$ defined as:

$$[\mathrm{A}]_{if} = \int_{-\infty}^{\infty} \psi_{\mathrm{f}}^{*} |\mathrm{A}| \psi_i \mathrm{d}q \tag{3.1}$$

q being the internal coordinate of the system and A, the operator involved in the phenomenon of interaction. In the particular case of the absorption phenomenon, the associated operator is the dipolar moment (μ) of the molecule. If the moment of transition equals zero, the probability of transition is zero as well. The transition is permitted if $[\mu]_{i,j} \neq 0$ when the molecule has a permanent dipolar moment ($\mu \neq 0$) or also if there is a variation of the dipolar moment associated with the vibration of the molecule ($\delta\mu/\delta q \neq 0$).

3.3.3.3 Quantitative Analysis (Absorption Spectrometry)

The well known Bouguert-Beer-Lambert's law is a relation between the intensity of IR absorption bands and the concentration of the absorbent molecule. Consider a parallel and monochromatic light beam (wavelength ν) going through an homogeneous sample composed of a cell of thickness "e" with parallel sides (Fig. 3.4).

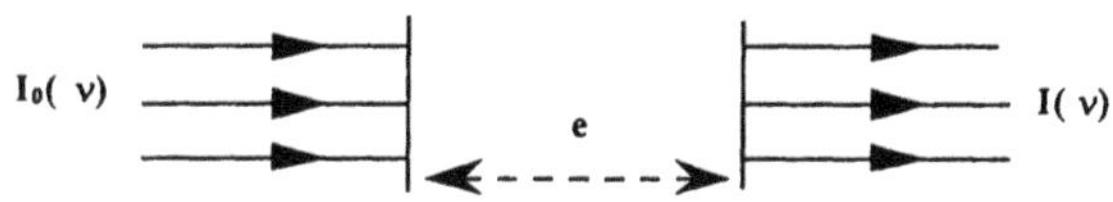

Figure 3.4. Schematic diagram used for Beer-Lambert's law.

$I_0(\nu)$ is the light energy per unit of time and surface and $I(\nu)$ the transmitted energy. The attenuation of the beam due to the absorption of the sample is given according to Beer-Lambert law:

$$(I_0/I)_\nu = \exp(-K(\nu)\cdot C\cdot e) = T \text{ (transmittance) at } \nu \text{ wavelength.}$$

C is the concentration of absorbent species, $K(\nu)$ is the molar absorption coefficient in $mole^{-1} \cdot l \cdot cm^{-1}$ in the case of liquids and in $atm^{-1} \cdot cm^{-1}$ in case of gases. An equivalent formalism is often preferred:

$$\log_{10} (1/T) = A(\nu) = \log (I_0/I)_\nu = \varepsilon(\nu)\cdot C\cdot e = \text{Absorbance.}$$

$$\varepsilon(\nu) = K(\nu) / 2.3 = \text{extinction coefficient.}$$

This is the basic equation of quantitative measurements in Infrared spectrometry.

Example of the experimental apparatus is in the following. From the Beer-Lambert law follows that the pressure will determine the path length necessary for the detection of species. The higher the path length the higher will be the absorbance. For pressure around 50 mTorr and a path length of few centimeters will be enough, but for pressure around 10 mTorr or less it will be necessary to increase the path length with an optical system such a White cell [59]. This is an optical cavity composed of 3 mirrors allowing multiple reflections on the mirrors and so, multiple pass in the plasma which absorb more light. Examples of experimental set-ups are shown on Figs. 5a and 5b. In such experimental systems, the unwanted coating of optical windows and mirrors due to plasma deposition remains the main problem. It can be solved or attenuated with some precautions. Windows and/or mirrors have to be far from the plasma source. An inert gas flow in front of the optical components can protect them, but in some cases, a modification of the plasma due to this gas can occur. In all cases to take into account the deposit on the optical components it is necessary to record an IR spectra of the reactor under vacuum, before and after each measurement

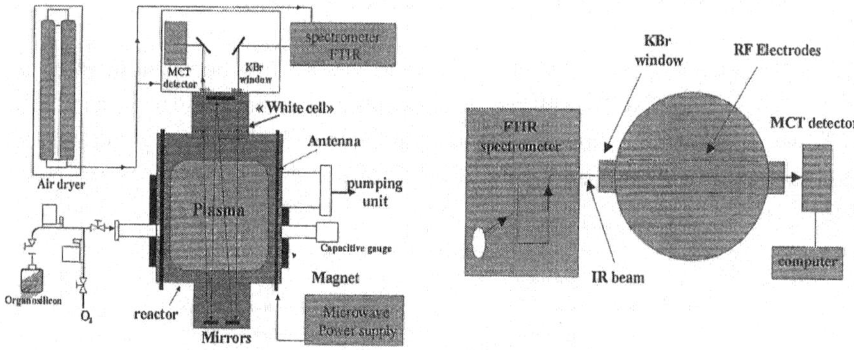

Figure 3.5. Experimental Set-up for (a) low (left) and (b) high pressure (right) plasma FTIR analysis.

under plasma condition. Moreover for transmittance or absorbance measurement, "Io" can be the intensity detected after one pass (high pressure plasma) or multipass (low pressure plasma) in a plasma reactor under vacuum or empty of gas and "I" the intensity detected after one or multipass in the reactor under plasma phase condition. By this way it will be possible to detect depletion of monomers, creation of new species and the other phenomena such as scattering.

3.3.3.4 Infrared Diagnostics of Organosilicon Plasma

Raynaud et al. [60] studied the dissociation of 3 organosilicon monomers in a low pressure Multipolar Microwave Plasma (MMP) reactor excited by Distributed Electron Cyclotron Resonance (DECR) reactor (see Fig. 3.5a) by infrared absorption. The first one being HMDSO [$O(Si(CH_3)_3)_2$] which is the only one to present a SiOSi structure. The second one is Tetramethylsilane (TMS) [$Si(CH_3)_4$] which is the only one without oxygen and finally the tetramethoxysilane [$Si(OCH_3)_4$]. The first result of this study is the high efficiency of the dissociation of the parent molecules in pure organosilicon plasmas. A second result is the appearance of organic molecules. Concerning HMDSO and TMS plasmas, differences have been detected between spectra recorded with and without oxygen. For the first time, the IR spectroscopy has allowed the detection of the main groups of vibration (SiO, CH, SiC,etc.) of the monomers such as HMDSO (Fig. 3.6). An example of decomposition in a pure organosilicon plasma has been analyzed in Fig. 3.7 and shows that even at the minimum power needed for plasma ignition (50 W), the IR absorption by

HMDSO, for example, has decreased drastically. Above 100 W, the fragmentation of all these monomers is virtually complete. The easier total dissociation of TEOS and TMS compare to HMDSO has also been shown. This reflects the stronger chemical bonding in HMDSO. Previous work [61] corroborates the easy destruction of HMDSO and TEOS in DECR plasmas: the mass spectrometric analysis of the neutral species in those plasmas shows an equally fast decrease of the mass peaks representative of the parent HMDSO and TEOS molecules.

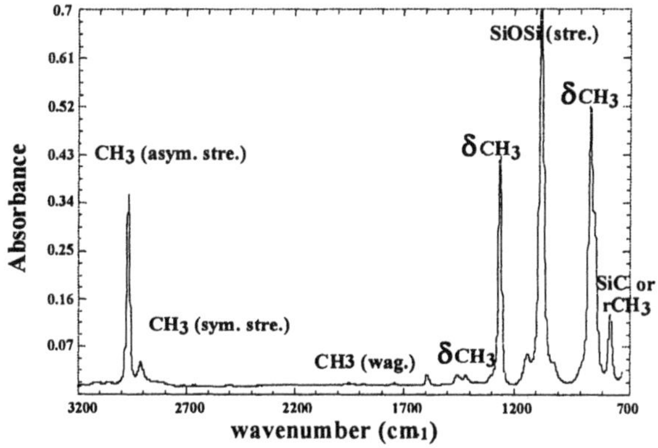

Figure 3.6. IR spectra of HMDSO gas phase. Pressure = 3.2 mTorr, power = 50 Watts.

The detection of new species in a pure organosilicon plasma, is shown in the IR absorption spectra of TEOS between 4000 cm^{-1} and 700 cm^{-1} (Fig. 3.8) during the plasma process. An interesting feature is the appearance of small structures. These absorption bands [62] are due to the fundamental ν_3 of C_2H_2 centered at 3288 cm^{-1}, the fundamental ν_3 of CH_4 centered at 3017 cm^{-1}, CO_2 (2350 cm^{-1}), CO (2147 cm^{-1}), the fundamental ν_4 of CH_4 centered at 1300 cm^{-1}, the fundamental ν_7 of C_2H_4 at 949 cm^{-1} and the fundamental ν_5 of C_2H_2 at 729 cm^{-1}.

The overall production rate of these new stable species also depends on the power. If the power increases up to 100 W and over, these species are recombined in the plasma. The new peaks (CH_4, C_2H_2, C_2H_4), characteristic of the

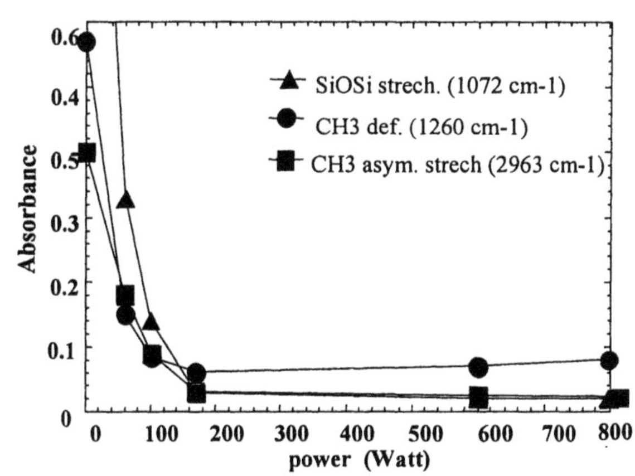

Figure 3.7. Absorbance vs. microwave power for three absorption peaks of HMDSO monomer.

organic molecules, are detected for the 3 precursors. On the contrary, free CO, CO_2 and the new PQR structure of the fundamental ν_2 of OCH_2 at 1749 cm^{-1} are detected only for TEOS. It seems that when CO bond is present in the organosilicon molecule, free CO, CO_2 and OCH_2 are detected in the plasma phase, which is not the case with TMS and HMDSO where CO doesn't exist. A new structure centered at 2125 cm^{-1} is probably due to SiH stretching mode in a big fragment of the parent molecule and appears only for the HMDSO plasma. A mixture: 50 % O_2 - 50 % organosilicon monomer has also been studied. The 50 % oxygen mixture does not produce significant differences in the plasma phase spectra compared to the spectra recorded for pure organosilicon plasma. But now, all the spectra (including HMDSO and TMS) show the characteristic structures of CO and free CO_2 coming essentially from the etching and the recombinations on the wall. It should also be noted that in every case there is a very high increase of pressure in the plasma phase which confirms the very strong dissociation. Experiments achieved with only pure oxygen in the reactor previously coated by $SiO_xC_yH_z$ films allowed the detection of CO, CO_2 and OH in the plasma phase and and showed the the decrease of CH peaks of the film (deposited on the KBr windows and mirrors). This confirms the etching effect of carbon by O_2 plasma and the appearance of CO and CO_2 in the O_2/HMDSO and O_2/TMS plasmas.

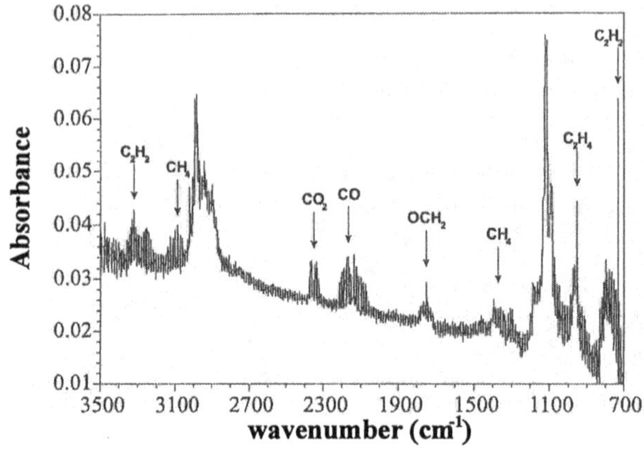

Figure 3.8. IR spectra of TEOS plasma phase. Pressure = 3.2 mTorr, power = 50 Watts.

M. Creatore et al. from d'Agostino's group in Bari Italy [63] and C. Courteille et al. [64] investigated the species produced in a high pressure RF plasma of HMDSO/O_2. Experiments have been achieved at constant pressure of 100mTorr and constant ratio mixture [HMDSO]/[HMDSO + O_2] = 5 % in a reactor similar to that one described in Fig. 3.5. They clearly show by FTIR Absorption Spectroscopy the presence of CO, CO_2 and water in the plasma phase attributed to combustion reactions between hydrocarbon radicals of HMDSO molecule and oxygen. Moreover formic acid (CO_2H_2) coming from the

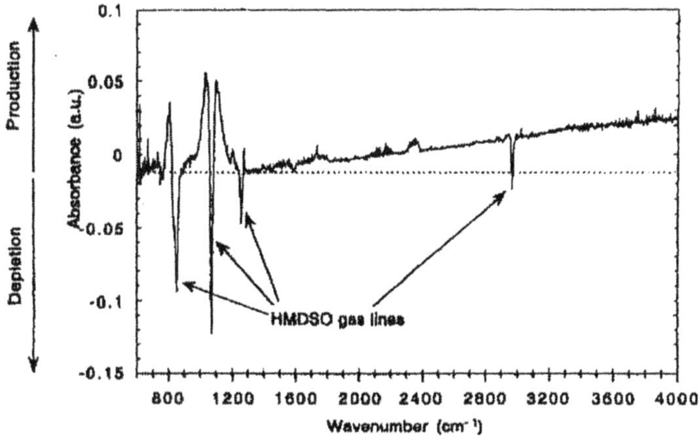

Figure 3.9. IR absorption spectrum of HMDSO-O_2 diluted with helium plasma (5.5 sccm HMDSO, 13 sccm O_2, pressure 0.1 mbar, RF power 100 W).

oxidation of methyl radical has also been detected. This molecule is probably easier created at high pressure than at low pressure. The combination of FTIR and mass spectroscopy measurement allowed us to show that high mass radicals $SiO_xC_yH_z$ diffuse to the surface and then carbon is etched by oxygen to form CO and CO_2. This fact confirms results in the DECR reactor with pure oxygen plasma (see above) which etch the surfaces to produce CO and CO_2. Another kind of species detected by FTIR is hydrocarbon (CH_4, C_2H_2), product from CH_3. They also found species such as $SiO(CH_3)_2$ or $SiCH_2$ product from the direct decomposition of the monomer.

Hollenstein [65] shows the efficiency of FTIR absorption spectroscopy in the detection of molecular components but also how to collect information on powder and radical formation. He shows the formation of SiO_x powder within an HMDSO RF discharge.

Figure 3.9 presents the absorbance spectra obtained from the IR gas phase signal (as a reference) and IR plasma phase signal. By this way, positive peaks indicate formation of new gaseous and/or solid species, negative peaks a dissociation of the monomer. That clearly shows formation of SiO_x powder (overlap between the 1070 cm^{-1} negative peak and positive broad "peaks"). Moreover, the elastic scattering due to the particles (powder) in the plasma induce a deviation of the base line. It's then possible to calculate the particle size and density.

Other FTIR diagnostics in HMDSO/Fluorocarbon mixture RF plasma revealed formation of particles in the plasma and changes in the plasma composition depending on the fluorocarbon precursor [66].

3.4. Open Questions and Conclusions

3.4.1 Species in the Gas Phase

Many papers deal with the detection of table species created in the gas phase as seen from OES or mass spectroscopy. Qualitatively speaking, it is believed all detectable species have been detected. There is a lack of quantitative information.

Stable species are not necessarily the most important ones from the point of view of the film deposition. In particular, short lived species like highly reactive radicals are difficult to detect and even more to quantify. Infrared laser diode measurements or LIF for light species could provide an interesting way to improve our knowledge.

The search for the real precursors leading to the growth of silicon oxide or nitride film is still in progress. This search involves a better knowledge of the chemistry in the gas phase. Infrared range probably contains many signatures of this chemistry. However, limitations come from detection thresholds. Ultra fast and out of equilibrium chemistry in these plasmas is still largely unknown.

An open question is about the origin of some species detected in the plasma (typically, hydrocarbonated partially oxidized compouds). There are two main pathways: the first one only involves homogeneous (gas phase) reactions the other one involves reactions between gas phase species and adsorbed species leading to the desorption of volatile groups.

Besides the nature of the species, some applications involving large substrate areas require to take into account the transport of material (mass) in large volume reactors. Chemical reactions of the species during the transport to also have to be considered. Fluid dynamics provide powerful tools for the transport of matter (mass) in the gas phase and it is now rather usual to find some simulations given by commercial codes like "Fluent". Our problem here is the coupling of the pure fluid mechanics approach to chemical reactions in one hand and electrical models which are the base for EEDF (Electron energy distribution function) calculations which, in turn, give the generation rate of species created by electron impact. A pioneering approach was recently published by Despax et al. [67- 68]; the output of the simulation was not only the spatial distribution of the growth rate but also predictive results on the composition of the deposited films as a function of the location of the wafer in the reactor (e.g. ratio silicon/oxygen). This work was done first on SiH_4/N_2O mixtures because the knowledge of the chemistry was good enough. Same approach has yet to be made for much more complex precursors like $TEOS/O_2$ or $HMDSO/O_2$. The main difficulties are a correct description of the chemistry and to find a set of reaction rates reliable because most of the excitation, dissociation or ionization cross sections are unknown and should be treated as adjustable parameters. Some attempts however to get a more argumented approach were recently published by Basner et al. [39] from mass spectroscopy analysis and Vallée et al. [69] from the development of a simple model which describes the experimental change of the deposition rate as a function of distance from the TEOS inlet to the substrate. One of the questions is the respective roles of electron impact and chemical reactions with atomic oxygen in the organosilicon dissociation. The answer is not unique and depends on the plasma parameters, specially the pressure. At high pressure neutral-neutral reactions in the gas phase can occur whereas electron impact is supposedly predominant at low pressure (typically 1mbar or less). However, these statements should be supported by careful calculations based on gas kinetics and electron impact dissociation cross section. Calculations for $TEOS/O_2$ helicon plasma mixtures from Vallée et al. [69] show that in this case and using their estimation the dissociation coefficient $k_e = 1.82 \cdot 10^{-7}$, electron impact effect is 10^3 more important than oxygen reaction. This result will be probably not true if the pressure was higher than 100 mBar.

Another difficult task is to take into account a non homogeneous distribution of the temperature. This aspect, very important for high power industrial reactors, has almost never been considered so far.

3.4.2. Interface Plasma-Surface; Growing Mechanism

The plasma species, those identified by the gas phase diagnostics and all the other ones, reach the surface at a given flow and with a given energy. Then, they can desorb immediately or after a given life time during which they move on the surface. This random walk can sometimes lead to a chemical reaction between moving species or between a particular site on the surface and the moving species. All these processes are sometimes reduced to a single coefficient called "sticking coefficient". This scenario is the basis of some models (see Stout and Kushner [70] for example) but it doesn't take into account the role of the ion bombardment.

A rather generally accepted scheme for the deposition mechanism is a reaction between two silanol adsorbed on the surface [70,71,72]:

$$\equiv \text{Si-OH} + \text{OH-Si} \equiv \;\rightarrow\; \equiv \text{Si -O- Si} \equiv + H_2O$$

Silanol production can be due to electron impact on organosilicon (a) or oxygen reactions (b).

(a) $e + (RO)_n\, Si\, (OH)_{4-n} \rightarrow (RO)_{n-1}\, Si\, (OH)_{4-n+1} + C_xH_y + e \quad n = 1 \text{ to } 4$

(b) $O + (RO)_n\, Si\, (OH)_{4-n} \rightarrow (RO)_{n-1}\, Si\, (OH)_{4-n+1} + C_xH_yO \quad n = 1 \text{ to } 4$

C_xH_y and C_xH_yO species have been detected by IR absorption spectroscopy [60].

Polycondensation of silanol groups allows a modelling of film growth rate [70,71]. The contribution of ions was ignored in these models. Raupp et al. [73] take this contribution into account in a very simple way. The oxygen positive ion flux Φ_+ was calculated from a sheath model [74] and then the contribution of the ions to the growth was supposed to be $\sigma_+ \cdot \Phi_+$, where σ_+ is the sticking coefficient for oxygen ions. The role of ions is probably much more complicated, however, models based only on neutrals are able to represent growth rates, step coverage and thermal activation energy. Then the role of ions was often neglected. It should be pointed out that the lack of basic data makes the models very adjustable; then a good fit does not necessarily mean that the physical basis of the model is absolutely correct. There are some experimental results (for example microwave plasmas plus RF induced DC self bias) suggesting that the energy provided by the ion flux changes the density of the films. More work is needed to clarify the role of ions.

Another important point both from one experiment to the other can be from a technological and academic point of view, is the role of the substrate on the early stage of the deposition (film growth). Historically, only silicon was considered; extending the domain of applications, other materials have to be considered. Growing processes on polymeric surfaces (PET, Kapton and PP) was investigated by Da Silva Sobrinho [75] and Dennler et al. [76]. Evidence was found to propose a layer by layer mechanism rather than an island coalescence. Obviously there is no evidence that the early stage are the same for other materials like metals or semiconductors. One can expect the surface mobility of adsorbed species and their lifetime at the surface and the density of chemisorption sites are substrate dependent. However, direct measurements seems very difficult and, to our knowledge, no attempt has been made so far.

In this chapter almost nothing has been said about the deposited film itself. Many papers were published on films compositions generally obtained through XPS or Infrared spectroscopy, as a function of gas mixtures, power, etc. A very general trend is a decay of the carbonated groups in the film as the oxygen to organosilicon ratio in the gas increases. Although some differences are observed, it can be concluded that no carbon was detected if the oxygen content in the gas phase is over 90 %. Some aspects dealing with the structure and properties of the films received less attention. First, the questions about morphology and densities. Many results give a growth rate expressed as nanometer per minute; the other alternative is μg per minute. Of course the density of the films is the key parameter to correlate these two approaches. Density is also directly related to some properties like barrier to oxygen or water vapor; also, the challenge to reduce the electrical permitivity deals with the density of plasma deposited organosilicon films. Practically, the value of the density ranged from values close to 1 for films deposited without oxygen to 2.2 or even more for films deposited from oxygen rich mixtures. Mechanical properties of plasma deposited films are also a field in which more work is needed. Hardness or internal stress are important parameters directly related to the durability of the coatings. Very few papers give the keys to correlate (or better to predict) these properties to the plasma parameters. The role of the structure and composition of the substrate is probably non negligible but there is no clear result on this question.

References

1. Inagaki N., Kondo S., Irata M. and Urushibata H., *J. Appl. Polym. Sc.*, **30** (1990) 3385.
2. Fayet P., Holland C., Jaccoud B. and Roullin A., *Soc. Vac. Coaters Proc. 38th Annu. Conf.*, 1995, 15.
3. Chatham H., *Surf. Coat. Technol.*, **78** (1996) 1.
4. Okuhara T. and White J. M., *Appl. Surf. Sc.*, **29** (1987) 223.
5. Fracassi F., d'Agostino R., Buccolieri A., Colaprico V., Angelini E., Grassini S. and Rosalbino F., *15th ISPC Proc. Orléans*, July 2001, 115.

6. Koller A., Hofer M. and Zimmermann H., *Proc. 11th ISPC*, Loughborough, UK 22627 August 1993, 1005.
7. Janocha B., Hegemann D., Oehr C., Brunner H. and Rupp F., *J. Geis-Gerstofer surf. Coating Technol.*, **142-144** (2001) 1051.
8. Schwarz J., Schmidt M. and Ohl A., *Coat. Techn.*, **98** (1998) 859.
9. Da Silva Sobrinho A. S., Latreche M., Czeremuszkin G., Sapieha J.E., Wertheimer M. R., *J. Vac. Sci. Technol.*, **A16** (1998) 3190.
10. D. Hegemann, U. Vohrer, C. Oehr, R. Riedel, Surf. Coat. Techn. **116-119** (1999) 1033.
11. Tachibana K., Nishida N., Harima H., Urano Y., *J. Phys. D Appl. Phys.*, **17** (1984) 1727.
12. Wrobel A. M. and Wertheimer M. R., *Plasma-polymerized Organosilicones and organometallics*, In R. d'Agostino Editor, Plasma deposition and treatments of polymers, Academic press, New York, 95.
13. Segui Y. and Bui A., *J. Appl. Polym. Sci.*, **20** (1976) 1611.
14. Wrobel A. M., Wertheimer M. R., Dib J., Schreiber H. P., *J. Macrom. Sci. Chem.*, **A14** (1980) 321.
15. Tajima I. and Yamamoto M., *J. Polym. Sci. - Polym. Chem. Ed.*, **23** (1985) 615.
16. Aovali G. and Hasiri N., *J. Appl. Polym. Sci.*, **29** (1984) 2617.
17. Sachdev K.G. and Sachdev H.S., *Thin Solid Films*, **107** (1983) 245.
18. Mukherjee S. P. and Evans P.E., *Thin Solid Films*, **14** (1972) 105.
19. Pai C. S., Miner J. F. and Foo P. D., *J. Electrochem Soc.*, **139** (1992) 850.
20. Selamoglu N., Mucha J. A., Ibbotson D. E. and Flamm D. L., *J. Vac. Sci. Technol.*, **B7** (1989) 1345.
21. Pai C. S. and Chang C. P., *J. Appl. Phys.*, **68** (1990) 793.
22. Ishii K., Ohki Y. and Nishikawa H., *J. Appl. Phys.*, **76** (1994) 5418.
23. Inagaki N., Kondo S. and Murukami T., *J. Appl. Polym. Sci.*, **29** (1984) 3595.
24. Kulisch W. and Rau C., *Proc. of the 12th ISPC Loughborough*, August pp1010 and *Thin solid Films*, **249** (1994) 28.
25. Nguyen V. S., Underhill J., Fridman S. and Pan P., *J. Electrochem. Soc.*, **132** (1985) 1925.
26. Fracassi F., d'Agostino R., Favia P. and van Sambeek M., *Plasma Sources Sci. Technol.*, **2** (1983) 106.
27. Boudreau M., Boumerzoug M., Masher P. and Jessop P.E., *Appl. Phys. Lett.*, **63** (1993) 3014.
28. Maruyama T. and Shirai T., *Appl. Phys. Lett.*, **63** (1993) 611.

29. Grebowicz J., Pakula T., Wrobel A.M. and Kryszewski M., *Thin Solid Films*, **65** (1980) 154.
30. Ishikawa M., Egashira Y. and Komiyama H., *J. Appl. Phys.*, **85** (1997) 2655.
31. Benissad N., Boisse-Laporte C., Vallée C., Granier A. and Goullet A., *Surf. and coating technol.*, **116-119** (1999) 868.
32. Aumaille K., Vallée C., Granier A., Goulet A., Gaboriau F. and Turban G., *Thin Solid Films*, **359** (2000) 188.
33. Tocitani G., Shimozuma M. and Tagashira H., *J. Vac. Sci. Technol.*, **A11** (1993) 400.
34. Selamoglu N., Mucha J.A., Ibbotson D.E. and Flamm D.L., *J. Vac. Sci. Technol.*, **B7** (1989) 1345.
35. Lamendola R. and d'Agostino R., *Plasma Processing of polymers NATO serie E: Applied Science*, R. D'Agostino et al. Ed., 321.
36. Pai P. G., Chao S. S. and Takagi Y., *J. Vac. Sc. Techn.*, **A4**(3) 689, 1985.
37. Schmidt M., Foest R., Bassner R. and Hanneman M., *Acta Physica Comenianae*, **35** (1994) 217.
38. Ropcke J., Ohl A., Schmidt M., *J. Anal. at. Spectrom.*, **8** (1993) 803.
39. Basner R., Foest R., Schmidt M., Hempel F. and Becker K., *Proc. XXIII ICPIG, Toulouse France*, July 17-22 1997, Vol IV p. 196.
40. Basner R., Schmidt M., Becker K. and Deutsch H., *Advance in atomic, molecular and optical physics*, Vol. 43 Academic Press.
41. Aumaille K., PhD. Thesis, Nantes, Dec 2000.
42. Fracassi F., d'Agostino R., and Favia P., *J. Electrochem. Soc.*, **139** (1992) 2636.
43. Latreche M., Segui Y., Delsol R. and Raynaud P., Proc. 11th ISPC, Loughborough, 22-27 August 1993, p. 1284.
44. Latreche M., PhD. thésis, Toulouse, March 1993.
45. Horii N. M., Okimura K. and Shibata A., *Thin Solid Films*, 343-344 (1999) 148.
46. Nothe N. and Bolt H., *Surf. Coating Technol.*, **131** (2000) 102.
47. Aumaille K., Granier A., Grolleau B. and Turban G., *J. Appl. Phys.*, **89** (2001) 5227.
48. http://physics.nist.gov/PysRevData/ionization/EII_table.html
49. Kroesen G. M. W., den Boer J. H. W. G., Bouffendi L., Vivet F., Khouli M., Bouchoule H. and de Hoog F. J., *J. Vac Sci. Technol*, **A14** (1996) 546.

50. Courteille C., Magni D., Howling A. A., Nosenko V., Fayet P., Hollenstein Ch., *Proc. 40th annual technical conference, Society of the vacuum coaters*, 304-308 (1997).
51. Segui Y. and Raynaud P., *Plasma processing of polymers*, NATO ASI series, Kluwer academic publishers dordrecht 346, 81 (1997).
52. Campargue A., Chenevier M., Lafayette L., Marcus B., Mermoux M. and Ross A. J., *Appl. Phys. Lett.*, **62** (1993) 134.
53. Raynaud P., Segui Y., Balmer G., Boulanger A. and Burke R., *J. Phys. D: Applied Phys.*, **30** (1997) L23.
54. Deschenaux Ch., *J. Phys. D: Appl. Phys.*, **32** (1999) 1876.
55. O'Neill J. A., Singh J. and Gifford G.G., *J. Vac. Sci. Technol.*, **A8**, (1990) 1716.
56. Magane M., Itabashi N., Nishiwaki N., Goto T., Yamada C. and Hirota E., *Jap. J. Appl. Phys.*, **29** (1990) L 829.
57. Hollas J. M., *Modern spectroscopy*, J. Wiley& Sons ed. (1987).
58. Griffiths P. R. and Haseth J. A., *Fourier Transform Infrared Spectroscopy*, Chemical analysis 83, J. Wiley & Sons Ed (1986).
59. White J. U., *J. Opt. Soc. Am.*, **32** (1942) 285.
60. Raynaud P., Amilis T. and Segui Y., *Appl. Surf. Sci.*, **138-139** (1999) 285.
61. Latreche M., Segui Y., Delsol R. and Raynaud P., *11th. ISPC*, Loughborough, John Harry Eds., 1284-1289 (1993).
62. Herzberg G., *Infrared and raman spectra of polyatomic molecules*, D. van Nostrand, Princeton, New Jersey (1968).
63. Creatore M., Palumbo F., d'Agostino R. and Fayet P., *7th international conference on Plasma surface Engineering* (2000).
64. Courteille C., Magni D., Deschenaux Ch., Fayet P., Howling A. A. and Hollenstein Ch., *Proc. 41th annual technical conference, Society of the vacuum coaters*, 327-332 (1998).
65. Hollenstein Ch., *Plasma and Polymers*, **3**(4) 249 (1999).
66. Shirafuji T., Mivasaki Y., Havashi Y. and Nishino S., *Plasma and Polymers*, **4** (1999) 57.
67. Caquineau H. and Despax B., *Chem. Eng. Sci.*, **17** (1997) 2901.
68. Date L., Radouane K., Despax B., Yousfi M., Caquineau H. and Hennad A., *J. Phys. D. Appl. Phys.*, **32** (1999) 1478.
69. Vallée C., Rhallabi A., Granier A., Goulet A. and Turban G., *J. Vac. Sci. Technol.*, **A18** (2000) 2728.
70. Stout P. J. and Kushner M. J., *J. Vac Sci. Technol.*, **11** (1993) 2562.
71. Kim M. T., *Thin Solid Films*, **360** (2000) 60.

72. A. Rhallabi and G. Turban J. Vac. Sci. Technol **A19** (2001) 743.
73. Raupp G. B., Cale T. S. and Hey H. P.W., *J. Vac. Sci. Technol.*, B (1992) 37.
74. Economou D. J., Evans D. R. and Alkire R. C., *J. Electrochem. Soc.*, **135** (1988) 756.
75. da Silva Sobrinho A. S., Czeremuszkin G., Latrèche M. and Wertheimer M. R., *J. Vac. Sci. Technol.*, **A18** (2000) 149.
76. Dennler G., Houdayer A., Segui Y. and Wertheimer M. R., *J. Vac. Sci. Technol.*, **A19** (2001) 2320.

CHAPTER 4

CORPUSCULAR DIAGNOSTICS OF PLASMA POLYMERIZATION PROCESSES

MICHAEL ZEUNER

Chemnitz, Germany

4.1. Corpuscular Diagnostics in Plasma Processing

In thin film technology, during the last few years, the demands on thin film quality and process reproducibility have dramatically increased. Highly sophisticated analytical equipment is now required for the necessary control of the basic discharge mechanisms. Especially in case of plasma surface processing, the information about the particle and energy flux to the surface is of essential importance for the understanding of the extremely complex plasma surface interaction processes.

However, this information especially remains hidden for standard bulk plasma diagnostics. The reasons why most methods fail, are the strong electric fields in the plasma sheath covering each surface in contact with the plasma, and consequently the large concentration gradients and the anisotropic velocity distributions of the charge carriers reaching orders of magnitude higher energies than in the bulk plasma region.

Optical diagnostic methods give very selective information about the species composition in the bulk plasma. But resolving the concentration within the plasma sheath is a highly dedicated experimental task, and no detailed statement about the velocity distribution could be derived (Chapt. 2.). A similar statement is true for Langmuir probe diagnostics, which is a useful tool for the bulk plasma diagnostics. But close to any surface the plasma sheaths of probe and surface affect each other, and the probe signal could not be evaluated correctly [1].

Mass spectrometry and electrostatic energy analyzers sampling directly at the substrate or electrode surface, respectively, turned out to be the most versatile tools in corpuscular diagnostics. Of course these methods could also be applied without any general restriction in bulk plasma diagnostics. Mass spectrometry combined with energy analysis provides detailed and impressive information on the energy distribution and flux of neutral particles [2-4], radicals [5-8], positive [9-14] and negative ions [15-18]. Particle flux, energy and angle distribution play an important role in plasma etching and sputtering as well as in plasma deposition processes. The energy transfer by incident particles is known to determine, together with other factors, the etch rate and anisotropy [19-23], deposition rate, and surface layer composition and structure [24-28]. Due to the plasma and sheath parameters and the features of the particles such as their charge state, mass and cross sections for collisions and chemical reactions, the transport mechanisms of these particles towards the surfaces are quite different. Hence the particle flux, energy distribution and chemical behaviour at the surface may alter dramatically.

In Chapt. 4 we explain the basic features of the bulk plasma, plasma sheath, and surface interaction processes which are essentially necessary to understand the experimental data. The description of experimental equipment used for corpuscular diagnostics is followed by exemplary results on the flux analysis of neutral species. Finally, we demonstrate the meaningful direct detection of positive and negative ions from the discharge. The discussion of the applications is confined to a few examples illustrating the information that may be obtained, the typical experimental equipment and important effects which have to be taken into consideration.

However, one should always take into account that plasma polymerization processes are difficult to analyze because of the high deposition rates of electrically insulating plasma polymer films on the (sampling) electrodes, walls, etc. in contact with the plasma. For this reason, in the experimental sections additionally some results from non polymerizing plasmas are presented in order to explain and illustrate phenomena happening in the same manner in the polymerization processes.

4.2. Basic Processes in Particle Interaction

The particle flux, its composition and particle energy distribution in any discharge arrangement are influenced by the three important factors. Within the discharge plasma, a particle composition is created which is affected by the collision frequency between the different species, their collisional cross sections, and their velocity distribution. Diffusion and the strong electrical field in the plasma sheaths are responsible for particle transportation to the surfaces. In the interaction of the surface with the particle flux the surface properties get changed, and secondary particles are re-emitted into the discharge.

Because these basic processes are very important for the interpretation and valuation of the corpuscular diagnostics, a brief introduction explains the basic plasma-chemical phenomena. However, the scope of this section gives only area for some basic explanations, detailed information may be found elsewhere.

4.2.1. Gas Discharge Phenomena

Within the discharge plasma, a typical species composition consisting of neutral background gas, electronically excited and metastabile species, radicals, positive and negative ions, electrons and photons is created [22,23,29,30].

In accordance with the Boltzmann equation, for each species, a spatial and a velocity distribution function results with the collisional operators describing the spatial and momentum scattering of the particles. The collisional operators in turn are determined by the velocity dependent collisional cross sections, σ_{ik}, and the density and velocity distribution function of the corresponding species. Fig. 4.1 recalls the central position of the Boltzmann-equation within the mathematical system describing the gas phase discharge phenomena. However, there exists a multiplicity of different collisional processes which are more or less important for the kinetics of the discharge.

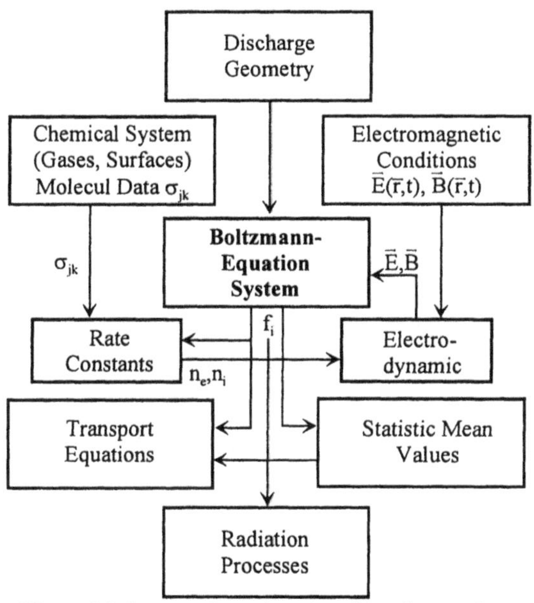

Figure 4.1. System of equations describing the gas phase discharge phenomena.

Within the bulk plasma, electrons play the most important role in the discharge kinetics. Because of the large mass difference in comparison with the background gas, the elastic collisions are of minor importance, but the electron velocity distribution shows energies high enough to drive a variety of inelastic collision events. The most important electron collisions are [31,32]:

Excitation:	$X_2 + e^- \rightarrow X_2^* + e^-$
Dissociation:	$X_2 + e^- \rightarrow 2X + e^-$
Ionization:	$X_2 + e^- \rightarrow X_2^+ + 2e^-$
Dissociative Ionization:	$X_2 + e^- \rightarrow X^+ + X + 2e^-$
Electron Attachment:	$X_2 + e^- \rightarrow X_2^-$
Dissociative Attachment:	$X_2 + e^- \rightarrow X^- + X$

The electron ionization and attachment is of basic interest for the presentation of the experimental data in Sec. 4.5.2 and 4.5.3, and for this reason the information about these processes is presented in the following.

However, collisions of the ions with the neutral particles have also a significant influence on the discharge phenomena. Especially in the plasma sheath with orders of magnitude higher energies of the ions and high energy neutrals, these collisions may play the important role for the particle flux (Sec. 4.3.2, 4.2.3) [33].

Fig. 4.2 presents typical absolute electron impact ionization cross sections of different gases. These cross sections are of fundamental importance for the post-ionization of neutral species in the ion source of the mass spectrometer (Sec. 4.4) and for the composition of the direct ion flux to the surfaces (Sec. 4.5). Exceeding the ionization energy, the cross sections are characterized

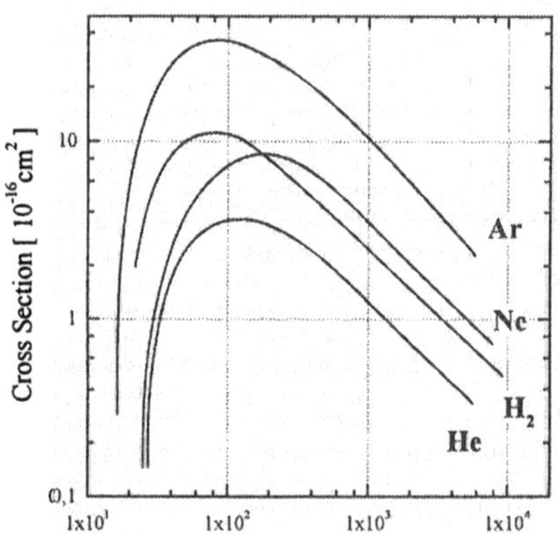

Figure 4.2. Electron impact ionization cross sectior of different gases [32] (with permission).

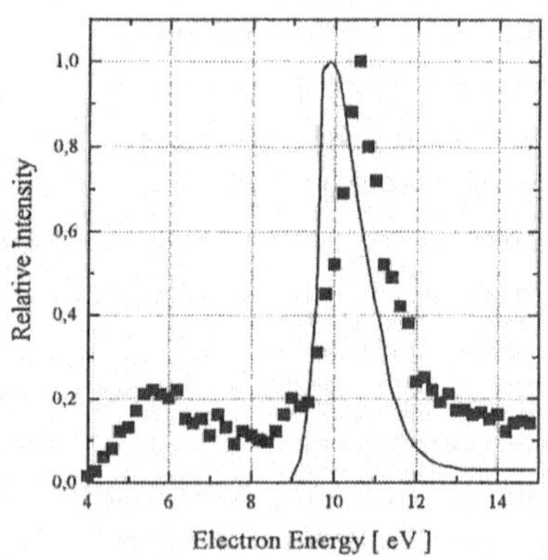

Figure 4.3. Electron attachment signal for the O^- ion generated in $CO + e^- \rightarrow O^- + C$ ($\sigma = 2x10^{-19}$ cm^2, Dots: Exp., Solid: Lit.) [34] (with permission).

by a broad energy dependence peaked at about 50–100 eV. As can be seen, changes in the electron energy distribution may alter the ion composition produced in the discharge or in the mass spectrometer ion source dramatically.

For electronegative gases, the electron attachment cross sections presented in Fig. 4.3 are of importance. In contrary, the typical energy dependence of a dissociative attachment cross section indicates a resonant character of the process with cross section peaking at low energy with a narrow width and a absolute value about 2 to 3 orders of magnitude lower than for positive ion production. However, the resonant attachment cross section is very sensitive to the electron energy distribution and could cause high negative ion production rates in the afterglow and in electron attachment mass spectrometry.

4.2.2. Plasma Sheath Effects

As a result of the interaction between a plasma and any surface, a plasma sheath always appears. The plasma sheath screens the discharge plasma against any potential disturbance caused by the surface. Within the plasma sheath, a space charge is present creating a high electrical field which in turn determines the transport phenomena and energy distribution of the charge carriers in this region.

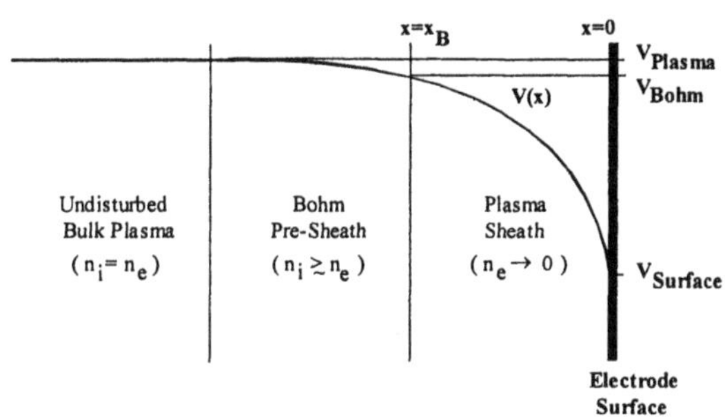

Figure 4.4. Potential distribution across a plasma sheath.

The origin of the plasma sheath is based on the different mobility of the hot electrons as the most important negative charge carriers and the low energy positive ions in the bulk plasma. Because of the different thermal fluxes of electrons and ions towards the surface, the potential distribution near the surface gets changed, and the surface initially receives a higher electron current. Depending on whether the surface is insulated or at a defined potential, it charges negatively or the plasma potential gets positive relative to the surface potential. The remaining positive space charge in the plasma sheath ($n_e \rightarrow 0$) drives an additional field current which, together with the drift current, finally compensates the fluxes of positive and negative charge carriers one to another (Fig. 4.4) [22,23].

The electrical field in the plasma sheath generated by the positive space charge of the ions $n_i(\vec{r},t)$ and the low negative one of the electrons $n_e(\vec{r},t)$ generates an electrical field which can be calculated using the Poisson equation:

$$\frac{d^2V(\vec{r},t)}{d\vec{r}^2} = -\frac{e\left(n_i(\vec{r},t) - n_e(\vec{r},t)\right)}{\varepsilon_0} \quad (4.1)$$

Solving the Poisson equation and taking the Debye-length λ_D and a mobility term for the charge carriers into account, the sheath thickness d_{sh} may be calculated in dependence on the sheath voltage. With the Boltzmann constant k_B and the electron temperature T_e for a stationary collision free plasma sheath the sheath thickness can be expressed as follows:

$$d_{sh} = \frac{2\sqrt{2}}{3}\lambda_D\left(\frac{e\,V_{sh}}{k_B\,T_e}\right)^{\frac{3}{4}} \quad (4.2)$$

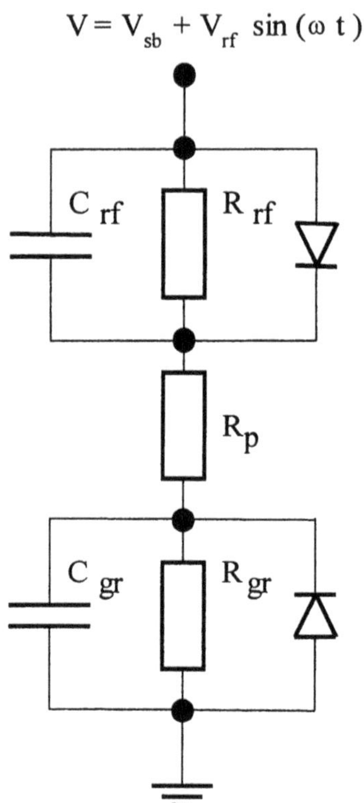

Figure 4.5. Replacement circuit diagram of a capacitively coupled RF discharge.

Between undisturbed plasma ($n_{pl}=n_i=n_e$) and the plasma sheath ($n_e \rightarrow 0$) the Bohm pre-sheath is located which results from the demand of monotony form of the electrical potential (Fig. 4.4) [35,36]. Within the Bohm region, the quasi-neutrality of the plasma is only a little disturbed with a small potential drop of $k_BT_e/2$ occurring across the pre-sheath [37-39]. Derived from the potential drop the Bohm criterion, a statement about the ion velocity at the plasma sheath boundary at $x=x_B$, could be given:

$$v_i(x_B) \geq \sqrt{\frac{k_B\,T_e}{m_i}} \quad (4.3)$$

Using the sheath thickness (4.2) and the Bohm criterion (4.3), the Child-Langmuir-equation that describes the current density in the plasma sheath can be obtained. For a collision and source free plasma sheath it may be written as follows:

$$j_i = \frac{4}{9}\varepsilon_0 \sqrt{\frac{2e}{m_i}} \frac{V_{sh}^{\frac{3}{2}}}{d_{sh}^{2}} \tag{4.4}$$

However, these basic formulas describing the plasma sheath characteristics may additionally be influenced by elastic collisions and ionization in the sheath, by the time modulated electron density in the sheath, and by the presence of negative ions. A detailed discussion can be found elsewhere [40-45].

Especially in RF excited discharges, the sheath characteristics are affected by the time varying potential. The replacement circuit diagram in Fig. 4.5 represents a typical capacitively coupled RF discharge with each plasma sheath represented by a capacitor, a resistor and a diode. In practice, the RF voltage is supplied via a coupling capacitor that causes that a DC self-bias voltage, V_{sb}, forms on the powered RF electrode. The diodes represent the high electron mobility and cause that the plasma potential is positive against any electrode potential. In the typical case of dominating capacitive sheath characteristics $1/(i\omega C_x) >> R_x$ and low plasma resistance $R_p \rightarrow 0$, the arrangement acts as a capacitive voltage divider, and a harmonic wave form modulation of the plasma potential is observed.

As can be seen from the replacement circuit diagram, the modulation amplitude of the plasma potential depends on the ratio of the sheath capacities, and also the self-bias voltage is determined by theses capacities:

$$V_{sb} = \left(\frac{C_{rf} - C_{gr}}{C_{rf} + C_{gr}} \right) V_{rf} \tag{4.5}$$

The exact value of the sheath capacitance is determined by the space charge located in the plasma sheath and the geometrical electrode area. In most cases the grounded electrode area drastically exceeds the RF electrode area due to the grounded chamber walls. Consequently, the self-bias, V_{sb}, gets negative up to nearly half of the peak-to-peak RF voltage, and we find a large RF potential drop in front of the RF electrode and only a small in front of the grounded one (Sec. 4.5.1) [9,46-48].

An additional important sheath effect is caused by the typical relation of the RF excitation frequency ω_{rf} and the plasma frequencies of ions, $\omega_{pl\text{-}i}$, and electrons $\omega_{pl\text{-}e}$ with $\omega_{pl\text{-}i} < \omega_{ex} << \omega_{pl\text{-}e}$. This relation explains why the ion movement through the plasma sheath takes several RF cycles, and hence is determined by the mean sheath voltage. In opposite, the electrons are able to respond instantaneously to the modulation of the sheath potential. Depending on the actual sheath voltage, an electron front penetrates into the sheath and compensates the positive space charge behind the electron front (Sec. 4.5.1). At the minimum sheath voltage, a short-term high electron current compensates the ion current flowing during the whole RF period [10].

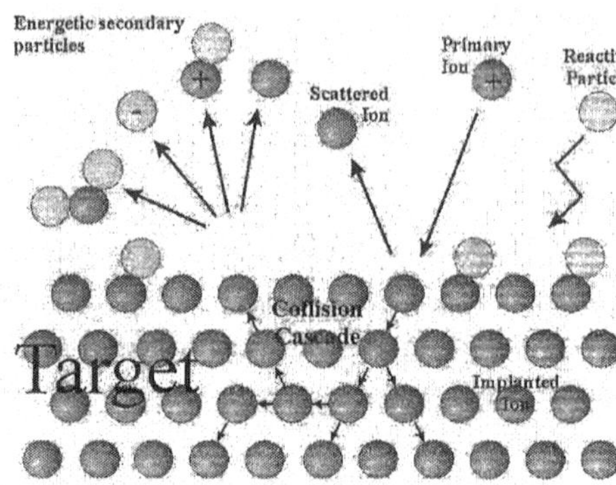

Figure 4.6. Schematic drawing of the surface interaction of high energy ions and reactive species.

4.2.3. Surface Interaction

The interaction of the gas discharge with the surface is determined by the received flux of neutral particles, excited species, radicals and ions. As explained in Sec. 4.2.2. these particles, especially the ions, may acquire large energies during passing the plasma sheath. Depending on the chemical character of the species and their energy, one can observe different interaction processes with the surface (Fig. 4.6) [26-28]:

- Direct adsorption and deposition of the incoming particles,
- Reflection of high energy particles,
- Implantation of high energy particles,
- Changes in surface stoichiometry and chemical reactions,
- Production of lattice defects,
- Particle flux induced desorption and
- Production of secondary particles (Sample and film sputtering, Secondary electrons).

Further processes may be driven by the particle flux from the plasma when a reactive background atmosphere is present:

- Influence on the adsorption characteristics of reactive species through surface activation and

- Particle flux assisted chemical reactions between reactive species and surface.

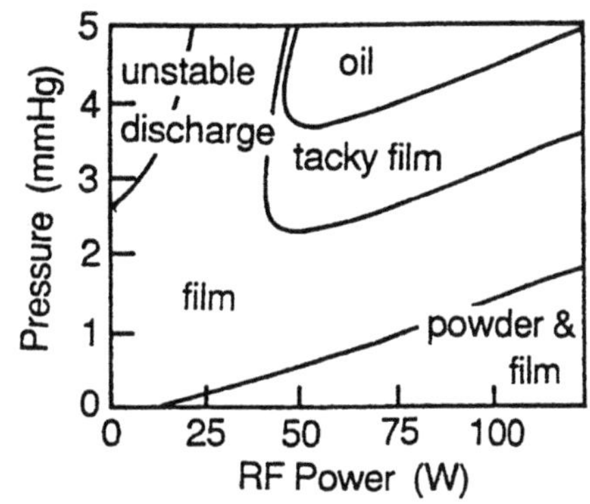

Figure 4.7. Appearance of plasma polymer films from ethylene dependent on the deposition conditions [49] (with permission).

In real surface processing, we usually observe a superposition of these different mechanisms, which may contrarily affect specific process characteristics. The complexity of the surface reactions is also a reason why these are only incompletely understood until now. In addition, lacking experimental data about the interaction of the incident particles have complicated the enlightenment of the phenomena. Actually powerful tools in the particle flux and energy analysis at the surfaces grant a deeper access to the understanding of these complex mechanisms. Subsequently, only some essential information should be presented about the interaction of high energy and reactive particles with the surface.

Fig. 4.7 presents, how the appearance of polymer films is typically affected by the deposition conditions. The deposition rate itself is mainly determined by the flux of activated monomer. The most important factor describing this activation effect is the specific energy, p_{el}, supplied per monomer gas unit (P_{el} - Total power, R - Gas constant, T - Temperature, Q_{mon} - Monomer flux, N_a - Avogadro number) [50]:

$$p_{el} = \frac{P_{el}\,RT}{Q_{mon}\,N_a} \qquad (4.6)$$

Below a minimum specific energy, a non optimum deposition rate results because of inadequate monomer activation, above the optimum energy, the deposition rate lowers because of an excessive monomer degradation.

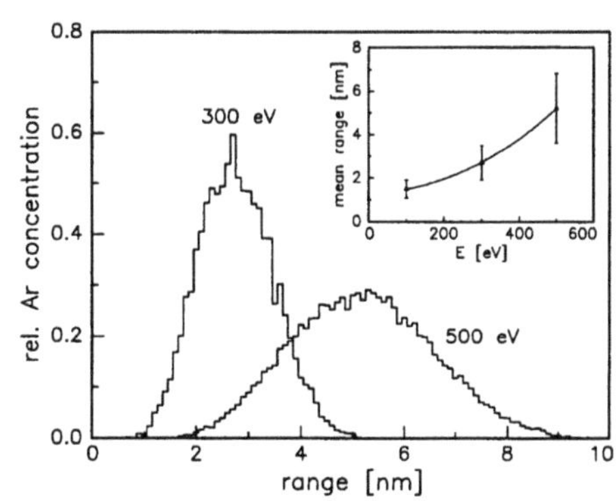

Figure 4.8. TRIM-calculated implantation profiles of Ar^+-ions in polyethylene (Parameter: Ion energy of Ar^+) [25] (with permission).

However, besides the gas phase activation of the monomer (Sec. 4.2.1), the appearance of the growing film may be affected by additional particle bombardment (Fig. 4.7). Due to high kinetic energies, these incident species could be implanted in a some nm thick region of the growing film where a modification of the growing film occurs (Fig. 4.8). Within this thin surface region, the chemical character of a deposited film may be dramatically altered, and densification and crosslinking are initiated. In a plasma polymerization process, the surface region always gets covered by new deposited material. This way the modified region migrates through the plasma polymer film, and the whole film is affected by implantation effects of high energy particles.

The pulse transfer in a collision cascade caused by a high energy particle may lead to the emission of secondary particles from the film. This sputter effect may be used in plasma etching and sputter deposition, and it happens always simultaneously in the most plasma polymerization processes.

The specific sputter yield, which is the ratio of emitted secondary particles per incident ion depends on different factors, with the reduced mass between incident and target atoms, the incident particle energy and angle, and the target surface binding energy being the most important ones. The incident ion energy in Fig. 4.9 (top) increases the sputter yield simultaneously, and for an optimum reduced mass a larger yield results. The sputter yield has been observed,

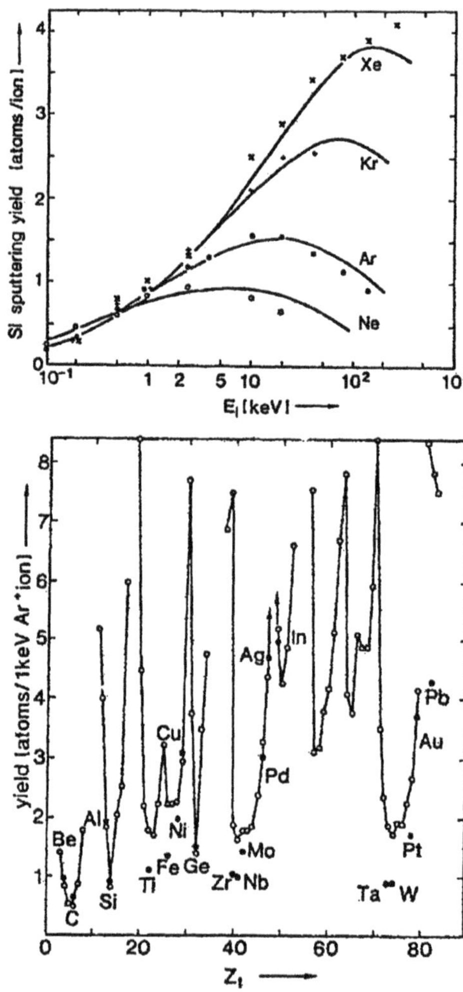

Figure 4.9. Energy dependence of the Si-yield using different noble gases (top) [51,52] and total sputter yields of the elements during 1 kV Ar bombardment (bottom) [53] (with permission).

additionally, to depend strongly on the atomic number of the target as shown in Fig. 4.9 (bottom).

However, sputtering in plasma polymerization is especially important, because it may substantially affect the resulting growth rate and, because of preferential sputtering, it may change the stoichiometry of the growing film. Although sputtering is a surface phenomenon, the whole plasma polymer film is affected by these processes during the film growth.

4.3. Experimental Equipment for Corpuscular Diagnostics

In corpuscular diagnostics, two essential experimental techniques are applied as follows: methods detecting the integral particle flux via a film growth, and methods analyzing the flux species selectively via electrical measurements. However, only techniques for direct particle flux measurement are considered in the following discussion. Optical methods (Chapt. 2) as well as Langmuir probes measurements are not included in this chapter.

Detecting the particle flux by the film growth is a low cost method. The possible conclusions are rather limited since no species selection takes place, neither by the charge state, nor by the mass or energy. Contrary to the restricted performance, these methods are of basic interest, because equipment effects are of minor importance, quantitative measurements could easily be performed, and a calibration standard for the complex methods is available.

The complex systems generally use electrical methods for particle separation and detection, and therefore ions from the process could be analyzed directly, whereas neutral species must be post-ionized. Quadrupole mass spectrometers are commonly applied in mass separation, and different types of energy filters are used. Both types, mass and energy analyzers, could be combined in a joined equipment, but exist also as stand alone versions. For this reason, mass and energy analyzers are discussed in separate sections.

In order to achieve qualitatively good results, negligible interaction with the analyzer background is necessary, and hence the particle mean free path must clearly exceed the typical analyzer dimensions. Consequently, most of the complex analyzers sample from the discharge via a pinhole aperture and are differentially pumped.

With the following categorization of the most important methods in corpuscular diagnostics, we present the hardware bases with the most important features and discuss important equipment effects.

4.3.1. Common Particle Collection by Film Deposition

Within this section, different methods allowing statements about the particle flux by measurements of the deposited film thickness are summarized.

First and foremost, common standard methods based on gravimetrical measurements by quartz crystal or beam micro-balances are discussed, which can be easily established and interpreted. Although these methods enable an ordinary information only about the growth or etch rate, respectively, their basic meaning lies in the simplicity and quantification of the measurements. For this reason, the gravimetrical measurements could be utilized establishing a calibration standard for the more complex methods discussed in Sec. 4.3.2. and 4.3.3.

In quartz crystal micro balances, the film deposition directly occurs on a quartz crystal (AT cut) that drives the oscillator which is tuned to resonance. With the growing film the resonance frequency, f_q, is reduced by Δf because of the additional mass load on the quartz surface. The thickness of the deposited film d_f can be expressed as (ρ - Density, N_{AT}=1.66 kHzm - Frequency constant):

$$d_f = \Delta f \frac{\rho_q N_{AT}}{\rho_f f_q^2} \qquad (4.7.)$$

For the evaluation of the sensor characteristics, the Z-match balances treat the two films as a system of coupled oscillators which allows for correct measurement of thick film systems [54].

For plasma polymerization processes with a wide spread of film properties and structure, the film density and acoustic parameters are often not adequately known and complicate the quantification of the results.

Less than quartz crystal balances, beam micro-balances with an electrical mass compensation are applied. However, these beam micro-balances are mechanically difficult to integrate into a discharge arrangement, but using massive samples, the full complexity of surface reactions by particle interaction could be observed, adsorption or desorption as well as etching, modification or deposition. Of course the application of a beam micro-balance requires carefully taking any additional forces at the sample into account, as there may be electrostatic forces or the aerodynamic lift.

Fig. 4.10 presents an example of a plasma modification experiment at a massive polyethylene sample treated in a CF_4 RF plasma. In the experiment, the polymer is placed on the balance beam within a parallel plate arrangement. Initially, fluorine incorporation in the polymer surface causes the mass increase.

Simultaneously, this modified surface film is etched on the plasma exposed side. After the short initial period the modification penetrates into the bulk polymer and etching on the plasma side succeeds with the same rate, and the modified surface film shifts with a stationary thickness into the bulk polymer.

Besides these basic gravimetrical experiments, sophisticated methods are applied in order to get complex information.

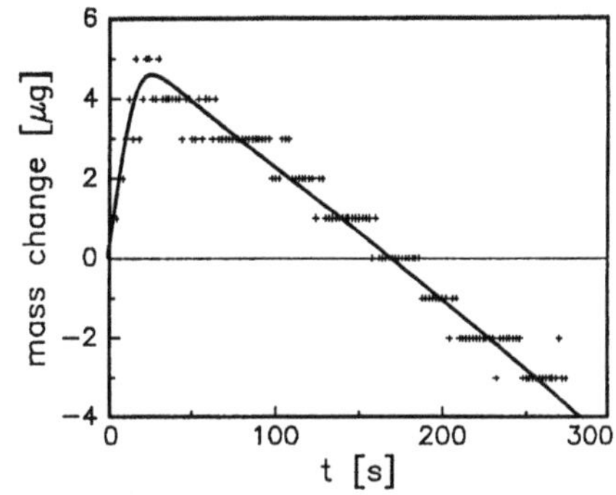

Figure 4.10. Mass change of a polyethylene sample during CF_4 RF plasma modification recorded by a beam micro-balance [25].

Pinhole arrangements are used as a technique for an angle resolved particle flux analysis. Depending on the discharge parameters, there may be an angle distribution of the particles created in the plasma sheath in front of the pinhole, and it may additionally reflect the spatial origin of the species within the discharge. The particles entering the collector through the pinhole cause a film formation on a spherically shaped collector (Fig. 4.11), and with the thickness data of the film, one can calculate the angle resolved particle flux into the collector. Due to the geometrical projection of the pinhole on the collector, an angle dependent correction of the effective pinhole area has to be taken into consideration.

Some basic conditions must be met for quantitative measurements. Sticking of the species at the collector should be reliable, and species re-emission by desorption or sputtering unlikely. Collisions with the background gas in the analyzer must be avoided and therefore the particle mean free path, λ_{sp}, must exceed the collector radius, R. For all pinhole experiments, the diameter, D, of the pinhole should be small against the collector diameter, R, but large enough to secure an adequate flux and deposition rate. An ideal aperture with the thickness, d, small in comparison with the diameter, D, ensures a low probability of interaction processes with the aperture walls and reduced shading effects (Fig. 4.11).

Pinhole techniques are also used for angle resolved measurements in combination with energy analyzers and mass spectrometers [11,12,56,57].

However, the sticking coefficients for the different species have primary importance in the interpretation of the experimental results. The simple gravimetrical methods represent, due to the effective growth rate, only the flux of the particles sticking at the surface. Contrarily, mass spectrometrical methods and retarding field analyzers mostly sample the whole particle flux and are therefore not suited for the analysis of the adsorption-desorption-kinetics. For the interpretation of the measurements and for the understanding of the growth mechanisms, knowledge about the kinetics of surface adsorption, reactions and desorption is essentially necessary.

Unfortunately, sticking coefficients for many species, especially radicals, are not adequately known. The reasons for this lack are the limited capabilities of the measurement techniques and the strong dependencies of the surface phenomena on discharge parameters, especially surface temperature and particle bombardment [58]. However, pinhole-cavity techniques could be used for analyzing the surface kinetics and completing the other experimental information [59-61].

The surface loss probability can be measured considering the transport of the reactive species inside a cavity (Fig. 4.12) which consists of a closed volume with a small entrance slit through which reactive species can enter. The surface loss probability is equal to the sum of the sticking coefficient and the reaction probability of the species to form a non-volatile product at the surface. Due to the surface loss and re-emission characteristics of the species a typical film thickness distribution is found at the bottom and top of the cavity. Assuming a cosine-type angle dependent re-emission characteristics, one can calculate, from the film thickness, the surface loss probabilities. A high loss probability produces a sharp film profile opposite of the entrance slit, whereas low loss probabilities tend to form a homogeneous film on both the bottom and top cavity inside.

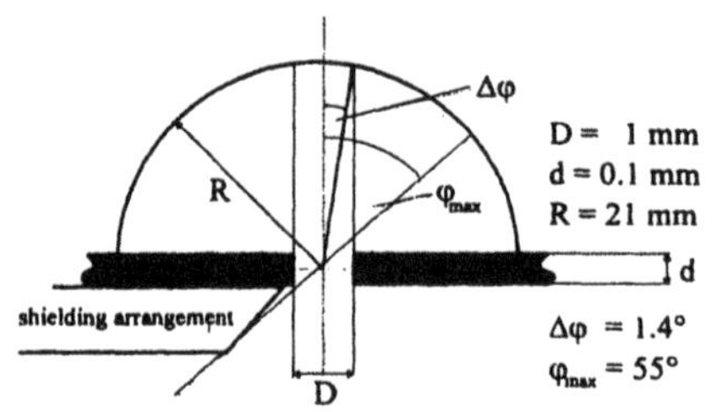

Figure 4.11. Schematic drawing of a semicircular pinhole collector for angle resolved flux measurements [55] (with permission).

However, the information provided by the cavity method is not species selective. Therefore determining species specific loss probabilities, the flux composition should be proved by mass spectrometrical measurements. As an important advantage, the temperature dependence of the surface reactions could be reproduced by this method. Nevertheless, other factors important for the real surface processes, like the flux of high energy particles, remain unconsidered by this technique.

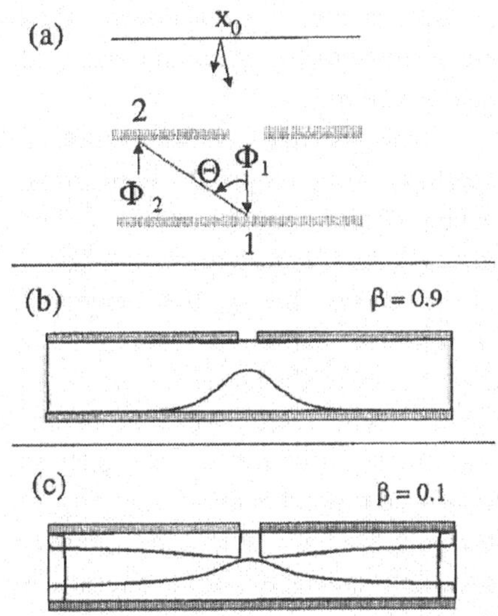

Figure 4.12. Schematic diagram of a cavity experiment with the particles entering at the top slit (a). Resulting film thickness for a surface loss probability of β=0.9 (b) and β =0.1 (c) [59] (with permission).

4.3.2. Mass Spectrometrical Methods

Quadrupole mass spectrometers represent the equipment standard for mass filtering in corpuscular plasma diagnostics. The main reasons for the dominating role of quadrupole mass spectrometers are the relatively small size, easy mounting to the process chamber, and the weaker dependence of the filtering characteristics on the kinetic energy of the particles. However, the performance quality increases for mono-energy ions, and for this reason, especially in plasma ion analysis, energy filters are inserted in front of the mass filter [1]. However, mass spectrometer systems could be used for direct line-of-sight analysis of ions or neutral species from the discharge plasma as well as for partial pressure analysis. There are special features adapting the spectrometer layout for the specific measurement problem.

Fig. 4.13. presents typical analytical equipment designed for particle flux analysis directly impinging on the discharge electrode. The complete mass spectrometer system consists of an aperture separating the process from the analytical equipment, an ion optical system for the transmission of the process ions, an electron impact ion source for post-ionization of neutral species, an energy filter, a quadrupole mass filter, and an ion detector. There are different requirements in the layout of each element, and care should be taken that stabile, quantitatively reproducible and meaningful results may be acquired.

For the interconnection between process and the mass spectrometer, a pinhole aperture should be used, which is essentially necessary for the line-of-sight based plasma particle detection. Also for an exact quantitative partial pressure analysis, an aperture offers clear advantages in comparison to needle valves.

The design of the entrance aperture should fulfil two basic geometrical requirements. The diameter D of the aperture is required to be below the Debye length λ_D ($D<\lambda_D$) in order to prevent any disturbance of the potentials at the ion optical mass spectrometer elements and their influence on the plasma sheath conditions. One should always bear in mind that the energy distribution detected at the aperture is mainly formed in the plasma sheath in front of the surface, and also the flux composition may be altered within this region. For this reason, any disturbance of the sheath conditions should be avoided, if possible, and sampling through an electrode or wall surface appears to be the optimum solution. Biasing the aperture electrically, and/or sampling from the bulk plasma always creates specific sheath conditions which have to be considered in the discussion of the results.

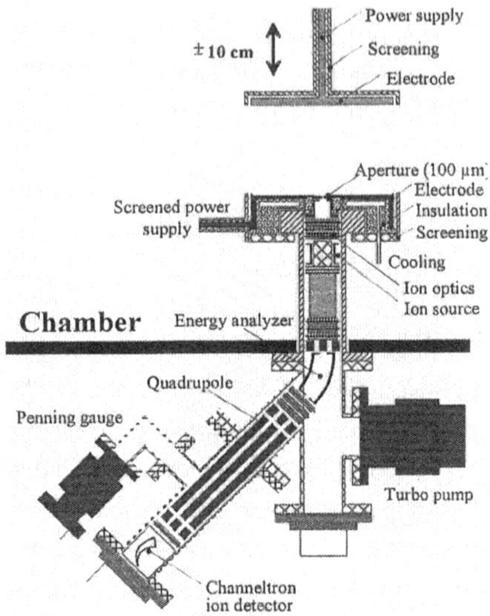

Figure 4.13. Drawing of an energy resolved mass spectrometer arranged for sampling through a discharge electrode [13].

As a second important condition, the diameter, D, of the aperture should also be smaller than the species mean free path, λ_{sp} ($D<\lambda_{sp}$). This condition guarantees a molecular flux regime within the aperture, a low probability for reactions within the aperture region, and a pressure independent gas flow conductivity of the aperture which is important for quantitative measurements. For typical discharge conditions, aperture diameters D between 50 µm and some 100 µm are sufficient. The simultaneously required low aperture thickness, d, in comparison with the diameter, D, of $d<D$ is mostly realised using metal foils.

After the particles have passed the aperture, the particle mean free path is required to be larger than the axial analyzer dimension to prevent significant collisions with the background gas. This condition is always met by a differential pumping of the analyzer system. Turbo pumps are mostly used to guarantee a gas load independent pump speed. Diffusion pumps are an unfavourable choice because of the back-diffusion of pump oils. The required

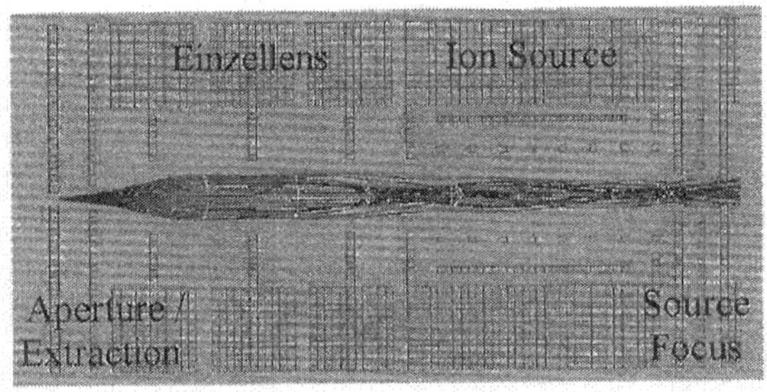

Figure 4.14. Calculation of ion trajectories in the mass spectrometer front region for different incident angles [62].

pump speed has to be determined from the process pressure and the gas flow conductance of the aperture and the mass spectrometer elements. At process pressures exceeding some 100 Pa an additional intermediate pumping stage is necessary which requires a much more sophisticated vacuum design of the analyzer system.

When designing the pumping system, it should be taken into consideration that the effective pumping speed is dramatically reduced by the components placed inside the spectrometer housing. For most applications, effective pumping speeds of some 10 l/s are required at least directly at the aperture which corresponds to nominal pumping speeds of some 100 l/s.

For the detection of plasma ions, ion optical elements directly behind the aperture are responsible for an effective transmission of these species (Fig. 4.14). An extraction electrode accelerates the ions after passing the aperture, and an einzellens focuses the ions to the entrance of the following elements. The ion optical elements generally cause angle and energy dependent discrimination effects. It is obvious that the system has a limitation in the acceptance angle of the ions in the order of several degrees. This maximum acceptance angle depends in turn on the energy of the ions, e.g. the acceptance angle decreases with increasing energy [13]. As will be explained in Sec. 4.3.3, the equipment should be favourably operated in the constant pass energy mode. All stabile ions are accelerated or decelerated, respectively, within the front region of the mass spectrometer to the pass energy. The focus length of the ion optical elements gets stabilized, and consequently the energy dependent discrimination effects are reduced.

For quantitative measurements, the operator should always have an idea on the acceptance angle of the mass spectrometer. In turn, the limited acceptance angle may be utilized for angle resolved ion analysis by tilting the aperture plate and the mass spectrometer with the ion optical system one against each other [11,12].

For the direct flux analysis of neutral plasma species and for partial pressure measurements a post-ionization must be realised. Different types of electron impact ion sources are used in quadrupole mass spectrometry, while cylinder symmetrical open sources are the standard solution. These sources consist of a cylindrical anode grid with mostly two filaments arranged at the outside (Fig. 4.14). The electrons emitted from the filament are accelerated towards the anode and ionize the neutral species within the cylindrical grid. This type of electron impact source represents a satisfying compromise between a good ionization rate and sufficient control of the electron energy.

Due to the maximum of the electron impact cross sections (Fig. 4.2) the sources are driven at an electron collision energy of about 70 eV - 100 eV. Lower electron energies are used in special applications like threshold ionization and electron attachment mass spectrometry and require a superior control performance of the source. These techniques are discussed in detail in Sec. 4.4.2 and 4.4.3.

Especially for the plasma flux analysis of neutral species, an optimum performance of the ionization is required. Geometrically, the ion source should be arranged directly behind the pinhole aperture, and the aperture diameter should be as large as possible. Biasing the ion source or applying a retarding potential in front prevents plasma ions from overlaying with the signal of the post-ionised neutrals. However, depending on the operating conditions, the ionization probability for the neutral flux is typically in the order of about 0.1 %, and especially for high energy neutrals this probability is further reduced. For this reason, the results from direct neutral flux analysis are much rarer than from ion flux analyzis.

The electron impact ion source is followed by an optional energy filter. The energy filter is, of course, needed for energy resolved measurements, but is also necessary for mass analysis of high energy species. The performance of a quadrupole mass filter is less sensitive against a small variation of the entrance energy. However, increasing energy reduces the mass resolution and at an energy of some 10 eV, the filter function of the quadrupole is lost. Because of this, a proper mass filter performance in plasma ion analysis requires a prefiltering of the ion energy. The functions and layout of energy filters applied in corpuscular diagnostics are widely spread and discussed in the following Sec. 4.3.3.

The quadrupole mass filter acts as a band pass filter due to the mass to charge ratio m/e of the ions. Applying a combination of RF and DC voltages to the quadrupole rods conditions (voltage combinations) resulting in stabile ion trajectories within the quadrupole may be obtained as follows from the Mathieu differential equations. Driving the quadrupole with a defined ratio of RF and DC voltage allows for a direct scan of the m/e ratio.

However, each quadrupole responds sensitively to the residence time of the ions within. Driving the quadrupole at a constant injection energy causes individual ion velocities for the different masses and results in a discrimination of the heavier ions. In the most modern mass spectrometers, the ratio between RF and DC voltage in the quadrupole may be varied dependent on the ion mass. This way resolution and sensitivity may be adapted to the experimental requirements, and the mass dependent discriminations effects could be compensated for at the expense of resolution.

An important parameter for the quality of the quadrupole, e.g. the mass resolution and the transmission characteristics, is the diameter of the quadrupole system. A small quadrupole with a rod diameter up to about 5 mm could be used in a mass range of not more than 200 amu. Heavier ion masses require larger quadrupole dimensions in order to achieve a sufficient transmission and resolution in the high mass range [63].

Triple quadrupole mass filters are often used to improve the performance of smaller quadrupole systems. In such an arrangement a geometrically short pre- and post-filter, which are driven by the RF voltage only, are arranged in front and behind the quadrupole. The transmission characteristics could clearly be improved without reaching the performance of larger rod systems.

Finally, the ion flux is detected by a Faraday collector or a secondary electron multiplier (SEM), favourably the spectrometer should contain both detector types. The Faraday collector has got a poorer sensitivity but offers some important advantages against a SEM. The Faraday collector is resistant against excessive current flow and unfavourable vacuum conditions, whereas the multiplier requires an operating pressure better than 10^{-4} Pa. Further more the Faraday collector provides an excellent signal linearity and could be used for calibration of the SEM signal.

Different types of SEM, with cascade or continuous dynode, are used in mass spectrometry and are optimized to be driven in an analogue or digital ion counting mode. The analogue mode delivers a sensitivity which is, by the amplification factor, better than a Faraday collector, whereas counting electronics allows qualitatively new operation modes. Ion counting gives a much higher dynamic range of about 7 decades with simultaneously better basic sensitivity. Additionally, gating and synchronizing the counting electronics by an external signal allows time resolved measurements typically up to about 100 kHz.

The operation characteristics of each SEM depends strongly on the surface conditions of the dynode coating and has to be periodically checked. The sensitivity clearly drops in the initial days of operation and has to be compensated by a higher multiplier voltage. It stabilizes in the later operation period, but unfavourable operating conditions like high pressure or reactive environment may affect the SEM performance, and the sensitivity should be checked at regular intervals. Besides the sensitivity care should be taken to prove the signal linearity, and to avoid the detector being driven in the saturation regime [64].

An absolute quantification of the mass spectrometrical measurements without any calibration standard is nearly impossible. However, direct particle flux measurement is much more difficult to quantify than partial pressure analysis.

The mass spectrometer layout for partial pressure analysis can be obviously simpler. No ion optics at the entrance, no energy analyzer, and only a relatively simple ion source are required for this purpose. It is very important that the pressure reduction between discharge and spectrometer is realised by a molecular flux connection which guarantees a linear pressure correlation and which avoids any changes in the relative gas admixture.

For exact partial pressure analysis the specific sensitivity of the mass spectrometer should be experimentally determined for each of the process gases that are expected to be supplied or to be formed in the discharge. The sensitivity should be determined directly related to the partial pressure in the discharge chamber, taking into account all sensitivity factors of the equipment including the pressure reduction, the post-ionization, the transmission, the quadrupole and the detector sensitivity.

The experimental determination of the sensitivity factors should be preferred wherever possible. While the chamber is filled up with the corresponding gas, and the total pressure is sensed with a pressure gauge with the sensitivity not depending on the kind of gas, the whole mass spectrum of each gas must be acquired. Subsequently absolute and relative sensitivity factors are calculated and give the constants for the partial pressure analysis.

With the calibration procedure performed, the partial pressures are calculated solving a system of linear equations:

$$I(MZ_i) = E_{ref} \sum_{j \in n} E_j e_{ij} p_j \qquad j=1...n \qquad (4.8)$$

where I is the ion current at mass MZ_i.

It is essentially necessary to check the absolute sensitivity, E_{ref}, of the equipment before each measurement cycle in order to compensate any possible system drift. This global sensitivity is reasonably determined using the process gas or a rear gas before the discharge operation. All other gas specific sensitivities are related as relative values to this absolute sensitivity, E_{ref}, and are only measured during the calibration procedure. The sensitivity of the most intense base peak of the gas j (related to the reference gas) is given by the relative sensitivity, E_j. The factors e_{ij} are related to the base peak in each mass spectrum and symbolize the fragmentation pattern of the gas j which contributes to an ion current at mass i. With the measured ion current, I, at selected masses MZ_i, the partial pressures, p_j, could be determined by solving the Equ. (4.8).

Especially the structure of the sensitivity matrix, e_{ij}, will influence the quality of the partial pressure measurement. The peaks in the mass spectra used for the calculation should be chosen so that most of the matrix elements, e_{ij}, remain empty, and that the intensity of the included peaks is in the same order of magnitude. However, in practice this condition is hard to achieve because often homologue series of gases produce the same dissociation fragments and cause a larger measurement insufficiency. However, the partial pressures should always be checked against the absolute pressure.

In contrary to the partial pressure analysis, the direct flux measurements are very difficult to quantify, and often only qualitative results are presented. This deficiency is caused by the mostly unknown post-ionization probability, the incident angle, mass and energy dependent transmission effects. Especially measurements of the ion flux using a retarding field analyzer could provide a reasonable quantitative standard (Sec. 4.3.3) [1].

4.3.3. Energy Filter Techniques

Energy filters are applied in different configurations and in combinations with mass spectrometrical methods as explained in the previous section. Besides the primary function of energy selection, band pass energy filters enable a sufficient performance of quadrupole mass filters by monochromating the particle flux, especially in analyzing high energy species. However, all energy analyzers use electrical fields, and for this reason only ions and post-ionised neutrals may be analyzed.

According to the function principle, two important types of energy filters are applied in corpuscular plasma diagnostics: retarding field and band pass energy filters. Retarding field analyzers (RFA) act as a high energy pass filter by applying a potential barrier for the low energy ions. In contrary, by means of an electrostatic deviation, the hemispherical energy filters analyse an energy band from the distribution function.

Retarding field analyzers are based on grid and lens systems. Retarding lens systems are actually replaced in mass spectrometry applications by hemispherical analyzers, and therefore only the grid type retarding field analyzers will be discussed in the following.

Grid type retarding field analyzers mostly consist of two or three plane or spherical grid systems and a particle collector (Fig. 4.15). The general working principle is similar to Langmuir probes by discriminating the low energy part of the distribution function. The particle velocity distribution of the ions with the mass, m_i, is calculated from the deviation of the particle current, $I(\varphi_r)$, versus the retarding potential, φ_r:

$$f(v) = -\frac{m_i}{e^2}\frac{dI(\varphi_r)}{d\varphi_r} \tag{4.9}$$

The retarding field analyzer senses the velocity component of the distribution function into the direction of the analyzer axis, and hence is able to analyse anisotrop distribution functions [65-67]. By means of the voltage setting, the analyzer could be tuned for positive or negative charge carriers, and a spherical grid arrangement together with a segmented collector allows angle resolved measurements of the velocity distribution [67]. Finally, geometrically miniaturized analyzers could be used for spatially resolved diagnostics [68].

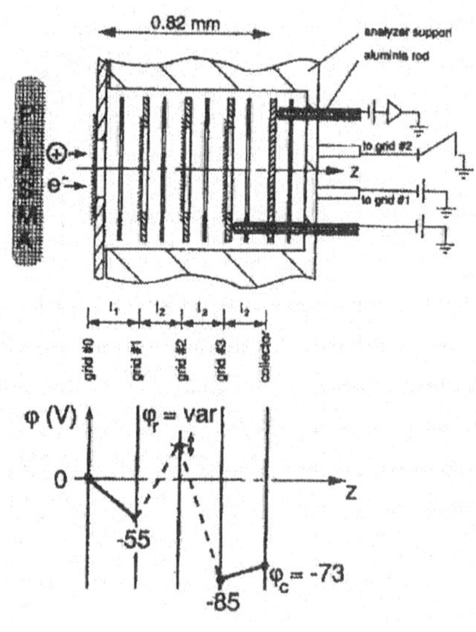

Figure 4.15. Schematic drawing of a retarding field analyzer and the potential distribution across the grid system [65] (with permission).

A schematic drawing of a retarding field analyzer together with the potential structure for positive ion analysis is shown in Fig. 4.15. The layout of each analyzer due to the number of grids and the electronic structure may differ but the basic functions remain the same. The particle flux enters through a pinhole aperture or an entrance grid biased on a value of the electrode or wall potential. The next grid suppresses the plasma electrons and is followed by the variable potential retarding field grid. A conceivable grid in front of the collector repels secondary electrons before the ions hit the collector.

Using retarding field analyzers, various experimental problems should be solved:

- The grid potentials must exactly be matched, and any insulation films on the grids would disturb the analyzer function. Therefore an application of a retarding field analyzers in plasma polymerization requires intense equipment control and maintenance.
- A mesh size of the grid smaller than the Debye length guarantees homogeneous equipotential lines.
- The axial analyzer dimension should be shorter than the ion mean free path. For most of the retarding field analyzers, no differential pumping is applied, and under these circumstances the maximum useful operation pressure is some 10 Pa.
- Transmission effects of the analyzer, due to the limited grid transparency, have to be taken into consideration.
- For small incident angles, the grid holes may act as channels, and an incident angle dependent discrimination may occur.
- Charged secondary particles produced at the grids may affect the measurement results.

Comparing the grid type retarding field analyzers with the hemispherical band pass analyzers discussed in the following one has to conclude that the retarding field analysers are clearly less powerful; however, they offer, on the other hand, some important features.

Because of the high energy pass characteristics, a combination of retarding field analyzers with mass spectrometers delivers only a poor system performance. For this reason, most of the retarding field analyzers are constructed as stand alone systems for energy analysis only. Finally, a poorer energy resolution and a clearly lower dynamic signal range characterizes the retarding field analyzers in comparison with the hemispherical analyzers. However, the most

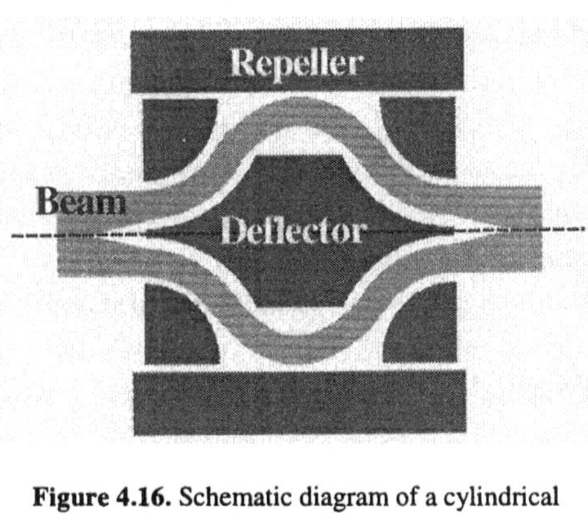

Figure 4.16. Schematic diagram of a cylindrical mirror analyzer.

important advantage of a stand alone retarding field analyzer is that the discrimination effects are easier to control, and exact quantitative measurements can be performed. Because of this the retarding field analyzers could provide a calibration standard for the more complex methods.

The optimum method for energy analyzers are represented by hemispherical analyzers which act as an energy band pass filter. In practice the two important types of hemispherical analyzers are used, the electrostatic cylinder analyser (Fig. 4.13) and the cylindrical mirror analyser CMA (Fig. 4.16).

The pass energy E of an electrostatic cylinder analyzer is determined by the potential difference φ_r applied between the cylinder plates and their radii r_1 and r_2:

$$E = \frac{z\,e\,\varphi_r}{\ln\left(r_2 / r_1\right)} \tag{4.10}$$

The energy resolution in turn is affected by the beam entrance angle into the analyzer, the diameter of the defining exit aperture, and the mean analyzer radius [1].

Whereas the electrostatic energy analyzer represents an off-axis analyzer, the cylindrical mirror analyzer, CMA, realises an in-axis version. The CMA analyzer is a serial arrangement of 3 miniaturized hemispherical analyzers, the first spreading the beam out of the axis, the second reflecting the ions, and the third directing the beam back into the axis. The general operation characteristics of both analyzer types is the same. The CMA allows a compact arrangement of experimental equipment, but delivers a poorer filter performance. Because of the small size, the filter characteristics are not affected so much by the adjustment problems and coatings at the filter elements. But the most important disadvantage of the CMA filter is the lower filter transmission because the central part of the beam hits the deflector in the arrangement.

The hemispherical filters may be driven in two operation modes. Scanning the voltage on the analyzer elements results in a variable exit energy which is unacceptable for a following mass spectrometer operation. Because of this the hemispherical analyzers are mostly driven in a constant pass energy mode, and the ions are accelerated or decelerated to the pass energy before entering the energy analyzer. An energy scan gets performed by shifting the reference potential of the whole equipment including a mass spectrometer together (Fig. 4.17). In practice the analyzer is driven with a pass energy of some 10 eV, and an FWHM of the pass energy in the order of 1 eV is achieved. After leaving the energy analyzer, the stabile ions are decelerated to the low kinetic energy required in the quadrupole.

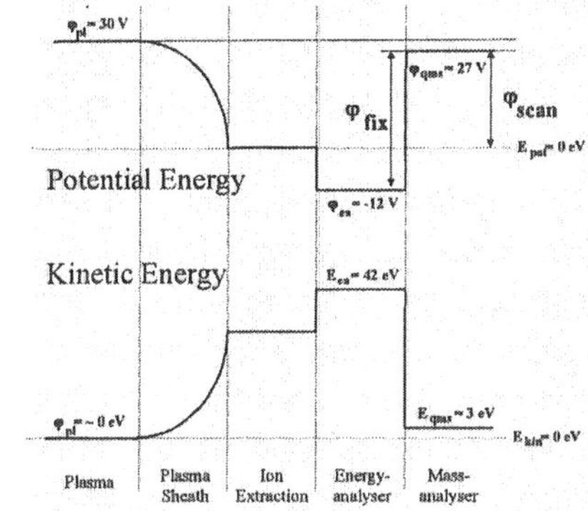

Figure 4.17. Potential structure and ion energy within a combination of hemispherical energy analyzer and mass spectrometer.

4.4. Neutral Species Detection

4.4.1. Neutral Flux Detection and Partial Pressure Analysis

An enormous number of scientific work is concerned with analytical investigations of plasma polymerization processes. Most activities focus on the analysis of the film properties, but a majority of contributions is devoted to the understanding of the deposition process where mass spectrometric techniques are mostly applied for gas and discharge diagnostics. The corpuscular diagnostics focuses on the deposition of the film systems:

- C-H – hydrocarbon films from a variety of monomer gases [69-71],
- C-F – perfluorinated films deposited from C_XF_Y-monomers [72-73] and
- Si-O-C-H – silicon-organic films mostly deposited from hexamethyldisiloxane (HMDSO), hexamethyldisilazane (HMDSN) and tetraethyloxysilane (TEOS) [74-78].

In most of the contributions, particle fluxes and gas composition are only analyzed qualitatively as complementary to other investigations. In addition, due to the often complex monomer substances, it is extraordinarily difficult in the mass spectrometrical measurements to draw conclusions about the original species because of their complex dissociation pattern which is formed in the discharge and in the electron impact source of the mass spectrometer. The following Sec. 4.4.2 and 4.2.3 reveal special methods which partially clarify the difficulties resulting from the fragmentation processes. However, in this section we present some selected examples on standard partial pressure and species analysis.

At the beginning, some fundamental statements are presented about the deposition of polymethylmethacrylate (PMMA) films from an RF discharge in methacrylacidmethylester (MMA) [79]. The film deposition was performed operating a parallel plate RF discharge in a sealed (no flow of monomer) vacuum chamber of 120 l volume.

Fig. 4.18 shows the time dependent partial pressures of the most important gases in the discharge measured by common mass spectrometry. The initial MMA monomer is decomposed over a period of a few minutes. The resultant low molecular weight hydrocarbons, especially ethane and propane, are generated by recombination of methyl- and ethyl-groups. Higher molecular weight alkanes up to pentane are detected, but in much lower concentrations. Hydrogen, carbon monoxide and a smaller amount of methane and carbon dioxide are generated as the final and most plasma stabile products. The production of the low molecular fragments including reactions with residual water in the chamber causes an increase of the total pressure by a factor of about 3.

The gas composition in the discharge is in turn the most important factor determining the observed film growth rate (Subplot Fig. 4.18). The initial maximum in the polymer growth rate is caused by plasma activated MMA, and only during this limited discharge period the typical (plasma polymer)MMA films are deposited. In spite of the quickly disappearing MMA concentration, a smaller decrease in the growth rate is observed in the later discharge period. The film growth after the maximum is continued by activated species produced from the low molecular weight hydrocarbons, and therefore, the chemical character of the layer changes from PMMA to amorphous plasma polymer hydrocarbon films, as observed by surface analysis for other monomer substances [74]. During this later discharge period the growth rate decreases proportionally with the partial pressure of ethane and propane.

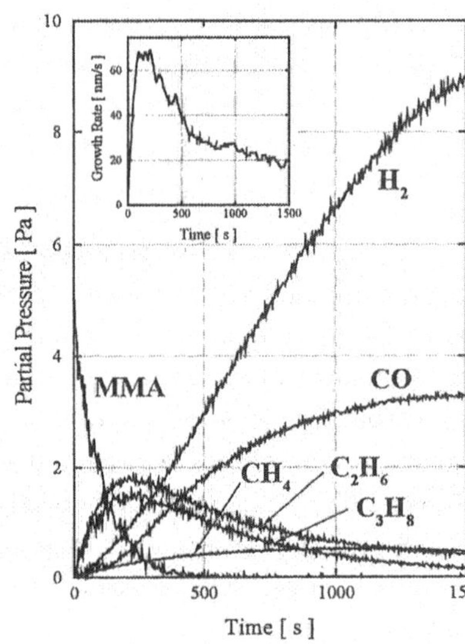

Figure 4.18. Time dependent partial pressures of the most important gases and corresponding film growth rate in a MMA discharge (p_{tot}=5–15 Pa, P_{rf}=160 W) [79].

However, any film deposition in a sealed vacuum chamber causes varying film properties due to the changes in the gas composition and therefore should not be applied in practical deposition processes. But analyzing the chemical behaviour in the sealed (static regime) process allows an overview and understanding of the complex reaction behaviour and the determination of the operation conditions for practical film deposition.

The discharge conditions in Fig. 4.18 require an optimum operation point with a gas residence time, τ, in the discharge of about τ=100 s for the deposition of PMMA-like films at a simultaneously high deposition rate. This gas residence time has to be adjusted by an adequate monomer gas flow and pump speed.

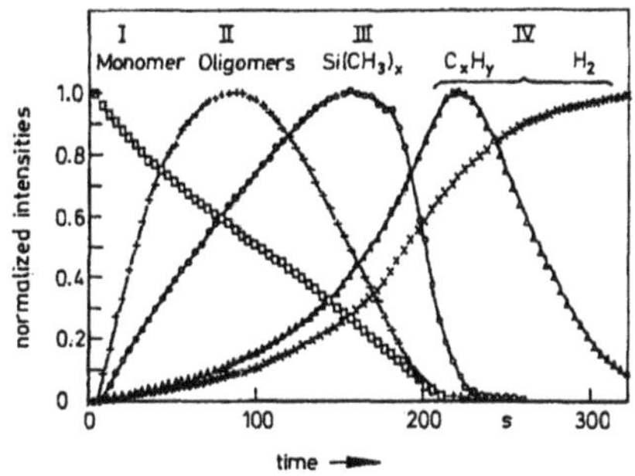

Figure 4.19. Plasma chemical gas conversion of HMDSO in the sealed chamber ($p_{tot}(t_0)$=15 Pa, P_{rf}=30 W) [74] (with permission).

The optimization of the operation point is especially important if several process gases are supplied simultaneously to the process. In order to avoid long initial reactions in the discharge, a gas admixture, that corresponds to the gas composition, which self-adjusts in the operation point favourably should be supplied to the process.

A process scaling is possible by applying the similarity law according to Equ. (4.6) which requires the same specific RF power supplied per gas particle. Consequently, a higher deposition rate at similar reaction conditions will be realized at higher RF power and an appropriate lower gas residence time. Of course the adjustment of the optimum operation conditions always requires a control of the important film parameters.

Using complex monomers, like silicon-organic substances, the gas composition is mostly discussed quantitatively, because sensitivity factors for all possible fragment gases are difficult to acquire. Fig. 4.19 presents an example of the plasma-chemical gas conversion in an RF discharge operated in an hexamethyldisiloxane (HMDSO) vapour [74]. As in the experiment explained before, the discharge is operated in a sealed vacuum chamber with a volume of 20 l, gaseous species are observed by common mass spectrometry, the deposition rate and the optical properties by in-situ and ex-situ ellipsometry, and the chemical structure by ex-situ FTIR-spectroscopy.

The monomer HMDSO is decomposed by diverse elementary processes. The unstable dissociation fragments are consumed by the growing polymer film or in the production of new stable gases (Fig. 4.19). Generally, the gaseous compounds generated in the discharge can be classified into four basic groups:

the initial monomer HMDSO (I), silicon-organic oligomers with a higher molecular weight than HMDSO (II), methylsilanes (III), and hydrocarbons and hydrogen as the finally dominating products (IV). The HMDSO content in the chamber decreases continuously, and new compounds are produced in the gas phase with different production rates. After the consumption of Si-O-containing gaseous compounds, only hydrocarbons and finally hydrogen are found.

The deposition rate of the polymer film changes in time as a result of the gas composition (Fig. 4.20.a). In the first period (A), the generation of polymerizing species from HMDSO leads to an increased deposition rate, in spite of the decreasing HMDSO concentration. With the further consumption of HMDSO and the other silicon containing reaction products, the deposition rate decreases. With the vanishing concentration of Si-O-containing gases, the remaining hydrocarbons drive a deposition process with clearly lower growth rate and changed film properties (B).

FTIR analysis demonstrates by the integral extinction of selected bands of the whole film that silicon containing groups reach a saturation level, whereas the CH-groups are continuously incorporated into the growing film (Fig. 4.20.b)

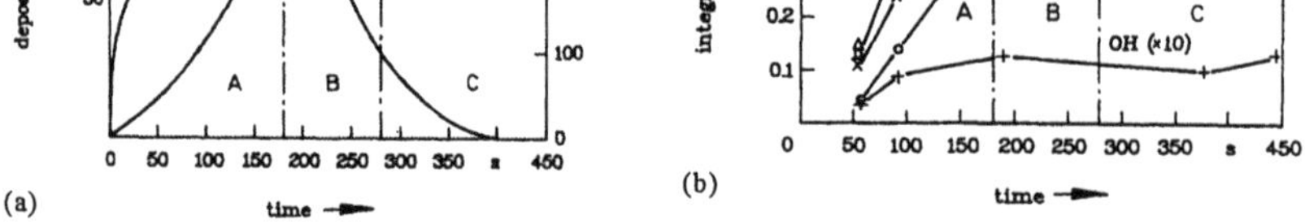

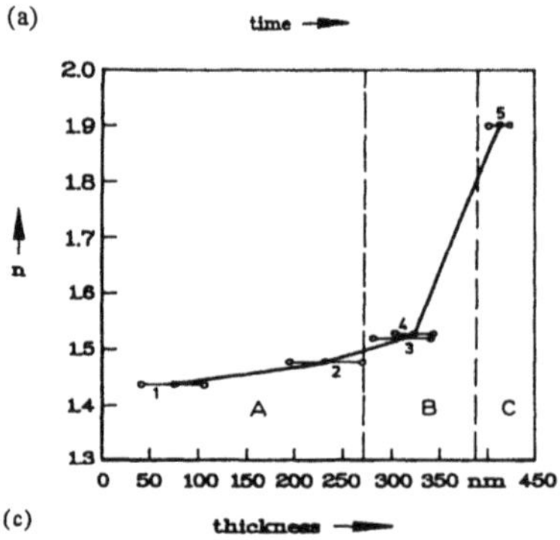

Figure 4.20. Relationship between plasma chemical gas conversion and structural properties of a plasma polymerized HMDSO film: (a) Deposition rate, (b) IR absorption, (c) Graded refractive index (Parameter: Sealed chamber, p_{tot}= 10-50 Pa, P_{rf}=30 W) [74] (with permission).

In the final period (C) of the deposition process, hard a-C:H films are deposited.

For the analysis of the refractive index films, which were deposited only in a limited time period using a shutter (labelled 1-5 in Fig. 4.20.c), spectroscopic ellipsometry was used. The variation of the film structure and chemical composition results in a graded refractive index across the film thickness from about 1.4 to 1.9 (Fig. 4.20.c).

4.4.2. Threshold Ionization Mass Spectrometry

Because of the very complex dissociation patterns of most organic substances used in plasma polymerization processes, an identification of the molecules is quite difficult. These problems result, on the one hand, from the multiplicity of fragments generated from each molecule and, on the other, from the superposition of the same fragments produced in the dissociation of different parent molecules. Especially radicals, which are very important for the plasma polymerization processes, could not easily be resolved from the contribution of the process monomer or other discharge gases. Threshold and electron attachment ionization are presented as methods partially overcoming these difficulties.

The threshold ionization technique (TIMS) is based on the principle of selective ionization of neutral species by means of the electron impact energy in the ion source of the mass spectrometer. A selected ion A^+ detected by the mass spectrometer could be formed in dissociative ionization of the parent molecule with the impact energy E exceeding the ionization potential E_{AB}:

$$AB + e^- \rightarrow A^+ + B + e^- \qquad E \geq E_{AB} \qquad (4.11)$$

In the discharge the radical A may be present which could be ionized by:

$$A + e^- \rightarrow A^+ + e^- \qquad E \geq E_A \qquad (4.12)$$

Since the dissociative ionization (4.11) requires the ionization and the dissociation energy to be transferred by the electron, the energy E_{AB} is about the binding energy of the molecule AB higher than the energy E_A required for the ionization of the radical A.

Tab. 4.1 gives the ionization potentials for producing fragment ions from different CH_X species. Applying an electron impact source with a sufficient low energy width of about $\Delta E<0.5$ eV allows a selective ionization of the different species. Exceeding the corresponding ionization potential, the spectrometer

Table 4.1. Ionization potentials of radicals from CH_4 [5].

Species	CH_4^+	CH_3^+	CH_2^+	CH^+	C^+
CH_4	12.6	14.3	15.1	22.2	25.0
CH_3		9.6	15.1	17.7	25.0
CH_2			10.3	17.4	20.2
CH				12.0	20.3
C					16.8

signal increases ideally step by step, which corresponds to each species getting ionized. The number of applications of the threshold ionization technique drastically increased with the availability of comfortable computer controlled mass spectrometers allowing automatic electron energy scans and discharge control operations.

In practice, the threshold ionization experiments are performed comparing the signals generated in the plasma off, where only the process gas may contribute to the signal intensity, and the plasma on state with the ionization of the radicals superimposed (Fig. 4.21). Due to the lower ionization potential of the radicals, a low energy shoulder in the electron energy dependent mass spectrometer signal identifies the radical contribution. For exact measurements, of course the contribution of the gas background has to be taken into account [1].

Threshold ionization mass spectrometry was successfully applied for the detection of different radical species in gas discharges:

- CH_X-radicals [5,81,82]
- CF_X-radicals, [80,83] and
- SiH_X-radicals [6,84].

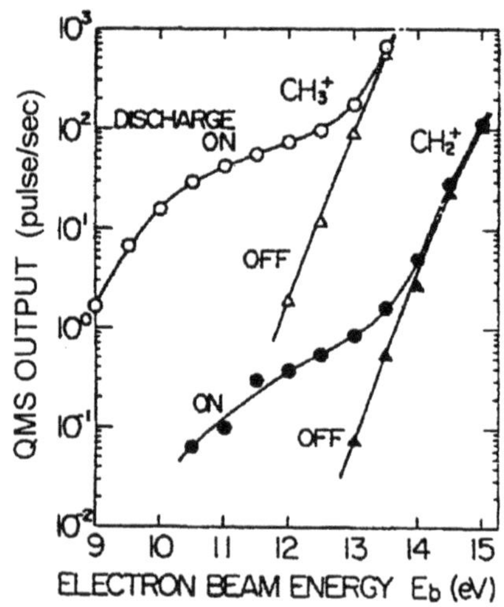

Figure 4.21. Mass spectrometer signal of CH_2 and CH_3 from a methane discharge dependent on the electron impact energy with the discharge switched on and off [5] (with permission).

A number of interfering factors should be considered for mass spectrometry of free radicals.

By applying retarding fields, ionized radicals and high energy electrons from the discharge must be prevented from entering the mass spectrometer ion source. The electron energy width should be reduced as much as possible by a low voltage drop across the source filament and by a good potential uniformity within the ion source. Radical formation by a thermal dissociation at the source filaments should be checked carefully. Finally, since the free radicals are extremely reactive, compounds may be formed and radicals may be consumed in surface reaction within the aperture and at any surface within the mass spectrometer as well as with the mass spectrometer residual gas at unfavourable vacuum conditions.

Precise measurements of the absolute radical flux are a difficult task. For the quantification, the dependence of the mass spectrometer ion current signal I_i is used:

$$I_i = \alpha\, \sigma(E_e)\, I_e\, n_n \tag{4.13}$$

α denotes the spectrometer specific loss and transmission factors, σ the impact cross section at an electron energy E_e, I_e the electron current and n_n the interesting species concentration in the spectrometer ion source. For the quantification of the radical measurements, mostly a similarity of the equipment effects between the detection of the radical and the parent molecule is assumed, and taking only the different impact cross sections into account, a quantitative calibration is achieved [5,81].

As an example, Fig. 4.22 presents axially resolved radical densities from an RF discharge in methane. With a variable position of the entrance aperture of the mass spectrometer, axially resolved measurements of the density profiles of the CH_2 radical between the discharge electrodes of a capacitively coupled RF discharge are performed. For each of the detected CH_3 and CH_2 radicals, specific density profiles are detected which depend in turn on the discharge conditions. At low discharge pressure a homogenous radical

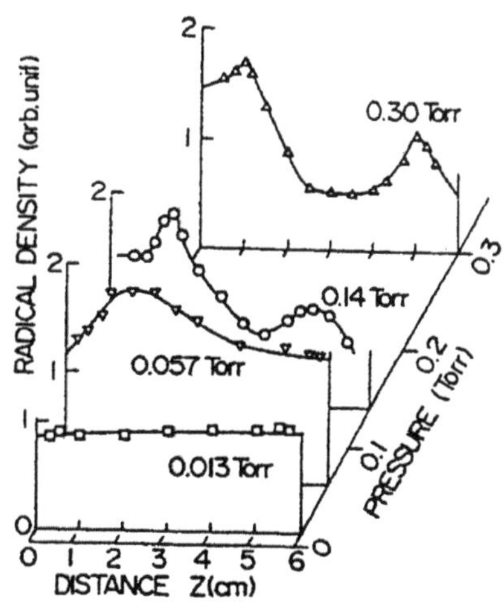

Figure 4.22. Axial density profile of the CH_2 radical in an RF discharge in methane (Parameter: Total pressure, P_{rf}=10 W, Powered electrode at z=0 cm) [5] (with permission).

generation rate is observed between the electrodes, however, at high pressure the radical production is located mostly in front of the powered RF electrode. The observed density profile is adjusted due to the radical specific diffusion and recombination coefficients.

4.4.3. Electron Attachment Mass Spectrometry

The complicated fragmentation patterns during electron impact ionization of larger molecules are an important handicap in distinguishing between the different constituents. Besides the threshold ionization mass spectrometry (TIMS), electron attachment mass spectrometry (EAMS) offers an elegant solution in the case of electronegative gases and an experimental complement for these problems.

Electron attachment mass spectrometry utilizes the production of negative ions from electronegative species (Sec. 4.2.1.) like oxygen, fluorine and chlorine containing process gases. The most important process in negative ion production for molecules and low energy electrons is the dissociative attachment:

$$XY + e^- \rightarrow X^- + Y \tag{4.14}$$

For the two important reasons the negative ion production is much more species selective than the impact ionization. Generally, the electron attachment is less destructive than the impact ionization, and large fragile molecules survive the attachment of a low energy electron rather than a high energy electron impact. The resonant character of the attachment cross sections is even more significant for the species selective character of this procedure. By setting the electron energy, particular molecules can be selected for the attachment event.

However, electron attachment cross sections for perflourcarbons are typically in the order of $5 \cdot 10^{-18}$-$5 \cdot 10^{-19}$ cm^2 [85-87], and hence 2-3 orders of magnitude lower than the electron impact ionization cross sections. For this reason, the resulting signals in electron attachment mass spectrometry are mostly of lower intensity.

Within the last few years the electron attachment mass spectrometry has been developed as a specialized but powerful tool in particular diagnostics [88,89]. The method requires a special layout of the electron attachment source and the mass spectrometer electronics that is nearly identical with the case of positive ion mass spectrometry. The attachment source is required to produce an electron beam with a narrow energy width in the range of nearly 0 eV up to about 10 eV, to adjust the electron energy to the resonant cross sections. Extracted from the attachment source, the negative ions are handled by a mass spectrometer electronics providing that all DC potentials are with the opposite polarity than in the case of positive ion detection. Even the most important attachment cross sections and their energy dependence are known from the literature, for quantitative analysis each mass spectrometer system requires gas specific check and calibration procedures.

Fig. 4.23 compares a positive ion mass spectrum obtained by common electron impact ionization (top) and a negative ion mass spectrum acquired by electron attachment (bottom). One can clearly see that in the electron impact ionization, the fluorine is removed from the parent molecule with a resultant variety of fragment ions. The electron attachment is much less destructive. Especially for the detection of large electronegative molecules, which are often found in plasma polymerization processes, this method represents a powerful instrument.

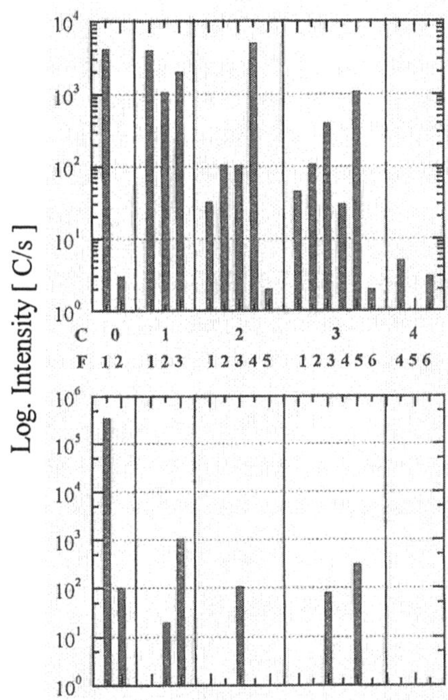

Figure 4.23. Mass spectra of positive ions produced by conventional mass spectrometry (top) and negative ions formed by electron attachment (bottom) (RF discharge in C_4F_8, P_{rf}=15 W, p_{tot}=1.3 Pa, Numbers between figures give ion composition $C_xF_y^{+/-}$, Intensities are integrated with respect to electron energy) [88] (with permission).

4.5. Process Ion Detection

The transport phenomena of the charge carriers to the surfaces, and hence flux densities and energy distributions, are strongly affected by the potential conditions in the plasma sheath facing any surface in contact with the discharge plasma. Besides the different mobility of electrons and ions the electrical excitation principle dominates the spatial and time varying electrical field in the plasma sheath. Due to the ratio of the excitation frequency ω_{ex} and the plasma frequency of the ions, $\omega_{pl\text{-}i}$, and electrons, $\omega_{pl\text{-}e}$, we distinguish regimes with a quite different behaviour.

DC and low frequency discharges with $\omega_{ex} << \omega_{pl-i} << \omega_{pl-e}$ are of minor importance for polymer processing. A dielectric shift current is essentially necessary for depositing or treating the mostly insulating polymer surfaces and plasma polymer films. Therefore, DC discharges are not considered in the following discussion. However, the behaviour of ions in DC plasma sheaths could easily be understood from the discussion of the capacitively coupled RF discharges.

The RF discharges with $\omega_{pl-i} < \omega_{ex} << \omega_{pl-e}$ are commonly used in polymer processing and offer the most interesting aspects with time modulated and high sheath potentials (Sec. 4.2.2).

Excitation frequencies above the plasma frequencies of the charge carriers $\omega_{ex} > \omega_{pl-e} >> \omega_{pl-i}$ exist in microwave discharges. Because the charge carriers are not able to follow instantaneously the excitation field, the sheath phenomena are dominated by the ambipolar diffusion and hence mobility effects. In practice plasma excitation and sample bias are often provided by different power supplies and different excitation frequencies. Because of this, a combination between the phenomena described in Sec. 4.5 takes place [22,23].

However, in polymer processing one should have in mind that a DC potential drop may occur across any insulating film. Therefore the surface potential of the polymer may differ from the electrode or substrate potential on which the film is deposited (Fig. 4.29).

Please, note that, contrary to Sec. 4.4, the formation of the ions treated in the following happens in the plasma itself. The composition of these ions is mainly dominated by the plasma specific ionization regime, and because no additional ionization in the mass spectrometer is applied, equipment effects are of lesser importance for the correct ion composition.

4.5.1. RF Plasma Processing

Using standard frequencies of 13.56 and 27.12 MHz RF excitation represents a standard tool in plasma polymerization. Generally, two basic modes of RF power coupling are applied, inductively and capacitively coupled discharges.

The inductively coupled plasma (ICP) excitation, or transformer coupled plasma (TCP) excitation, respectively, realizes an effective principle in high density plasma generation. A primary RF coil transfers the power to the plasma which acts as the secondary part of a transformer configuration. Taking only the

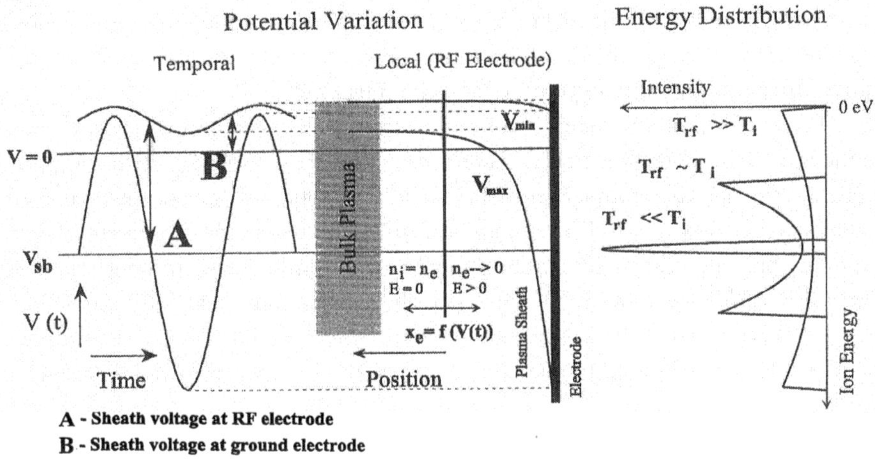

Figure 4.24 Spatial and temporal potential variation in an RF sheath and resulting ion energy distribution at different ratios between RF cycle time T_{rf} and ion residence time in the sheath T_i.

transformer mechanism into account, a low plasma sheath voltage results. However, in practice the high RF voltage at the primary coil may couple capacitively to the discharge plasma. In the case that no Faraday shield screens the electric RF field of the primary coil, a similar behaviour as for the direct RF coupling results [90-92].

Capacitively RF power coupling describes the power transfer directly provided by the electric RF field. Because of the dominating effects of the capacitively RF power coupling on the plasma sheath characteristics and the ion behaviour at the chamber walls, only this excitation regime is discussed in detail in the following.

The typical setup of a capacitively coupled RF discharge is represented by the parallel plate configuration with the powered RF electrode being in direct contact with the plasma (Sec. 4.2.2). Fig. 4.24 shows schematically the resulting time and spatial variation of the potential distribution within the discharge arrangement. Because of the coupling capacitor, a DC self-bias voltage may develop at the powered RF electrode of the order of nearly half of the RF peak-to-peak amplitude. Consequently we have a high RF modulated sheath voltage in front of the powered RF electrode (A), and a clearly lower sheath voltage in front of the grounded electrode (B). Besides the potential variation several additional factors determine the ion transport in the plasma sheath and the resulting flux density and energy distribution at the chamber walls and electrodes [9,10,22].

However, due to the different values of the sheath voltage we expect quite different ion energy distributions at the powered RF electrode and at the grounded or/and floating electrodes and chamber walls.

As an important factor, the ratio between RF cycle time T_{rf} and ion residence time in the sheath, T_i, determines how the ion energy distribution is affected by the RF modulation of the sheath potential. At a residence time of several RF cycles $T_{rf} \ll T_i$, the ions are influenced only by the mean sheath voltage, and the energy distribution reduces to a single peak. In contrary, at a very low residence time $T_i \ll T_{rf}$, the distribution function gets fully modulated by the RF frequency with a resulting saddle shape of the energy distribution (Fig. 4.24). The resulting modulation energy width, ΔE, scales with [67,93,94]

$$\Delta E \sim \frac{V_{sh}}{\omega_{rf}\, d_{sh}} \sqrt{\frac{e}{m_i}} \tag{4.15}$$

The ion residence time in the sheath depends on the ion mass m_i, on the sheath voltage, V_{sh}, and on the sheath thickness, d_{sh}, which in turn is affected by the sheath voltage, V_{sh}, and by the Debye length, λ_D. Inserting the Debye length in Equ. (4.2), it is obvious that the resulting sheath thickness is strongly affected by the sheath voltage, V_{sh}, and plasma density, n_{pl}, whereas the electron temperature, T_e, has a much smaller effect. Due to Equ. (4.2), we obtain the lowest sheath thickness, d_{sh}, and hence the largest modulation effect of the energy distribution at a low sheath voltage, V_{sh}, and a low Debye length, λ_D.

Besides the RF modulation the collisions of ions with the background gas during the passing of the plasma sheath may considerably affect the ion transport. The ratio between sheath thickness, d_{sh}, and ion mean free path, λ_i, describes the importance of collisional effects that could differ between the ion species due to the collisional cross sections. At an ion mean free path larger than the sheath thickness $\lambda_i \gg d_{sh}$, collisional processes are of minor importance, whereas at $\lambda_i \ll d_{sh}$, the ion energy distribution becomes completely thermalized, and the sheath parameters get in turn influenced by the collisional processes [10-13,67,95,96].

In Fig. 4.25, an example of the energy distributions of different ions at the powered electrode of an RF discharge is presented. For the ions with a small collisional cross section (ArH^+, Ar_2^+), the transit through the plasma sheath is almost collisionless. The resulting saddle shaped structure for these ions, at the high energy end of the distribution, reflects the RF modulated electric field across the sheath. The heavier the ion, the more RF cycles it experiences transiting the sheath, and the effect of RF modulation gets reduced.

The large charge transfer cross section [33] and the lower mean free path of the Ar^+ ion causes a collision induced low energy structure in the distribution function. Ions generated with thermal energy by charge transfer inside the sheath, and less so those produced by electron impact, experience only a part of the sheath voltage and generate this structure.

The multiple peaks are formed by the effect of the RF modulated electron density in the sheath and ions that are produced by charge transfer out of the actual electrical field at $x>x_e(t)$ (Fig. 4.24). During the increase of the RF voltage and spatial expansion of the electric field, all these ions are accelerated together and cause the peaks in the low energy part of the distribution. These multiple peaks overlay the saddle shaped structure [10].

Additionally, for the Ar^+ ion only, a low intensity high energy tail is detected. The low intensity high energy tail in the Ar^+ ion energy distribution (Subplot Fig. 4.25) is produced in a collision process by a high energy doubly charged ion Ar^{2+} generated in a charge exchange collision with a high energy singly charged Ar^+. Because of the higher energy of the doubly charged ions, the singly charged ions generated in this process reach up to double the energy of an Ar^+ ion coming directly from the bulk plasma.

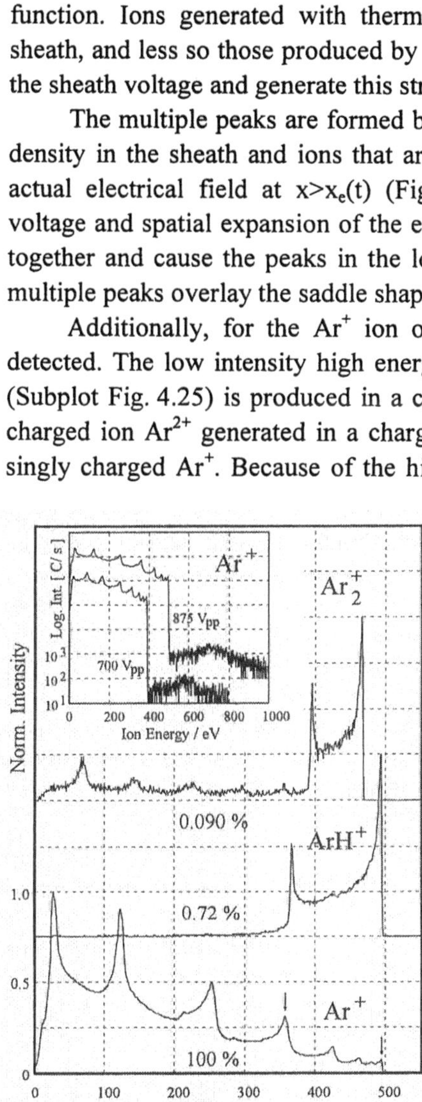

Figure 4.25. Normalized energy distributions of different ions at the powered RF electrode (V_{rf}=875 V_{pp}, p=5 Pa, Relative intensities in percent, Curves shifted, Subplot in log. scale) [13].

The voltage across the plasma sheaths in an RF discharge can be described by the capacitive sheath model (Sec. 4.2.2). In the capacitively coupled, unconfined and therefore asymmetrical RF discharge, the negative self bias of nearly half of the peak to peak RF voltage, is created at the powered electrode (Fig. 4.24). Changes in the RF amplitude mainly affect the self bias and have negligible influence on the plasma potential. As a result, the sheath voltage at the powered RF electrode increases with the RF voltage, whereas the sheath voltage at the grounded electrode remains almost constant [9].

Fig. 4.26 shows the resulting Ar^+

energy distributions at both discharge electrodes of a parallel plate electrode arrangement at different RF voltage. As expected from the potential structure, the maximum energy at the powered electrode is strongly coupled to the RF voltage by the DC self bias and is of the order of several hundred eV, whereas at the corresponding energy at the grounded electrode is only slightly increased and always below some 10 eV.

Clear differences in the shape of the energy distribution at the two electrodes are observed. The multiple peak structures at the RF electrode represent the large RF amplitude applied across this sheath. In contrast, at the grounded electrode, only a single peak at the high energy end is seen because the sheath voltage is low and less modulated (Fig. 4.26).

It is clearly seen from the intensity ratio between the collision free and collisional part in the energy distribution in Fig. 4.26 that, at the grounded electrode, the effect of collisions is less significant at higher RF voltage (right), whereas this effect is almost constant at the powered electrode (left). This phenomena result from the influence of the discharge parameters on the plasma sheath thickness according to Equ. (4.2). While at the grounded electrode, the sheath voltage remains nearly constant, the higher plasma density at higher

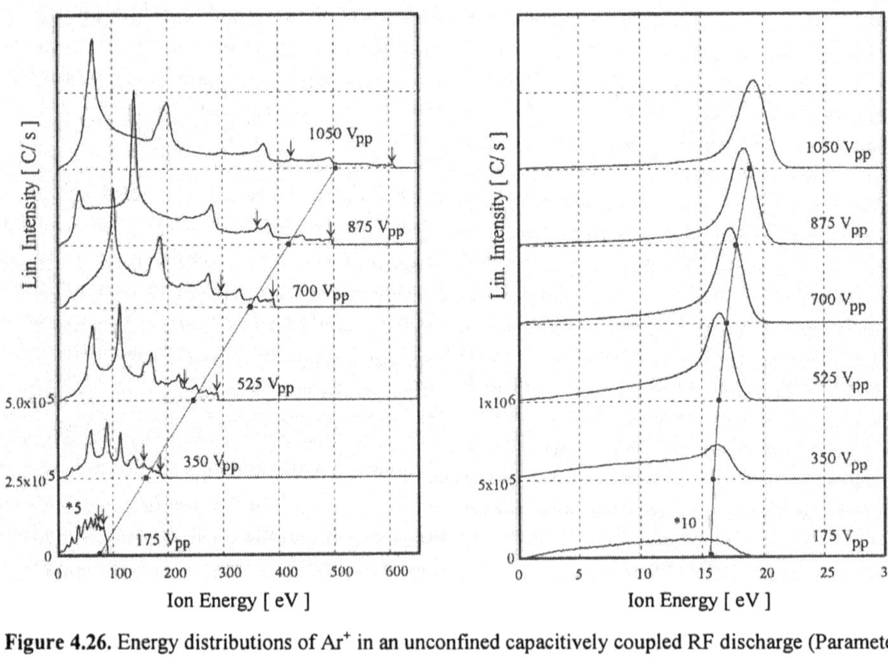

Figure 4.26. Energy distributions of Ar^+ in an unconfined capacitively coupled RF discharge (Parameter: $RF_{p\text{-}p}$ voltage, p_{tot}=5 Pa, Curves shifted, Mean sheath voltage by vertical line, Modulation width by arrows) [13].Left: Powered electrode. Right: Grounded electrode.

power causes the plasma sheath thickness to decrease. Consequently, the resulting energy distribution becomes less influenced by collisions. In contrast, at the powered electrode the higher sheath voltage and the higher plasma density nearly compensate their effect on the sheath thickness, and because of this, the collisional character of the energy distribution changes only a little.

As the final example for the ions, Fig. 4.27 illustrates the influence of the discharge pressure on the energy distribution at the powered RF electrode. Even at low pressure the Ar^+ energy distribution is strongly affected by collisions in the sheath, as seen from the multiple peaks in the low energy region. At a higher discharge pressure the collisional low energy part of the distribution becomes more marked because of the lower ratio between the ion mean free path and the sheath thickness, λ_i/d_{sh}. The ion mean free path is affected, however, to a lesser extent, the sheath thickness decreases with pressure as could be seen from the multiple structure in the low energy tail [10,67,94,97].

In collisions of the ions with the background gas high energy neutral species are formed. Because the generation rate and energy distribution of these high energy neutrals are strongly coupled to those of the ions, finally, the high energy neutrals should be discussed. However, these neutral species are hard to detect and their flux is difficult to quantify as explained in Sec. 4.3.2.

The generation rate of the high energy neutrals is determined by the collisional frequency of the ions. Consequently the energy transferred in a collision event depends on the original energy of the ion and the character of the collision event. Charge transfer and elastic collisions are of practical importance. The elastic collisions are affected by mass ratio of the collision partners and the collision geometry. The final energy distribution of these high energy neutrals at the surface is affected by further collisions with the background gas.

In Fig. 4.28, an example is presented that shows the energy

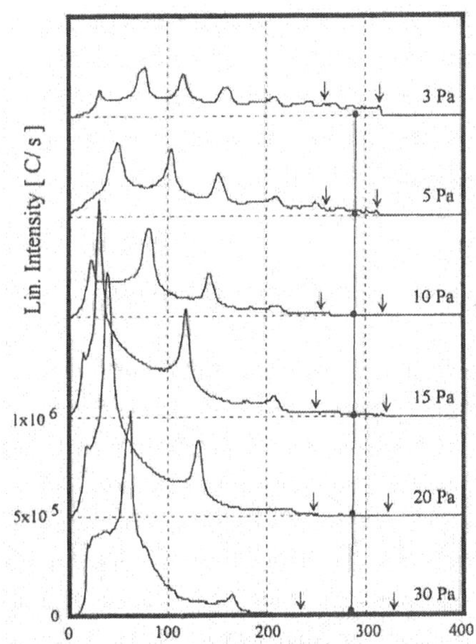

Figure 4.27. Energy distribution of Ar^+ at the RF electrode in a capacitively coupled discharge (Parameter: Total pressure, V_{rf}=570 V_{pp}, Curves shifted, Mean sheath voltage by vertical line, Modulation width by arrows) [97].

distributions of neutral species impinging on the powered electrode of an RF discharge. Clearly, the mean energy of the ions exceeds the energy of the neutral species, and the sharp structures in the ion energy distribution are not transferred to the neutrals.

Due to the higher collisional probability of the ions, the flux of the high energy neutrals increases with the discharge pressure. Under conditions of low sheath voltage and low pressure, most of the neutrals arrive with low energy. As sheath voltage and pressure increases, and with a dominating multiple scattering of the neutrals, a maximum in the energy distribution in the order of some 10 eV is observed [2,98].

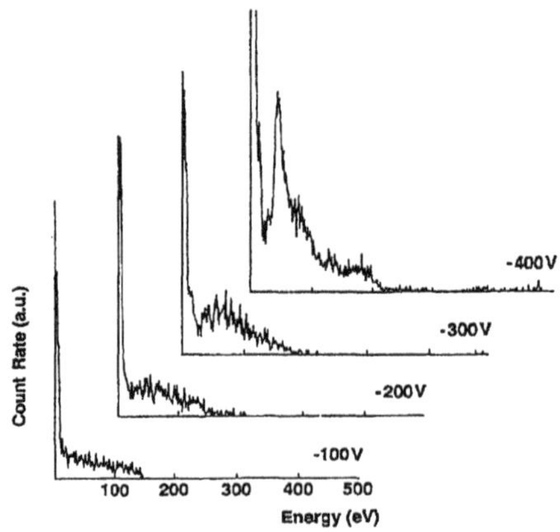

Figure 4.28. Energy distribution of neutral Ar atoms at the RF electrode of a capacitively coupled discharge (Parameter: DC-Self-bias, p_{tot}=6.65 Pa) [2] (with permission).

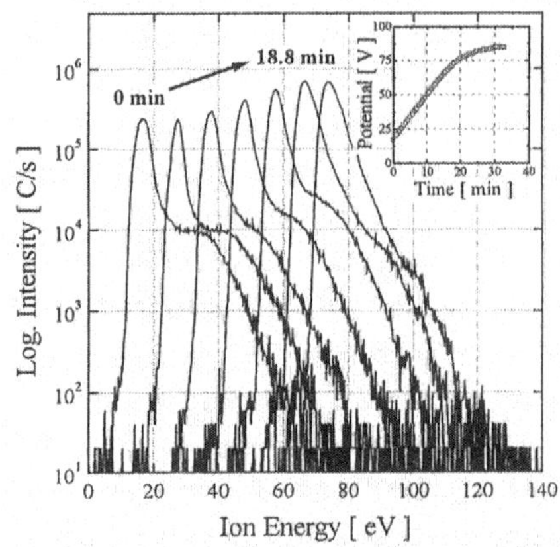

Figure 4.29. Time dependent F^+ energy distribution in PTFE sputtering starting with clean chamber walls (Q_{Ar}=8.5 Pa l/s, p_{tot}=0.3 Pa, P_{rf}=130 W) [72].

4.5.2. *RF Sputter Deposition*

Plasma polymerization of polymer films from a gaseous precursor is the common technique mostly used, but RF magnetron sputtering of a solid polymer target may also be applied for deposition of plasma polymers and this technique offers some advantages [72,99,100] (Please, see also Chapter 9). An important difference to plasma polymerization is the lower process pressure (0.1 Pa ~ 1 Pa) and the directed particle flow resulting in a more effective process performance. However, the sputter deposition process is obviously complex, with several reactions occurring in the plasma volume, at the substrate and at the target surface. These reactions become known using corpuscular diagnostics.

Analyzing the particle flow and energy at the sample position provides information about the potential conditions in the plasma sheath as already discussed in Sec. 4.5.1. Additionally, this method is a powerful tool for investigations of the kinetics of plasma-surface-interactions, for analyzing chemical reactions as well as collision phenomena. The RF sputtering of PTFE presented in the following was performed using a balanced RF magnetron equipped with a massive PTFE target. The target was initially sputtered in an argon atmosphere.

The sputter deposition process is known to be driven mainly by neutral species [101-104]. Due to the necessary post-ionization of the neutral species in the mass spectrometer, the sensitivity for neutrals is low (Sec. 4.3.2) [4]. Analysis of the ions directly produced in the discharge is much more sensitive and also delivers information about the total particle flux to the film.

Fig. 4.29 shows time resolved ion energy distributions of the F^+ ion starting at t=0 min with cleaned metallic chamber walls. A characteristic shift of the distribution function is observed which takes place in stationary conditions after about 30 min. The maximum of the energy distribution indicates the mean plasma potential (Subplot Fig. 4.29). The PTFE coating (fluorocarbon plasma polymer) of the chamber walls allows the plasma potential to float against the surface potential of the film with an additional potential drop occurring across the insulating PTFE film. The plasma sheath voltage of about 10...15 V remains nearly constant and can be deduced from the difference between the low energy foot point of the distribution and the maximum. This sheath voltage determines the energy of positive ions bombarding the growing film. Due to the fact that a defined reference potential is needed for the ion energy analysis the ions must pass the potential drop across the deposited PTFE film at the mass spectrometer aperture. However, this potential drop happens at such a short distance that the shape of the ion energy distribution is not affected by collisions across this distance, and the distribution function is only shifted on the energy scale (Fig. 4.29). This shift is corrected for all positive ion energy distributions presented in the following so that the distribution gives the correct energy of the particles bombarding the deposited PTFE film.

From the energy distributions (Fig. 4.30), conclusions about the origin of the ion species are possible. According to a capacitive sheath model in a typical RF discharge only a small RF voltage drops occurs across the sheath at the grounded electrode or the floating walls, respectively (Sec. 4.5.1). Besides the ion acceleration due to this small electrical field, the energy distributions in Fig. 4.30 are affected by the collision cascades in the sputter target. The original

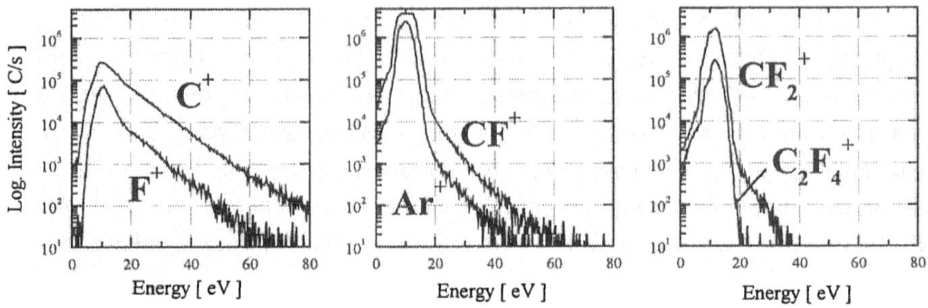

Figure 4.30. Typical energy distributions of ions from PTFE sputtering (p_{tot}=0.3 Pa, P_{rf}=130 W) [72].

energy distribution of sputtered particles is characterized by a peak at the surface binding energy and a $\sqrt{E}$ -varying intensity in the high energy tail [71,102,104].

The energy distributions in Fig. 4.30 are generally characterized by a low energy peak produced by ionization of thermal background gas and acceleration while passing the sheath voltage in front of the deposited PTFE film. For some ions, a more or less intense high energy tail is observed. High energy C^+, F^+, CF^+, CF_2^+ and CF_3^+ ions have their origin in the ionization of high energy neutrals produced in the sputter process. Note, that positive secondary ions are not able to overcome the target sheath, and, therefore the above high energy ions must be produced by plasma ionization of neutrals from the sputter process.

The high energy tail becomes less pronounced as heavier is the CF_y^+ ion. For all $C_xF_y^+$ ions with x≥2, we do not detect any high energy structure in the energy distribution. Because of this, the directed particle flux produced at the sputter target is mainly composed from C, F, CF, CF_2 and CF_3 molecules, and the heavier C_xF_y should be produced in secondary gas phase reactions but not directly in the sputter process. The high energy tail of Ar^+, which is used as the sputter gas, results from the collisions between background argon gas and high energy particles from the sputter process.

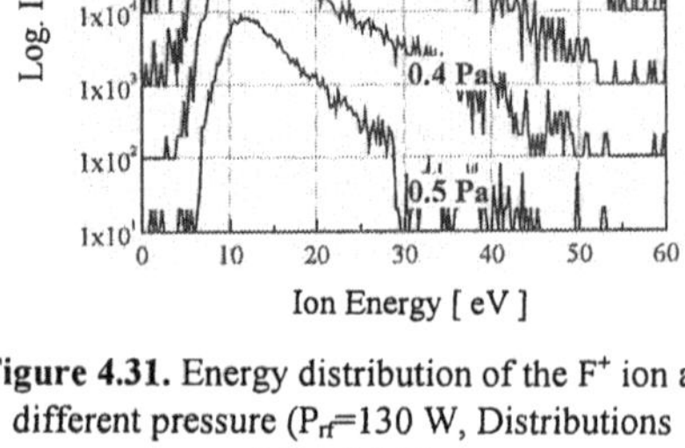

Figure 4.31. Energy distribution of the F^+ ion at different pressure (P_{rf}=130 W, Distributions shifted) [72].

In Fig. 4.31, the effect of the discharge pressure on the F^+ energy distribution is demonstrated. Especially collisions during the passage of the distance from the target towards the substrate clearly affect the shape of the distribution function. As higher the discharge pressure (Fig. 4.30) is and therefore as lower the ion mean free path is, as more the high energy tail of the distribution is lowered by collisions [71,105,106].

4.5.3. Microwave Plasma Processing

In comparison to RF discharges, ions emitted from a microwave excited discharge show a rather simpler behaviour. Most of the practically used microwave discharges are excited at a frequency of 2.45 GHz. A very effective power transfer is achieved utilizing an electron cyclotron resonance heating (ECR) in the microwave field. The introduction of a steady magnetic field, in which there is a resonance between the applied microwave frequency, ω_{mw}, and the electron cyclotron frequency, $\omega_{ce} = eB/m_e$, allows an operation at high density without the need of a cavity resonance. Because of the cyclotron resonance, the gyrating electrons rotate in phase with the right hand polarized wave, seeing a steady electric field over many gyro orbits. With the electron mean free path large enough, the ECR heating enables the most effective energy transfer and low plasma sheath potentials [22,23].

The typical microwave excitation frequency is orders of magnitude above the ion plasma frequency and a little above or within the order of the plasma frequency of the electrons $\omega_{ex}>\omega_{pl\text{-}e}>>\omega_{pl\text{-}i}$. This frequency relation causes all charge carriers being not or only less able to respond to the microwave modulation of the sheath potentials, and consequently behaving like in a DC plasma sheath. However, the plasma parameters are strongly affected by the electron velocity distribution and mobility which in turn is influenced by the strong magnetic fields in an ECR discharge. Because of this the magnetic fields may cause significant spatial inhomogeneities in behaviour of ECR discharges.

Fig. 4.32 shows representative ion energy distributions obtained at the sample position in an ECR microwave discharge. Due to the high excitation frequency, the ions experience the time averaged electrical field within the sheath resulting in a single peak in the ion energy distribution with the position determined by the mean sheath voltage. Because ECR discharges are typically driven at a pressure about 1 - 2 orders of magnitude lower than in an RF discharge, collisional processes in the plasma sheath are of minor importance, and the low energy structure in the energy distribution formed by collisional

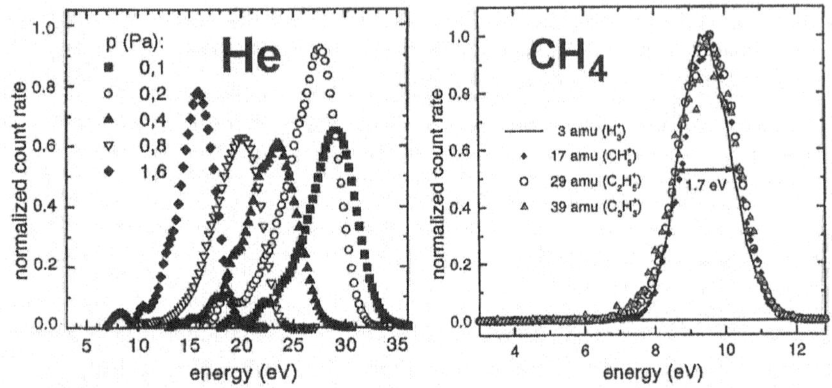

Figure 4.32. Energy distribution obtained at the grounded sample position of an ECR discharge. Left: He^+ energy distribution (Parameter: He pressure, P_{mw}=145 W) [107]. (with permission) Right: Different $C_XH_Y^+$-ions (Parameter: Ion mass, CH_4, p_{tot}=0.4 Pa, P_{mw}=160 W) [69] (with permission).

processes are not significantly observed. The mean sheath voltage in turn depends of course on the discharge parameters, especially on the electron velocity distribution. This velocity distribution is affected by the more or less resonant energy transfer to the electrons. At lower pressure and higher RF power, the ECR conditions could be easier realized with a clearly higher electron temperature and a consequently larger plasma potential and sheath voltage. Exactly this behaviour causes that the ion energy in Fig. 4.32 (left) is higher at lower pressure. A similar but less pronounced increase of the ion energy is observed at a higher microwave power supplied to the discharge.

In comparison between the helium ions (Fig. 4.32, left) and the hydrocarbon ions from the methane discharge (Fig. 4.32, right) a higher mean plasma potential for Helium resulting from a higher ionization potential of the process gas is obvious.

Within the same discharge, the energy distributions between different ions are nearly not affected by the ion specific characteristics (Fig. 4.32, right). Because only the mean sheath voltage dominates the ion transport, the mass dependent residence time of the ions within the sheath is unimportant. In addition, because of the low operation pressure of the ECR discharges and the low collisional probability, the ion specific differences in the collisional cross sections are not reflected in the ion energy distribution [107-109].

4.5.4. Negative Ion Detection

Negative ions are known to influence the electrical discharge characteristics in electronegative gases and to play a dominant role in powder formation during deposition processes [110-112,122]. In spite of the importance of negative ions in discharge kinetics, their detection by mass spectrometric methods is usually difficult because ions produced in the bulk plasma are confined by the plasma sheath potentials. Techniques for the sampling of negative ions by amplitude modulation of the RF voltage [15,16,113] or by positive biasing of the mass spectrometer aperture [114] have delivered useful results. These methods, however, influence the negative ion generation process or the plasma sheath characteristics in front of the sampling orifice, respectively, and provide only partial information on a steady state RF discharge. Observation of ions generated in the plasma sheath was applied as an alternative method to acquire knowledge about negative ion formation in a parallel plate RF discharge in oxygen discharge without any plasma disturbance [17-18].

Fig. 4.33 shows ion energy distributions for different negative ions from a parallel plate discharge in oxygen detected at the grounded electrode. Besides the expected negative oxygen ions (O^-, O_2^-), negative fluorine (F^-) and chlorine ($^{35}Cl^-$, $^{37}Cl^-$) ions are detected, the latter giving a maximum contribution to the negative ion current of less than 5 %. The absolute level of impurities is much lower but because of their high electronegativity the negative ion spectrum is very sensitive to them.

All ions in Fig. 4.33 are generated in the plasma sheath in front of the driven RF electrode or at its surface at a potential which is negative with respect to ground. The electrical field in this sheath accelerates the negative ions away from the RF electrode, and with their collision probability low enough, they pass through the bulk plasma. Because of the accumulated energy they are able

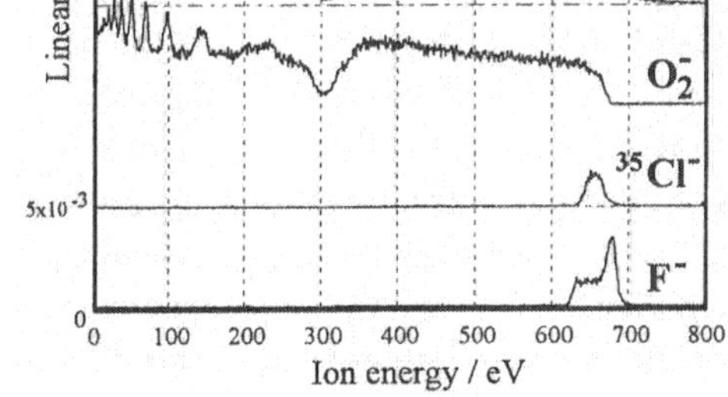

Figure 4.33. Negative ion energy distributions at the grounded electrode of a capacitively coupled Oxygen discharge (V_{rf}=1600 V_{pp}, V_{sb}=-650 V, p_{tot}=1.5 Pa, Electrode separation d=3 cm, Curves shifted) [17].

to overcome the small retarding field across the sheath in front of the grounded electrode (Subplot Fig. 4.33). The maximum energy on entering the mass spectrometer corresponds to the negative self-bias voltage of the driven electrode [9-13].

The fluorine and chlorine ion energy distribution consist only of a high energy peak. These ions are generated at the electrode surface, probably from an adsorption film. For both O^- and O_2^- ions there is, in addition to the high energy peak, a structured low energy section of the distribution. These ions are generated mainly by electron attachment at

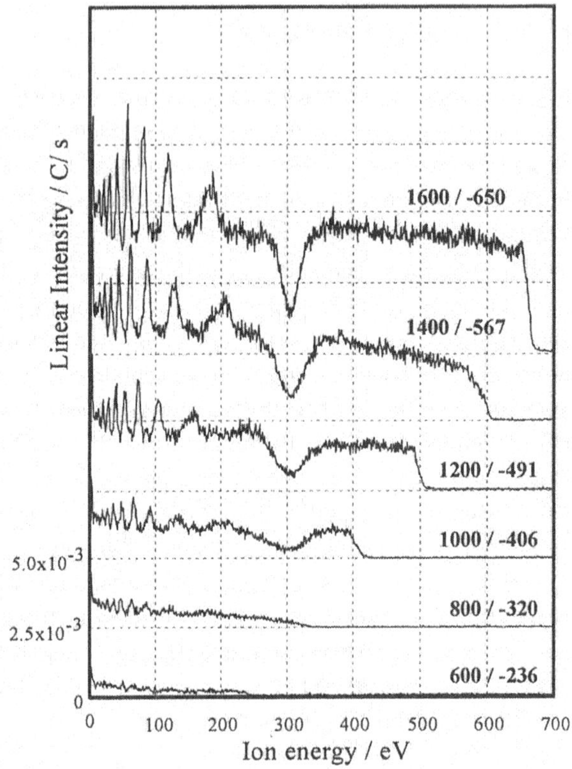

Figure 4.34. Energy distribution of O_2^- at different $RF_{p\text{-}p}$ voltage and self-bias (p_{tot}=1.5 Pa, Electrode separation d=3 cm, Curves shifted) [17].

different space potentials in the RF plasma sheath. Differences between the distribution functions for the O^- and O_2^- are caused by their different collision cross sections in the bulk plasma and, more important, by differences in the generation process. Dissociative attachment or pair production can deliver a considerable contribution to the O^- intensity [17,18,112].

Fig. 4.34 shows energy distributions for the O_2^- ion at different RF voltage amplitudes. The multiple peak structures seen in the low energy tail have their origin in the RF modulated electron density in the plasma that is known from works on positive ions (Sec. 4.5.1), and, hence, the phase coupling of generation and acceleration of the ions.

However, negative ions reaching the opposite electrode surfaces with high energy may contribute significantly to the energy flux. This negative ion bombardment is able to drive re-sputtering effects and may alter the properties of deposited films [115,116].

4.5.5. Pulsed Plasma Operation

Pulsed mode RF discharges are actually used as a promising tool controlling specific discharge parameters. By changes in the plasma density and the electron energy distribution, a controlled optimization of the process performance could be achieved [117,118]. Especially the negative ion induced chemistry is strongly affected in a pulsed mode RF discharge because of the abolished negative ion confinement during the afterglow period [15,16]. Contrary to the background of powder formation in dusty plasmas by negative ion reactions, this phenomenon has an immense practical importance [113,119].

In this section we show that additional information on the kinetics of a modulated discharge may be obtained from the measurement of the ion energy distribution at the electrode surface. The energy distribution directly indicates changes in the discharge potential conditions during the modulation cycle. In this example, the discharge is powered by a broad band amplifier capacitively coupled to the discharge by a fully tuneable π-type matching network. The amplifier is controlled by three frequency generators producing the sine wave RF frequency, f_{rf}=13.56 MHz, the modulation frequency, f_{mod}, and wave form, and TTL gate pulses controlling the mass spectrometer counting electronics. The following investigations used square wave modulation with 50 % duty cycles and 100 % modulation depth [120].

Fig. 4.35 schematically shows the time dependent voltage characteristics at the powered RF electrode. The square wave RF modulation supplied by the broad band amplifier is changed by the electrical properties of the matching network and the discharge itself. Additionally, the capacitive coupling of the RF voltage allows the DC self bias to be created at the RF electrode [9].

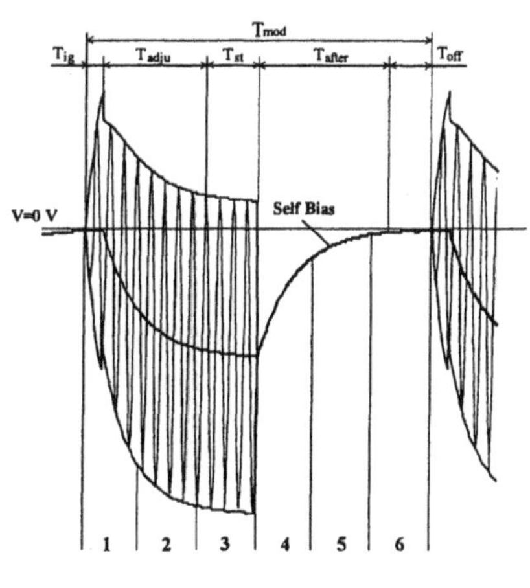

Figure 4.35. Schematic shape of the cover curve during a modulation period (RF and modulation frequency not true to scale, Bottom numbers indicate time periods for Fig. 4.37).

Each modulation cycle starts with the discharge ignition (T_{ig}). The ignition voltage depends on the Paschen curves and additionally on the modulation frequency, because a residual charge carrier density from the previous cycle eases the re-ignition. The following period (T_{adju}) is characterized by a further increase of the RF voltage superimposed with the creation of the negative self bias voltage. Finally we reach a stationary discharge regime (T_{st}) with a constant RF and bias voltage.

Switching the RF voltage off, the following afterglow period (T_{after}) is determined by the time constant of charge carrier recombination. Their density and the capacity of the coupling capacitor cause a delayed decay of the self bias voltage. At last, the arrangement is in a real off state with all discharge activities having died away (T_{off}).

However, the energy distribution of the positive ions at any electrode is determined by the sheath voltage in front of it. Because the plasma potential is positive with respect to each electrode in contact with the plasma, the plasma sheath voltage at the grounded electrode is formed by the upper part of the cover curve of the voltage modulation (Fig. 4.35). We find the highest voltage during the discharge ignition, which decreases while coming to the stationary discharge state. The sheath voltage instantaneously decays to low values when the RF voltage is switched off because the electron energy is immediately lowered [118].

According to the time scaling of these discharge periods, the sheath voltage characteristics depend on the modulation frequency, f_{mod}. The discharge ignition (T_{ig}), the adjustment period (T_{adju}) and the afterglow period (T_{after}) are nearly independent of the modulation time, and follow immediately each plasma on or off pulse, respectively. For this reason, at a low modulation frequency the sheath characteristics are dominated by the long stabile discharge (T_{st}) and the real off state (T_{off}). As the modulation time gets in the order of the initial (T_{ig}, T_{adju}) and the afterglow time (T_{after}), these periods get more and more dominating on the sheath voltage characteristics. A further increase of the modulation frequency causes the discharge system to be unable to react to the modulation and consequently a stationary discharge is found.

The positive ion energy distributions of Ar^+ from an argon discharge at different modulation frequencies in Fig. 4.36 clearly indicate these dependencies. At the lowest modulation frequency of 500 Hz, we find the same shape of the energy distribution as for the stationary discharge, but only with half of the intensity. At this frequency, the period for building up the stationary discharge and the plasma off period clearly surpass the initial periods

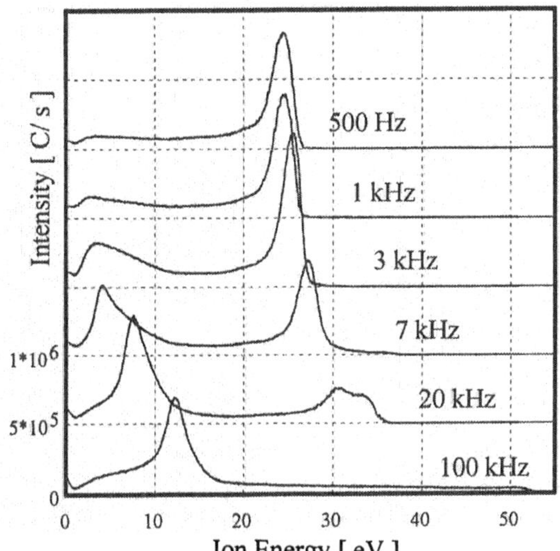

Figure 4.36. Ar^+ energy distribution dependent on square wave modulation frequency (Parameter: Modulation frequency V_{rf}=1200 V_{pp}, MHz, p=5 Pa, Curves shifted) [120].

(T_{st}>>T_{ig},T_{adju}; T_{off}>>T_{after}). This causes nearly a constant sheath voltage during the plasma on cycle which dominates the observed distribution function (Sec. 4.5.1) [10,13,67].

Increasing the modulation frequency, the energy distribution becomes more and more dominated by influences of the initial and afterglow periods. It is obvious that the higher sheath voltage during the initial discharge period (T_{adju}) becomes more important at higher frequency and shifts the upper end of the distribution towards higher energy. At a modulation frequency of 100 kHz no stationary discharge state is created any more.

At the same time, a low energy peak is formed which reflects the decaying sheath potential during the afterglow period. At a higher frequency, the new discharge on pulse breaks off the afterglow earlier with the sheath potential only less decreased, and the corresponding peak is located at higher energy.

The different superimposed discharge periods from Fig. 4.35 could be time resolved experimentally by synchronizing the modulation frequency generator and the channeltron counting electronics. The discharge periods marked in Fig. 4.35 contribute to the distribution with the same number in Fig. 4.37. During the ignition (No. 1), a residual low energy signal from the previous afterglow is seen. A high energy peak comes up with the energy a little larger than during the stationary discharge period (No. 2, 3). After switching the discharge off (No. 4), the high energy peak decays, and the collapsing sheath potential causes the low energy peak. This tendency continues in periods No. 5 and 6 with a lowered energy of the afterglow peak.

Because of the retarding potentials of a stationary plasma sheath, negative ions are dramatically affected by the modulation of the RF excitation. Negative ions generated in the bulk plasma region of a stationary RF discharge are confined, and only ions produced in the sheath in front of the RF powered electrode at negative potential against ground are able to gain enough kinetic energy to overcome the retarding field at the opposite electrode. (Sec. 4.5.4.).

Fig. 4.38 shows the O^- energy distributions from a pulsed mode RF discharge. Additional to the high energy ions produced in the RF sheath (Sec. 4.5.4), negative ions are now able to escape from the bulk plasma or are produced during the decay of the sheath voltage in the afterglow. The high energy part of the O^- energy distribution function is nearly not affected by the discharge modulation (Subplot Fig. 4.38). If the modulation time falls in the order of the afterglow time, additional negative ions can be detected at the electrode, as seen from the rapidly increasing intensity at low energy. The intensity of these low energy negative ions has a pronounced maximum over the modulation frequency [16]. If the frequency is too high the sheath does not decay. However for values too low the afterglow cycles are too long separated giving only a low time averaged intensity. These low energy negative ions must not necessarily be produced in the bulk plasma during the stationary discharge. Because of the dramatically changed electron energy distribution function in the afterglow [117] and the resonant electron attachment cross sections [85-87], the

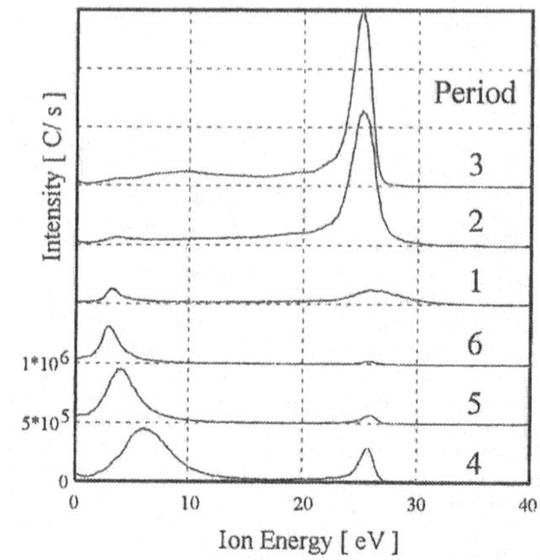

Figure 4.37. Time resolved Ar^+ energy distribution across the modulation period (Parameter: Time period indicated in Fig. 4.35, V_{rf}=1200 V_{pp}, f_{mod}=3 kHz, p=5 Pa, Curves shifted) [120].

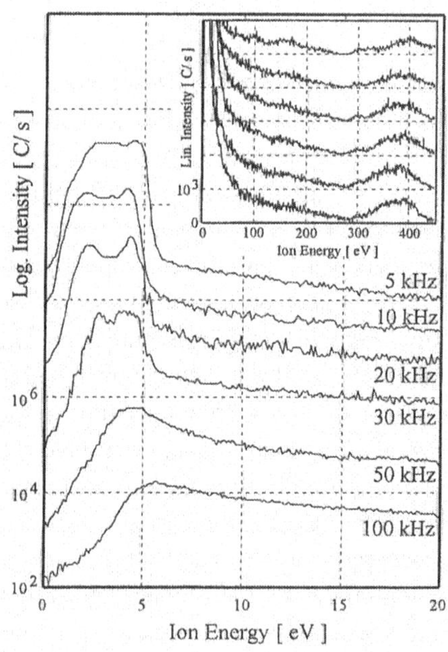

Figure 4.38. O^- energy distributions in a pulsed mode RF discharge (Parameter: Modulation frequency, V_{rf}=1000 V_{pp}, p=1.5 Pa O_2, Electrode separation d=3 cm, Curves shifted) [120].

negative ions can also be formed during the afterglow period. However, quantitatively the flux density of the negative ions increases at the optimum modulation frequency by 1 - 2 orders of magnitude.

4.5.6. Plasma Ion Mass Spectra

Beside the energy distributions and their effect on the surface bombardment, the ions play an important role for the bulk plasma chemistry, especially for the formation and growth of oligomers and powders [121]. Therefore in this section mass spectra of positive and negative plasma ions are discussed. Note that most of the mass spectrometers suited for plasma ion analysis acquire energy filtered mass spectra. Remembering the mass specific ion energy distributions in Sec. 4.5.1, one should always bear in mind a comparable energy distribution for the ions analyzed in a mass scan.

Fig. 4.39 compares the mass spectra of the fluxes of positive ions and thermal neutral species to the substrate from an ECR discharge in methane. Obviously the plasma ion mass spectrum extends to much larger masses than the mass spectrum of the neutrals. The ion mass spectrum structures containing up to 20 carbon atoms could be detected. The most intense ion signal is seen for the C_3H ion which clearly exceeds the intensity of the CH_X^+-ions produced from the direct ionization of the methane precursor. Up to the ions $C_{15}H_Y^+$ with the maximum masses in Fig. 4.39, the flux density decreases by a factor of about 1.000, but due to the large number of carbon atoms in these large molecular ions, their contribution to the total ionic fraction of the carbon

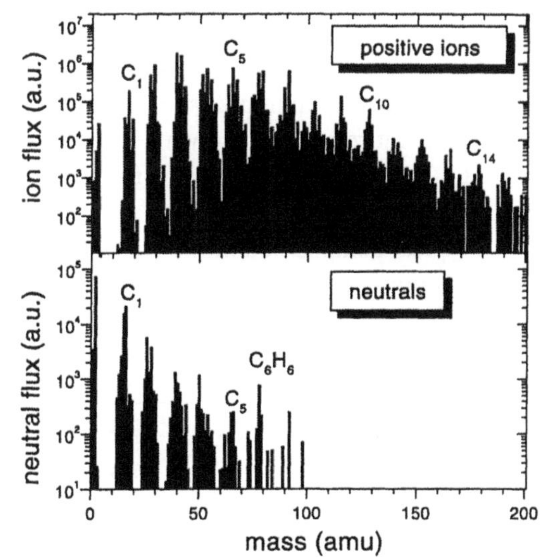

Fig. 4.39. Mass spectra of positive ion (top) and thermal neutrals (bottom) from an ECR methane plasma (p_{tot}=1.5 Pa, P_{mw}=160 W, Spectra corrected for mass dependent transmission) [69] (with permission).

flux is still important [69-71,79,122].

The lower part of Fig. 4.39 presents the corresponding distribution of the flux of thermal neutrals. The dominant neutral species formed in the plasma is H_2, which is a clear indication that the methane is significantly depleted under the discharge conditions. The generation of oligomer species already in the discharge plasma could also be deduced from the neutral gas spectrum, but the CH_X hydrocarbons are the dominant carbon species. In contrary to the ions, the highest neutral mass detected corresponds only to C_7H_Y molecule.

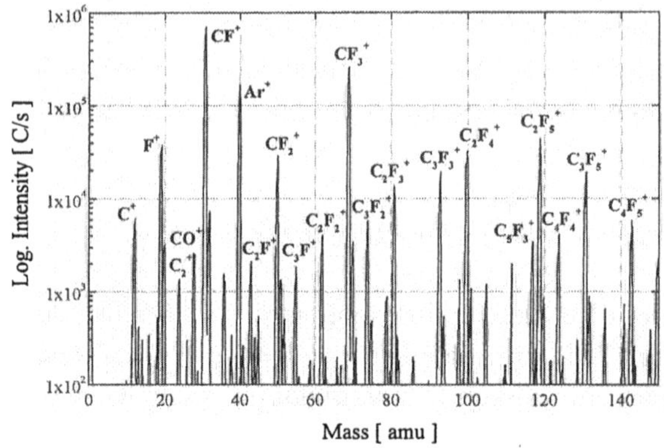

Fig. 4.40. Typical ion mass spectrum during PTFE sputtering (p_{tot}<=0.3 Pa, P_{rf}=130 W, E^+=12 eV) [72] (with permission).

The difference between the extent of the plasma ion and the neutral species mass spectrum is based on equipment and plasma chemical reaction processes. The post-ionization of the neutral species within the electron impact ion source of the mass spectrometer causes dissociation of higher molecular species favouring the observed low molecular fragments. In addition, it appears that at this relatively high pressures, ion-molecule reactions in the bulk plasma play a dominant role in the growth of positive ions. The small amount of CH_Y^+- and $C_2H_Y^+$-ions can be explained by the fact that these ions are consumed in the growth process.

A typical composition of the ion flux obtained in RF magnetron sputtering of PTFE is presented in Fig. 4.40. In the experiment, argon is used as the initial sputter gas with different fluorocarbon gases formed in further reactions. Consequently, beside the Ar^+ ions, the spectrum is dominated by various $C_XF_Y^+$

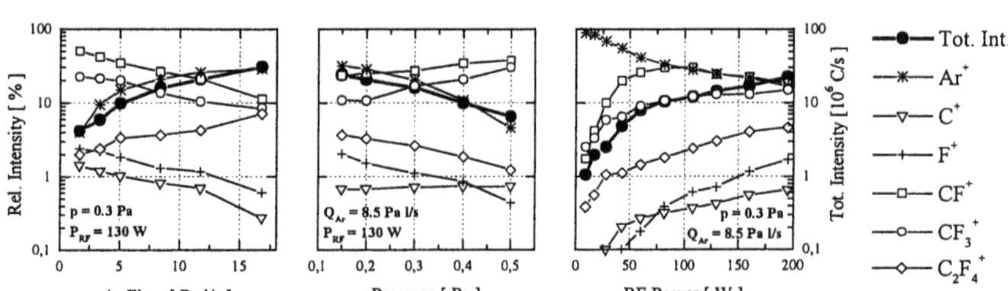

Fig. 4.41. Relative intensities and total ion signal dependent on the discharge parameters [72].

ions with up to X=4. Fig. 4.41 shows together the total signal intensity of all ions in the observed mass range and the signal tendency of characteristic ions. The total signal intensity gives an idea about the plasma density. In the observed parameter range, we detect a higher ion flux at high RF power and low pressure.

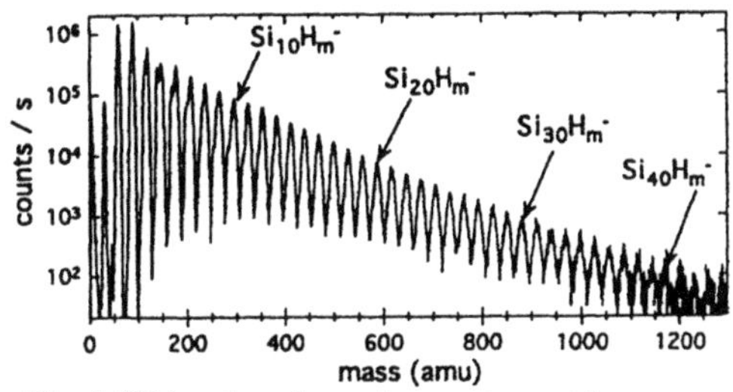

Fig. 4.42 Negative plasma ion spectrum at low mass resolution ($\Delta M \sim 20$ amu) obtained from an RF discharge in SiH_4 (p_{tot}=13 Pa, P_{rf}=8 W, f_{mod}=1 kHz) [122](with permission).

The higher flux at higher argon partial pressure results from the higher ionization cross section of argon compared to the C_XF_Y species [72].

The relative signal intensities of the ion species in Fig. 4.41 are produced by ionization of gaseous species in the discharge and hence are affected by the plasma density and the concentration of the corresponding neutral particle. However, it is difficult to identify one of these species directly with the deposition rate. But the dominating role of CF, CF_2 and CF_3 for the film deposition is clearly seen in Fig. 4.41.

The confinement of negative ions in a steady state RF or microwave discharge was already discussed in Sec. 4.5.4 and 4.5.5. Fig. 4.42 presents an example of a negative plasma ion spectrum from a silane discharge. Masses above 1.000 amu and a chemical composition with up to $Si_{40}H_Y^-$ have been detected. In contrast, the largest detected neutral counterparts reach up to 200 amu only.

These large negative ion clusters are correlated to be an important factor in powder formation within the bulk discharge plasma. Modulation of the discharge excitation, with a frequency low enough to allow the negative ions to escape from the bulk plasma, reduces both, the high molecular share of negative ions and the rate of powder formation [111,113,122].

References

1. Auciello O. and Flamm D.L., *Plasma Diagnostics* (Academic Press, Boston, 1989).
2. Janes J. and Börnig K., *J. Appl. Phys.* **73** (1993) 2724.

3. Franke F., Neumann H., Zeuner M., Frank W. and Bigl F., *Surf. Coat. Technol.* **97** (1996) 90.
4. Jenett H. and Hodoroaba V.-D., *J. Vac. Sci. Technol.* **A15** (1997) 3158.
5. Sugai H. and Toyoda H., *J. Vac. Sci. Technol.* **A10** (1992) 1193.
6. Kae-Nune P., Perrin J., Guillon J. and Jolly J., *Plasma Sources Sci. Technol.* **4** (1995) 250.
7. Singh H., Coburn J.W. and Graves D.B., *J. Vac. Sci. Technol.* **A17** (1999) 2447.
8. Schwarzbach W., Tserepi A., Derouard J. and Sadeghi N., *Jpn. J. Appl. Phys.* **36** (1997) 4644.
9. Köhler K., Coburn J.W., Horne D.E. and Kay E., *J. Appl. Phys.* **75** (1985) 59.
10. Wild C. and Koidl P., *J. Appl. Phys.* **69** (1991) 2909.
11. Janes J. and Huth C., *J. Vac. Sci. Technol.* **A10** (1992) 3522.
12. Janes J. and Huth C., *J. Vac. Sci. Technol.* **A10** (1992) 3086.
13. Zeuner M., Neumann H. and Meichsner J., *J. Appl. Phys.* **81** (1997) 2985.
14. Zeuner M., Neumann H. and Meichsner J., *Plasma Sources Sci. Technol.* 7 (1998) 252.
15. Overzet L.J., Lin Y. and Luo L., *J. Appl. Phys.* **72** (1992) 5579.
16. Overzet L.J., Beberman J.H. and Verdeyen J.T., *J. Appl. Phys.* **66** (1989) 1622.
17. Zeuner M., Meichsner J. and Rees J.A., *J. Appl. Phys.* **79** (1996) 9379.
18. Zeuner M., Neumann H., Zalman J. and Biederman H., *J. Appl. Phys.* **83** (1998) 5083.
19. Flamm D.L., *Solid State Technol.* **22** (1979) 109.
20. Jurgensen C.W. and Rammelsberg A., *J. Vac. Sci. Technol.* **A7** (1989) 3317.
21. Hope D.A.O., Hydes A.J., Cox T.I. and Deshmukh V.G.I., *SPIE* **1392** (1990) 185.
22. Chapman B., *Glow Discharge Processes* (John Wiley, New York, 1980).
23. Franz G., *Kalte Plasmen* (Springer, Berlin 1990).
24. Wild C., Wagner J. and Koidl P., *J. Vac. Sci. Technol.* **A5** (1987) 2227.
25. Meichsner J., Nitschke M., Rochotzki R. and Zeuner M., *Surf. Coat. Technol.* **74-75** (1995) 227.
26. Rossnagel S.M. and Cuomo J.J., *Vacuum* **38** (1988) 73.
27. Mohan S. and Krishna M.G., *Vacuum* **46** (1995) 645.
28. Ensinger W., *Nucl. Instr. Meth. Phys. Res.* **B127** (1997) 796.

29. Frey H., *Grundlagen der Plasmaphysik* in *Vakuumbeschichtung 1*, ed. by Frey H. (VDI, Düsseldorf, 1995).
30. Janzen G., *Plasmatechnik* (Hüthig, Heidelberg, 1992).
31. Märk T.D. and Dunn G.H., *Electron Impact Ionization* (Springer, Wien, 1985).
32. Engel A.v., *Ionized Gases* (Clarendon Press, Oxford 1955).
33. Robinson, R.S., *J. Vac. Sci. Technol.* **16** (1979) 165.
34. Wagner H.E., Meichsner J., Kroutilina V. and Lerch R., *Bull. Am. Phys. Soc.* **45** (2000) 34.
35. Liebermann M.A., *IEEE Trans. Plasma Sci.* **16** (1988) 638.
36. Sheridian T.E. and Goree J., *Phys. Fluids* **B3** (1991) 2796.
37. Riemann K.-U., *J. Phys. D: Appl. Phys.* **24** (1991)493.
38. Scheuer J.T. and Emmert G.A., *Phys. Fluids* **B2** (1990) 445.
39. Meijer P.M. and Goedheer W.J., *Phys. Fluids* **B3** (1991) 1804.
40. Lister, G.G., *J. Phys. D: Appl. Phys.* **25** (1992) 1649.
41. Gogolides E. and Sawin H.H., *J. Appl. Phys.* **72** (1992) 3971.
42. Gogolides E. and Sawin H.H., *J. Appl. Phys.* **72** (1992) 3988.
43. Graves D.B. and Jensen K.F., *IEEE Trans. Plasma Sci.* **PS-14** (1986) 78.
44. Bridsall C.K., *IEEE Trans. Plasma Sci.* **19** (1991) 65.
45. Trombley H.W., Terry F.L. and Elta M.E., *IEEE Trans. Plasma Sci.* **19** (1991) 158.
46. Koenig H.R. and Maissel L.I., *IBM J. Res. Develop.* **14** (1970) 168.
47. Gottscho R.A., Scheller G.R. and Stoneback D., *J. Appl. Phys.* **66** (1989) 492.
48. Köhler K., Horne D.E. and Coburn J.W., *J. Appl. Phys.* **58** (1985) 3350.
49. Iganaki N., *Plasma Surface modification and plasma polymerization* (Technomic, Lancaster, 1996).
50. Rutscher A. and Wagner H.-E., *Plasma Sources Sci. Technol.* **2** (1993) 279.
51. Blank P. and Wittmaack K., *J. Appl. Phys.* **50** (1979) 1519.
52. Zalm P.C., *J. Appl. Phys.* **54** (1983) 2660.
53. Gschneider K.A., *Sol. State Phys.* **16** (1964) 275.
54. Batella J., *Messungen an dünnen Schichten während des Beschichtungsprozesses* in *Vakuumbeschichtung 3*, ed. by Kienel G. (VDI, Düsseldorf, 1994)
55. Eisenmenger-Sitter C., Bergauer A., Bangert H. and Bauer W., *J. Vac. Sci. Technol.* **A12** (1994) 536.
56. Pellet C., Desgranges C., Schwebel C. and Aubert J., *Appl. Phys.* **A55** (1992) 359.

57. Stoffels E., Stoffels W.W., Tachibana K. and Imai S., *Jpn. J. Appl. Phys.* **36** (1997) 4632.
58. Kersten H. and Kroesen G.M.W., *J. Vac. Sci. Technol.* **A8** (1990) 38.
59. Hopf C., Schwarz-Selinger T., Jacob W. and Keudel A.v., *J. Appl. Phys.* **87** (2000) 2719.
60. Perrin J., Takeda Y., Hirano H., Takeuchi Y. and Matsuda A., *Surf. Sci.* **210** (1989) 114.
61. Nurrudin A., Doyle J. R. and Abelson J. R., *J. Appl. Phys.* **76** (1994) 3123.
62. SIMION V.4.02, Idaho Nat. Eng. Lab (1989).
63. Voo, A.C.C., Ng R., Tunstall J.J and Taylor S., *J. Vac. Sci. Technol.* **A15** (1997) 2276.
64. Dylla H.F. and Blanchard W.R., *J. Vac. Sci. Technol.* **A1** (1983) 1297.
65. Böhm C. and Perrin J., *Rev. Sci. Instr.* **64** (1993) 31.
66. Flender U. and Wiesemann K., *J. Phys. D: Appl. Phys.* **27** (1994) 509.
67. Liu J., Huppert G.L. and Sawin H.H., *J. Appl. Phys.* **68** (1990) 3916.
68. Nguyen P.L., Turkot R.B. and Ruzic D.N., *J. Vac. Sci. Technol.* **A9** (1991) 682.
69. Jacob W., *Thin Solid Films* **326** (1998) 1.
70. Deschenaux C., Affolter A., Magni D., Hollenstein C. and Fayet P., *J. Phys. D: Appl. Phys.* **32** (1999) 1876.
71. Zeuner M., Neumann H., Zalman J., Slavínská D. and Biederman H., *Vacuum* **51** (1998) 417.
72. Zeuner M., Hirsch D., Neumann H., Zalman J. and Biederman H, *Proc. 14th Int. Symp. on Plasma Chem. (ISPC)* (Prague, 1999) 1527.
73. d'Agostino R., Cramarossa F. and Illuzzi F., *J. Appl. Phys.* **61** (1987) 2754.
74. Poll H.-U., Meichsner J., Arzt M., Friedrich M., Rochotzki R. and Kreyßig E., *Surf. Coat. Technol.* **59** (1993) 365.
75. Granier A., Nicolazo F., Vallée C., Goullet A., Turban G. and Grolleau B., *Plasma Sources Sci. Technol.* **6** (1997) 147.
76. Charles C., Garcia P., Grolleau B. and Turban G., *J. Vac. Sci. Technol.* **A10** (1992) 1407.
77. Magni D., Deschenaux C., Hollenstein C., Creatore A. and Fayet P., *J. Phys. D: Appl. Phys.* **34** (2001) 87.
78. Alexander M.R., Jones F.R. and Short R.D., *Plasmas Polymers* **2** (1997) 277
79. Zeuner M. and Meichsner J., *Vacuum* **46** (1995) 27.
80. Pecher P. and Jacob W., *Appl. Phys. Lett.* **73** (1998) 31.

81. Schwarzenbach W., Tserepi, A., Derouard, J. and Sadeghi N., *Jpn. J. Appl. Phys.* **36** (1997) 4644.
82. Singh H., Coburn J.W. and Graves, D.B., *J. Vac. Sci. Technol.* **A17** (1999) 2447.
83. Vasile M.J. and Smolinski G., *Int. J. Mass Spectrom. Ion Phys.* **24** (1977) 171.
84. Robertson R. and Gallagher A., *J. Appl. Phys.* **59** (1986) 3402.
85. Christophorou L.G., Olthoff J.K. and Rao M.V.V.S., *J. Phys. Chem. Ref. Data* **25** (1996) 1341.
86. Christophorou L.G. and Olthoff J.K., *J. Phys. Chem. Ref. Data* **27** (1998) 1.
87. Christophorou L.G. and Olthoff J.K., *J. Phys. Chem. Ref. Data* **28** (1999) 967.
88. Stoffels E., Stoffels W.W. and Tachibana K., *Rev. Sci. Instr.* **69** (1998) 116.
89. Stoffels W.W., Stoffels E. and Tachibana K., *Jpn. J. Appl. Phys.* **36** (1997) 4638.
90. Kortshagen U. and Zethoff M., *Plasma Sources Sci. Technol.* **4** (1995) 541.
91. Hopewood J., *Appl. Phys. Lett.* **62** (1993) 940.
92. Lieberman M.A. and Gottscho R.A., *Design of High Density Plasma Sources for Materials Processing - Phys. Thin Films* **18** (Academic Press, New York, 1994).
93. Kuypers A.D. and Hopman H.J., *J. Appl. Phys.* **67** (1990) 1229.
94. Manenschijn A., Janssen G.C.A.M., Drift E.v.d. and Radelaar S., *J. Appl. Phys.* **69** (1991) 1253.
95. Kushner M.J., *J. Appl. Phys.* **58** (1985) 4025.
96. Zeuner M. and Meichsner J., *Vacuum* **46** (1995) 151.
97. Zeuner M., Neumann, H. and Meichsner J., *Jpn. J. Appl. Phys.* **36** (1997) 4711.
98. May P.W., Field, D. and Klemperer D.F., *J. Appl. Phys.* **71** (1992) 3721.
99. Hishmeh G.A., Barr T.L., Sklyarov A. and Hardcastle S., *J. Vac. Sci. Technol.* **A14** (1996) 1330.
100. Biederman H., Ojha S.M. and Holland L., *Thin Sol. Films* **41** (1997) 329.
101. Wucher A. and Oechsner H., *Surf. Sci.* **199** (1988) 567.
102. Siegmund P., *Nucl. Instr. Meth. Phys. Res.* **B27** (1987) 1.
103. Andersen H.H., *Nucl. Instr. Meth. Phys. Res.* **B33** (1988) 466.
104. Zalm P.C., *Surf. Interf. Analysis* **11** (1988) 1.
105. Ellmer K. and Lichtenberger D., *Surf. Coat. Technol.* **74-75** (1995) 586.
106. Wouters S., Kadlec S., Nesladek M., Quaeyhaegens C. and Stals L.M., *Surf. Coat. Technol.* **76-77** (1995) 135.

107.Pecher P., *Quantitative Determination of the Particle Fluxes Emanating from Methane ECR Plasmas* (IPP-Report, Garching, 1998).
108.Holber W.M. and Forster J., *J. Vac. Sci. Technol.* **A8** (1990) 3720.
109.Reinke P., Schelz S., Jacob W. and Möller W., *J. Vac. Sci. Technol.* **A10** (1992) 434.
110.Gottscho R.A., *Phys Rev.* **A36** (1987) 2233.
111.Howling A.A., Sansonnes L., Dorier J.-L. and Hollenstein C., *J. Phys D: Appl. Phys.* **26** (1993) 1003.
112.Stoffels E., Stoffels W.W., Vender D., Hagerlag M., Kroesen G.M.W. and de Hoog, F.J., *Contrib. Plasma Phys.* **35** (1995) 331.
113.Howling A.A., Sansonnens L., Dorier J.-L. and Hollenstein C., *J. Appl. Phys.* **75** (1994) 1340.
114.Lergon H.G., Venugopalan M. and Müller K.G., *Plasma Chem. Plasma Proc.* **4** (1984) 107.
115.Cuomo J.J., Gambino R.J., Harper J.M.E., Kuptsis J.D. and Webber J.C., *J. Vac. Sci. Technol.* **15** (1978) 281.
116.Kester D.J. and Messier R., *J. Vac. Sci. Technol.* **A4** (1986) 496.
117.Ashida S., Shim M.R. and Lieberman M.A., *J. Vac. Sci. Technol.* **A14** (1996) 391.
118.Samukawa S., *Jpn. J. Appl. Phys.* **33** (1994) 2133.
119.Smith B.A. and Overzet L.J., *J. Appl. Phys.* **78** (1995) 5195.
120.Zeuner, M., Neumann, H and Meichsner J., *Vacuum* **48** (1997) 443.
121.Garscadden A., Ganguly B.N., Haaland P.D. and Williams J., *Plasma Sources Sci. Technol.* **3** (1994) 239.
122.Hollenstein C., Schwarzenbach W., Howling A.A., Courteille C., Dorier J.-L. and Sansonnens L., *J. Vac. Sci. Technol.* **A14** (1996) 535.

CHAPTER 5

ELECTRICAL AND OPTICAL PROPERTIES OF PLASMA POLYMERS

JACEK TYCZKOWSKI
Technical University of Lodz
Faculty of Process and Environmental Engineering
Lodz, Poland

5.1. Introduction

Plasma polymers are high-molecular systems produced, as has been already mentioned in Chapter 1, in the consequence of chemical processes proceeding in the region of low-temperature, non-equilibrium plasma. The nature and properties of such materials are closely connected with a thin film form, in which they are generally produced. If a monomer (source precursor) is introduced to the plasma region, a thin film of the plasma polymer begins to arise on all surfaces in this region. By optimizing parameters of the process we can obtain solid, homogeneous thin films without macroscopic defects (such as pinholes, powders, cracks, porosity). Such films are just those which at present awake the great interest of science and high technology. In general, the thickness of the films is between a few nanometers and a few micrometers. Therefore very often the term *thin plasma film* is used as a synonym of the term *plasma polymer*.

A large variety of chemical species created in the plasma, their statistical combination towards high molecular structures and generally a high degree of crosslinking cause the structure of such a material to be very often much closer to that of covalent glasses than that of conventional polymers. Thus, the thin plasma films can be treated as a single amorphous macromolecule consisting of the three-dimensional basic molecular network with attached side units such as hydrogen atoms and sometimes also other functional groups.

As an example, a two-dimensional model of the molecular structure of a plasma polymerized hydrocarbon film (pp-hydrocarbon), also termed as hydrogenated carbon film (a-C:H), is shown in Fig.5.1. A theoretical amorphous

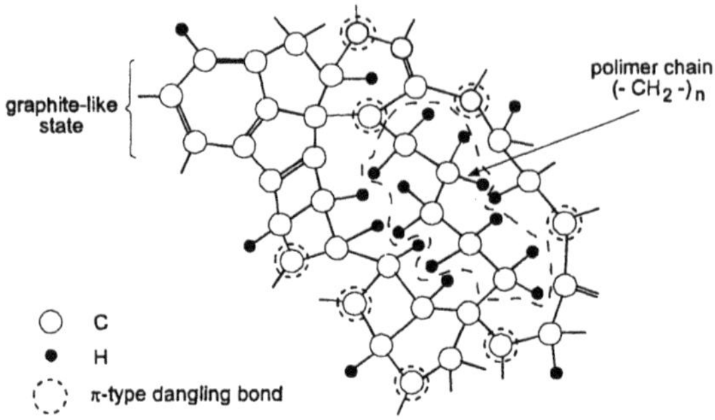

Figure 5.1. Two-dimensional model structure of a hydrogenated carbon film (a-C:H).

carbon structure is built only from quaternary carbons, and in other words, it is extremely deformed diamond crystal. In the presented structure, apart from the quaternary carbon framework, other chemical species can also exist, e.g. graphite-like states (double bonds), dangling bonds (in the case of a-C:H they are rather π-type, that differs from a-Si:H where σ-type dangling bonds dominate [1]), C–H bonds, and hydrocarbon chains. Of course, their contents and space distributions in the film closely depend on deposition conditions. If in the film there are atoms of other elements (oxygen, nitrogen, silicon, germanium, sulfur, metal atoms, etc.) the molecular structure is yet more complicated.

Although the precise description of such structures, from the molecular point of view, is incredibly difficult, one has to realize that every chemical unit in the film participates in the creation of its electronic structure and, in consequence, influences electronic (electrical and optical) properties of the film. If we want to tailor the electronic properties of thin plasma films according to our requirement we should find the correlation in the following sequence: deposition conditions – chemical structure – electronic structure – electronic properties. The control of the electronic properties by deposition conditions is one of the most important tasks currently undertaken in the scope of plasma polymers.

The last two fields of the sequence, namely the electronic structure and electronic properties of thin plasma films, as well as the correlation between them, are the main subjects of this chapter.

5.2. Electronic Structure

5.2.1. Theoretical Models

A high degree of crosslinking of solid plasma polymers and the resulting random network constitute the main reason why these materials are regarded as amorphous. The application of Sir Mott's [2] theory of electronic structure of amorphous solids to plasma polymerized thin films is, therefore, an entirely justified approach.

To understand the nature of the electronic structure of the amorphous state it is simplest to start from the electronic structure of a crystalline material. It is widely recognized that such a structure is described by the band model based on the periodicity of the crystal lattice. Schematically, a two-dimensional crystal model and a corresponding band model (energy distribution of density-of-states) are shown in Fig. 5.2(a). The black triangles represent a short-range order. A periodic array of the triangles gives the crystal lattice (a long-range order). In this case the band model, in its simplest version, is the following: two bands of extended states exist – the conduction and valence bands – and they are separated by the forbidden band (bandgap) E_{gap}. Electrons and holes can be transported through the crystal if their energies are in a range of the conduction or valence bands, respectively.

If we provoke a disorder of the periodicity simultaneously maintaining the short-range order, we transform the crystal structure to the amorphous state. We may imagine it taking into account a diamond crystal as an example [3]. In this crystal each carbon atom has four immediate neighbors arranged in a tetrahedral geometry. Introduction of a statistical distribution of bond angles about their average value of 109°28' causes the destruction of the tetrahedral geometry. If a maximum deviation of bond angles is not too high, in this case not higher than 20°, practically any bond is not broken. Each carbon atom is still bonded to four other atoms that satisfies its chemical valence. Now, we have, however, a new structure – the random network structure without the periodicity but with the short-range order (Fig. 5.2(b)). This is the idealized model of amorphous state. But how does the band structure model look for this state?

The basic concept in the theory describing the electronic structure of amorphous solids is the notion of a localized state. A criterion for electron localization in amorphous materials was first introduced by Anderson [4], who showed that, under specific conditions of disorder, the probability of diffusion of an electron to a certain distance drops to zero with the increase of this distance. Mott [2], who analyzed the localization of charge carriers in both the conduction and the valence bands in amorphous materials, proved that the Anderson's localization may occur only in certain fragments of both bands with the remaining part of each band consisting of extended states. This implies that, identically as for the case of crystalline state, a charge carrier excited to an energy level belonging to the range of extended states can freely move along the whole system, and its diffusion ability is limited only in the energy range of localized states. The energy values separating the localized from the extended states have been called the mobility edges.

Based on these considerations a band model was proposed for the amorphous state [2,5]. This model, known as the Mott-CFO model (Mott - Cohen, Fritzsche, Ovshinsky), is schematically presented in Fig. 5.2(b). Ranges of localized states neighboring the bandgap characterize both the conduction and the valence bands. The energies denoted by E_V and E_C constitute the borders between localized and extended states and correspond to the mobility edges. The distance between these two values is called the mobility gap E_g and is equivalent to the bandgap for crystalline materials. The distance E_AE_B is called the optical gap E_{OPT} and is determined by the optical absorption edge (see: *Sec.*

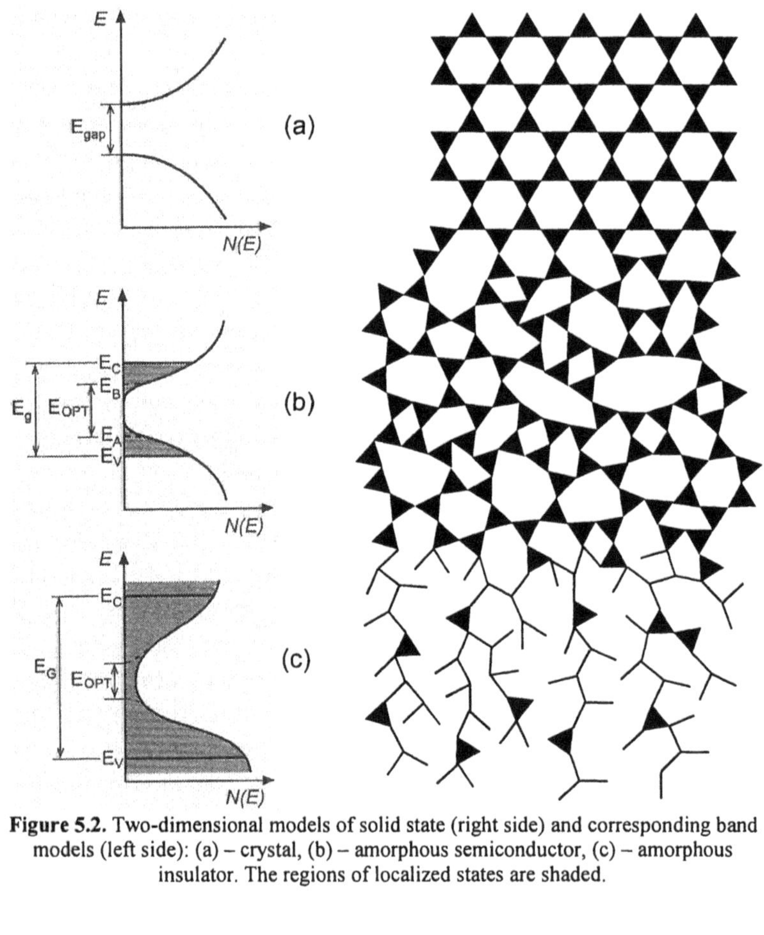

Figure 5.2. Two-dimensional models of solid state (right side) and corresponding band models (left side): (a) – crystal, (b) – amorphous semiconductor, (c) – amorphous insulator. The regions of localized states are shaded.

5.2.2.1). In that case the charge carrier (electron and hole) transport can proceeds via the extended states (as in a crystal) or by a hopping mechanism via the localized states (see: *Sec.* 5.3.4).

For a perfect amorphous material, the difference between E_g and E_{OPT} amounts to a few tenths of an electron volt and is exclusively due to the localized states resulting from the lack of the long-range order. However, real amorphous materials, such as thin plasma films, contain numerous defects in their structure (some examples are shown in Fig. 5.1) creating an additional number of localized states within the mobility gap. They may fill the gap out either partly or even completely thus bringing about a significant reduction of E_{OPT} compared to E_g. Since the appearance of the Mott-CFO model a number of specific models describing particular, real amorphous materials have been proposed. All of them, however, constitute its various modifications assuming only different distribution of density-of-states and maintaining the fundamental feature: the existence of both the localized and the extended states [5,6]. All materials of which the electronic structure is described by such a model are called **amorphous (a-) semiconductors**.

The a-semiconductor model was applied until quite recently, sometimes uncritically indeed, to most, if not all, amorphous materials, also those deposited by plasma processes [7]. On the basis of the theory of electron state localization one can predict, however, that with increasing disorder, for example as a result of the destruction of the short-range order (Fig. 5.2(c)), the mobility edges move towards the centers of the bands and above some critical strength of disorder all states are localized [8]. In that case a new category of materials originates, of which the electronic structure is qualitatively different from that generally accepted for a-semiconductors. To distinguish between both types of amorphous materials, those materials without the extended states in the conduction and valence bands have been called **amorphous (a-) insulators** [9].

It should be emphasized that a-semiconductor and a-insulator are qualitatively unlike materials in contrast with crystalline semiconductors and insulators differing only quantitatively from each other.

In Fig. 5.2(c) a typical band model of a-insulator is illustrated. All states are localized. It means that electrons and holes can be transported through the material only by hopping. This process proceeds above E_C and below E_V, respectively, where the density of states is sufficiently high. It has been shown that a step change in carrier mobility in the vicinity of E_C and E_V justifies the concept of the mobility gap but its meaning is different from that for a-semiconductors [10]. The distance E_CE_V is often called the transport gap E_G.

5.2.2. Experimental Methods

The investigations on the electronic structure of thin plasma films are focused mainly on two principal groups of problems. The first one comprises the questions concerning basic parameters of the band model, such as the transport gap E_G (or the mobility gap E_g), the optical gap E_{OPT}, the Fermi level energy E_F, the electron affinity χ, the ionization potential J. The second group of problems refers, in turn, to determination of the density-of-state distributions in the bands and in the transport gap. Various experimental methods, a lot of them adopted from studies performed on the crystalline solids [11,12], are used to solve these problems. The most useful experimental methods available to study the electronic structure of plasma polymers will be briefly presented in the further part of the section. It should be added, however, that the experimental results are at present enhanced by more and more sophisticated calculations based on more and more realistic atomic models [13-15].

5.2.2.1. Optical Spectra

Optical studies of plasma polymers have concentrated on absorption measurements, mainly of the absorption edges in the high absorption region (absorption coefficient $\alpha > 10^5\ m^{-1}$) which enable us to determine E_{OPT}. Although this parameter still has no immediate physical meaning and there is some ambiguity about its experimental determination, nevertheless it is treated as one of the most significant parameters describing the electronic structure of amorphous materials.

To define the optical gap, some researchers [16-18] use the energy E_{04} at which the absorption coefficient amounts to $10^6\ m^{-1}$. This value of the optical gap is identified with the mobility gap E_g in amorphous semiconductors. However, this is more complicated in the case of amorphous insulators. It was shown, for example, that values of E_{04} are evidently lower than the transport gaps E_G for plasma deposited insulating a-Si_XC_Y:H films [19].

Most frequently E_{OPT} is determined according to the Tauc relation [20]:

$$\alpha h\nu = B(h\nu - E_{OPT})^n, \tag{5.1}$$

where α is the absorption coefficient, $h\nu$ the energy of light, n a parameter connected with the density-of-states (DOS) distribution in the transport gap (in the band tails) and B a proportionality factor that also provides information about the DOS distribution. It is connected with the length of localized state tails. Smaller values of B indicate longer tails of these states [16,17,21].

The optical gap for the semiconducting thin plasma films is usually determined by assuming a parabolic function for the DOS distribution in the band tails ($n = 2$ in Eq. (5.1). In certain cases, however, (especially for amorphous insulators) a better description of the absorption edges was found using $n = 3$ [9,19,20,22-24]. This can be interpreted as a close-to-linear DOS distribution in these materials. Figure 5.3 presents an example of absorption edges measured for semiconducting and insulating a-Si_XC_Y:H films produced by plasma polymerization from methylosilanes [9]. It is obvious that in this case $n = 3$ is more adequate for the insulating film while the semiconducting film is better characterized by $n = 2$.

More information on the DOS distribution in the transport gap can be obtained from optical absorption spectra recorded in the low absorption region of the absorption edges ($\alpha \leq 10^5\ \mathrm{m}^{-1}$). These studies require, however, measurements of very small α values. A lot of techniques have been proposed to this end, each one having its individual advantages, disadvantages and limitations.

Over the last years especially two techniques, in particular PDS (Photothermal Deflection Spectroscopy) and CPM (Constant Photocurrent Method), have attracted considerable attention. PDS and CPM are in certain respects considered as complementary methods in the sense that the former is sensitive to surface/interface and bulk states, while the latter leads to information primarily on bulk properties. Their combination in principle allows one to determine both the surface/interface and bulk contributions.

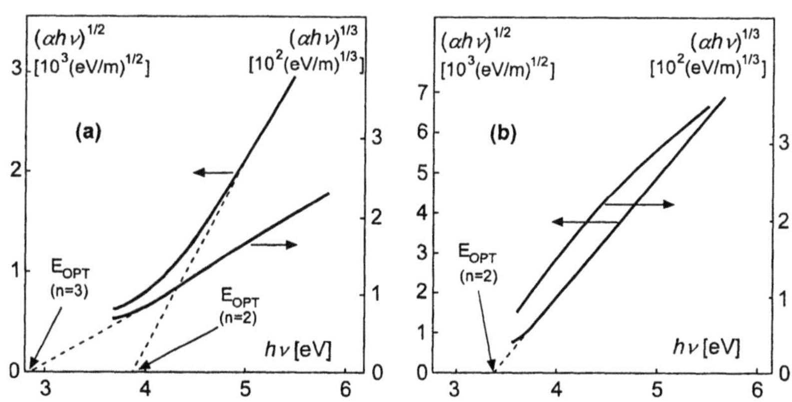

Figure 5.3. Absorption edges presented according to Eq. (5.1) for $n = 2$ and $n = 3$: (a) – pp-dimethylsilane (insulator), (b) – pp-methylsilane (semiconductor) [9].

The PDS, adopted for the plasma deposited thin films, is one of photothermally based spectroscopies that have been applied for 20 years to study the absorption behavior of solids, liquids and gases. The attractive features of this technique are an extremely high absorption sensitivity and the low sensitivity on scattering effects. Absorbance values as low as $\alpha d \approx 10^{-7}$ (d is the film thickness) have been measured. Fig. 5.4 shows a typical setup for PDS measurements. A sample of thin film on substrate is placed in a cell filled by nonabsorbing medium having a large $\partial n/\partial T$ (n – index of refraction, T – temperature), for example CCl_4. A monochromatic light penetrates CCl_4, traverses the investigated film (absorbing medium), and subsequently propagates through the substrate (non-absorbing medium). Part of the energy absorbed in the film will diffuse back into CCl_4 and establish a certain temperature distribution close to the film surface, which in turn entails a distribution of the index of refraction $n(T)$. A He-Ne laser beam probes the refraction index gradient and deflects by an angle $\Delta\varphi$ from its original straight path. This angle is related to the absorption coefficient [25].

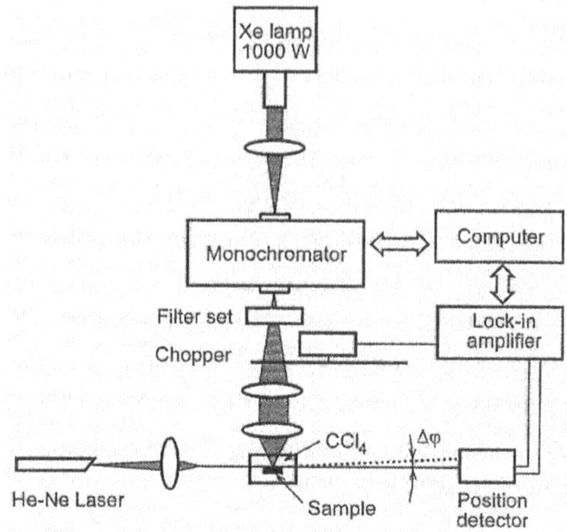

Figure 5.4. Experimental setup of the photothermal-deflection-spectroscopy [25] (with permission).

The PDS technique has been used to investigate the electronic structure among others of such plasma deposited amorphous hydrogenated films as: silicon (a-Si:H) [25,26], silicon-carbon (a-Si_XC_Y:H) [27,28], silicon-nitrogen (a-Si_XN_Y:H) [28], carbon (a-C:H) [29] (also for plasma deposited diamond films [30]), carbon-tin (a-C_XSn_Y:H) [31], carbon-sulfur (a-C_XS_Y:H) [32], boron-nitrogen (a-B_XN_Y:H) [33], germanium (a-Ge:H) [34,35], silicon-germanium (a-Si_XGe_Y:H) [36].

The second important technique, namely the CPM technique, is based on bulk photoconductivity measurements. In the bulk photoconductivity process, photogeneration of carriers (free electrons or holes) is connected with the optical absorption coefficient α and the quantum efficiency η of the generation. On the other hand, transport of these carriers is characterized by their mobility μ_0 and lifetime τ_l. If we want to find from the measured photocurrent the spectral dependence of α, we have to keep in mind the fact that also other material related parameters (η, μ_0, τ_l) are in principle spectrally dependent. In some cases, when additional independent information about the material is available and experimental conditions are carefully designed, the correct information about the absorption spectrum and DOS can be obtained. The basic CPM idea is to keep the photocurrent constant when changing the wavelength of light. In that case $\alpha(h\nu) \sim 1/N(h\nu)$, where $N(h\nu)$ is the number of photons for a given light energy.

Unfortunately, the CPM is limited to semiconducting films that reveal a relatively high photoconductivity and it has been used in practice only to a-Si:H in the group of plasma deposited films [37].

Among other absorption spectroscopy methods one should mention such as the transient photocapacitance spectroscopy [38], the modulated photocurrent spectroscopy (MPS) [39,40], the transient photocurrent measurements (TPC) [41], and the photoacoustic spectroscopy (PAS) [42]. However, all these techniques, similar to the CPM, are suitable only for the study of amorphous semiconductors, mainly a-Si:H, and are quite useless for amorphous insulators.

5.2.2.2. Internal Photoemission

Internal photoemission (photoinjection) has been used for more than 30 years as a tool to study the electronic structure of solids [11]. Not infrequently thin films of plasma polymers have also been a subject of such investigations [22,43-53]. The phenomenon of photoinjection is generally referred to as injection of electrons or holes from a solid into another solid, for example, from metal electrode to an investigated material, resulting from an interaction between photons and solids in their contact region. In that case measurements of photocurrent as a function of exciting photon energy enable us to estimate the contact barrier heights, and in consequence, the basic parameters of the band model.

The main problem in the photoinjection investigations, however, is to distinguish between the internal photoemission and bulk photogeneration processes. To eliminate this difficulty the photocurrent measurements should be performed for both the sandwich and coplanar geometry of the investigated film. It has been shown that the photocurrent measured under coplanar geometry gives information only on bulk processes [37]. On the other hand, for the sandwich type of sample the bulk and internal photoemission processes are observed. Thus, the "pure" photoinjection effect can be extracted [19,52].

For the photoinjection process the photoconductivity of a film is given, according to the Fowler theory [11], by the following relation:

$$Y \propto (h\nu - \Phi)^2, \tag{5.2}$$

where Y is the photocurrent quantum yield, $h\nu$ is the exciting photon energy, and Φ is the contact barrier height (for electrons or holes). Usually the exponent 2 is used in this relation but in some cases other values, as 2.5 or 3, have been proposed [47,49].

Figure 5.5 presents an example of photocurrent quantum yield spectra for a film of plasma polymer deposited from tetramethylsilane and equipped with two gold electrodes, in the first case in the sandwich geometry (Fig. 5.5(a)) and in the second one as the coplanar system (Fig. 5.5(b)). As one can see, the bulk photoconductivity is observed only at a photon energy higher than 4.75 eV. Thus, the photoconductivity recorded at lower photon energies (for the sandwich sample) must originate from photoinjection. Two photocurrent thresholds are visible in this case. It has been found that the lower threshold is attributed to the height of the contact barrier for hole injection from the positively polarized electrode, whereas the upper threshold is connected with the height of the barrier for electrons injected from the negatively polarized electrode. (In this case, the both electrodes are illuminated – the upper electrode directly and the rear one through the film).

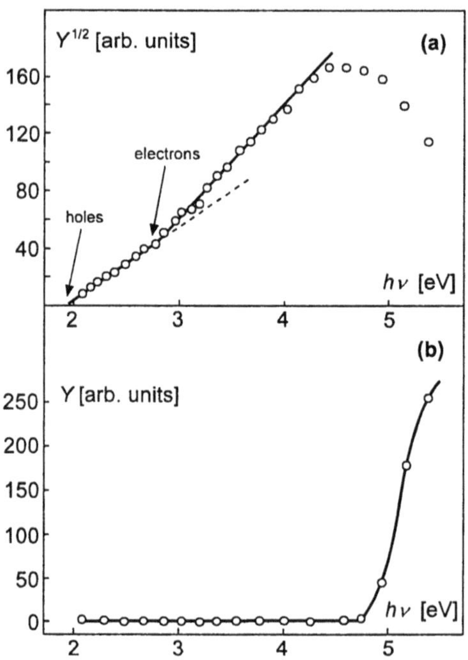

Figure 5.5. Photocurrent quantum yield spectra for pp-tetramethylsilane equipped with two Au electrodes: (a) – in the sandwich geometry, (b) – in the coplanar geometry [19].

Taking into consideration lowering of the barrier heights by the external voltage applied to a sample during the photoinjection measurement, the absolute heights of the barriers can be determined. For amorphous insulators the lowering is attributed to the Schottky effect [22,45,48,52], and is described by the general relation [11]:

$$\Phi = \Phi_0 - \beta(U/d)^{1/2}, \tag{5.3}$$

where Φ_0 is the absolute barrier height, U is the applied voltage, d is the film thickness, and β is the Schottky coefficient. A typical average value of the coefficient β is $2\text{–}3\cdot10^{-5}$ eV(m/V)$^{1/2}$.

A simple model of photoinjection from metal electrode to insulator, e.g. ref. [54], shows that a sum of the heights of contact barriers for electrons Φ_{0e} and holes Φ_{0h} gives the value of the transport gap E_G for the investigated insulator [22,52]:

$$\Phi_{0e} + \Phi_{0h} = E_G \tag{5.4}$$

The sum should be independent of the kind of electrodes employed and of the state of the contact, e.g. of the presence of surface states. These factors influence the shape of the energy bands at the contact and thus the heights of barriers, but their sum remains unchanged.

When the contact is practically free from surface states studies of the barrier heights for a metal–insulator contact enabled to calculate the electron affinity χ and the ionization potential J according to relations [22,52]:

$$\chi = W_M - \Phi_{0e}, \tag{5.5}$$

and

$$J = E_G + \chi = W_M + \Phi_{0h}, \tag{5.6}$$

where W_M is the work function of the metal.

For semiconducting films, contrary to insulators, generally a very weak dependence of the barrier height on the applied voltage is predicted [44,45,49], which results from a simply blocking contact model [11]. In some cases,

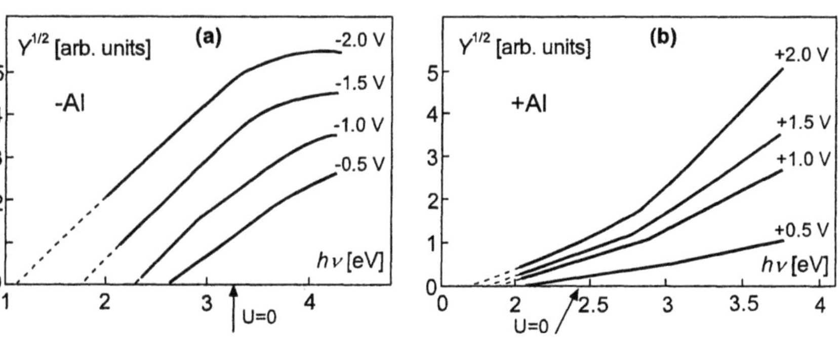

Figure 5.6. Fowler's diagrams for semiconducting a-Ge_XC_Y:H (pp-tetramethylgermanium): (a) – negative, and (b) – positive potential applied to the illuminated Al electrode [50].

however, namely for semiconducting a-Si_XC_Y:H and a-Ge_XC_Y:H films, unexpected dependences of the barrier height versus the applied voltage have been observed. In Fig. 5.6 photoinjection currents plotted in the Fowler's coordinate system (Eq. (5.2)) for a-Ge_XC_Y:H films are presented, both for negative and positive polarity of the illuminated electrode. The barrier heights for electrons and holes are determined from the intercept of the curves with the abscissa. For electrons (that are minority carriers in the films) a linear relation of $\Phi \propto U$ is found whereas for hole injection (injection of majority carriers) a relation of $\Phi \propto U^{1/2}$ appears. On this basis a model of electron and hole photoinjection from metal into amorphous semiconductors was proposed (Fig. 5.7) [50,53]. According to the model the transport gap E_G in such materials can be calculated as follows:

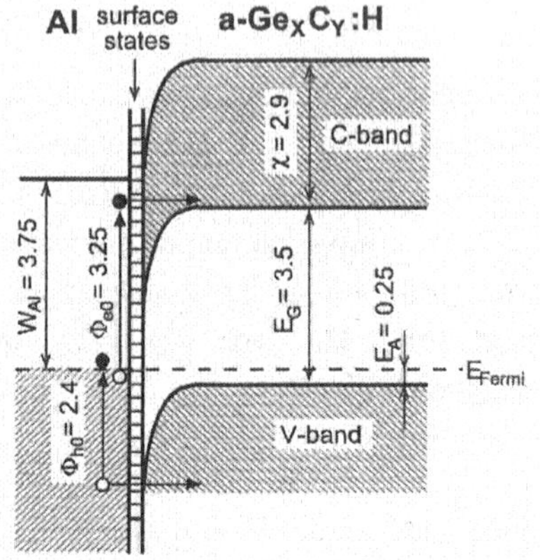

Figure 5.7. Energy level model explaining the electron and hole photoinjection from metal (Al.) into semiconducting a-Ge_XC_Y:H film. (Values are in eV) [50].

$$E_G = \Phi_{0e} + \Delta E_F, \tag{5.7}$$

where ΔE_F is the distance between the Fermi level position and the edge of the transport band. ($\Delta E_F \approx E_A$, where E_A is the activation energy of the bulk conductivity.)

Photoinjection measurements can also give information on the electronic structure of material surface [55]. It is possible to determine the density and type of surface states, the position of the Fermi level at the surface, and the work function of the material. Such investigations have been carried out for plasma polymers as well, for instance, for plasma deposited organosilicon and organogermanium films [46,52].

5.2.2.3. Photoelectron Spectroscopies

The most powerful experimental tools for probing the electronic structure of solids are to be found in the area of photoelectron spectroscopy. The chief reason for the widespread use of these techniques is that photoemission spectra can yield information on the absolute energy placement of an energy band. It is possible to examine the conduction and valence bands independently and to determine their absolute energy positions.

Using the photoelectron techniques, we should, however, keep in mind that their key feature is a very high sensitivity to surface characteristics. Typically, the penetration depth for these techniques is only several dozens of interatomic layers [12].

As far as plasma polymer films are concerned, UV-photoelectron spectroscopy (UPS) and x-ray photoelectron spectroscopy (XPS) are the most frequently used methods to determine their electronic structure. With both UPS and XPS, the mode of operation is to fix the incident photon energy and to measure, as a function of energy, the number of emitted photoelectrons. The only important difference between both methods is the photon energy, which spanned from low energy of UV photons that can only excite the outermost orbitals (the valence region) up to high-energy x-ray photons that can also ionize and probe the core levels. In general, the photoelectron-energy distribution function may be, at first approximation, regarded as replicas of the electronic structure of the solid, and most strictly speaking, the DOS (density-of-states) distribution function for the bands [56,57].

The UPS spectra and the low-energy side of XPS spectra (this is a region from 0 to 20–30 eV) are especially valuable to determine the valence band structure. Such investigations have been widely performed, for example, on plasma polymerized hydrocarbons (a-C:H) [1,58-62]. The detailed electronic structure of the a-C:H films is complex for at least three reasons: namely, the hybridization of C atoms may be sp^2 or sp^3, considerable amounts of incorporated H atoms may affect the structure, and the material may be heterogeneous due to preferential clustering of sp^2 C atoms. All these features are a strong function of deposition conditions, so changing the parameters of the plasma deposition process one can control the electronic structure of a-C:H from typical insulating polymer-like films to a semiconducting graphite-like material [63,64].

Figure 5.8 shows by way of example an XPS spectrum of a-C:H (r.f. plasma polymerized methane film). The relevant features of the spectrum are three main peaks: the first at about 18 eV connected with the C 2s states, the second at about 8 eV ascribed to the C 2pσ states, and third around 3 eV that can originate from the 2pπ states of carbon [60].

Analogous investigations, sometimes called the valence band spectroscopy, have also been performed on other plasma polymers, e.g. a-Si:H [65-67], a-Si_XC_Y:H [13,62,68,69], a-C_XN_Y:H [70], a-Si_XN_Y:H films [71,72]. All

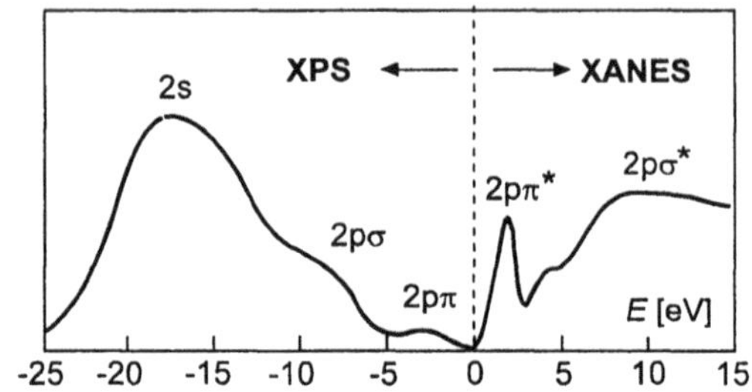

Figure 5.8. XPS and XANES spectra of a-C:H films [60,74].

these studies give us the important information on the valence region of the materials.

The next problem concerning the electronic structure of plasma deposited films is determination of the conduction band DOS. It has appeared that the x-ray near-edge spectra (XANES) are profitably used for this end. The XANES is a special case of more general technique, EXAFS, consisting in the study of x-ray absorption as a function of photon energy. The absorption intensity, which is related to the photoexcitation of electrons from deep core states (mostly from *K*-shell), exhibits oscillations known as the extended x-ray absorption fine structure (EXAFS) [57,73]. Although this technique is not a typical photoelectron spectroscopy (we measure the x-ray absorption but not the photoelectron energy), the XANES is mentioned here just because its primary physical process, namely the photoionization of a given atom by a x-ray photon, is the same as for the XPS.

In Fig. 5.8 the XANES spectrum for a-C:H is presented. It shows the first 15 eV of oscillations in the excitation spectrum of the carbon 1s core electrons and, at first approximation, reflects the density of unoccupied states, in this case – π^* and σ^* states of the conduction band [1,74].

Recently, in order to image the conduction band structure with high lateral resolution, a technique that combines XANES spectroscopy with a photoemission electron microscope has been proposed [75]. In this case the x-ray absorption spectra are recorded through measurements the secondary electron yield, which is directly proportional to the x-ray absorption. The electrons are collected and accelerated by a high electrostatic field between the sample and the first electrode of the immersion objective lens of the microscope. After magnification by a two-stage projective lens system and intensification via a multichannel-plate, the lateral distribution of electrons is made visible on a luminescent screen. The lateral resolution, which can be reached, is order of 130 nm (at $h\nu = 540$ eV). Using this technique the lateral distribution of the unoccupied density of states has been mapped for carbon thin films [75].

Although less common, the secondary electron spectroscopy appears to be another useful method for analysis of unoccupied states in the material. Fine structures superimposed on the "cascade slope" of a typical secondary electron spectrum can yield some information about the σ-electrons [76]. Such studies were performed among others for a-C:H films produced from methane [58]. It was found that the intense, narrow peak observed for crystalline and polycrystalline graphite at 7.6 eV above the Fermi level, which is related to a maximum in the conduction σ^*-band state density, is not detected in the hydrogenated carbon films. According to the authors, this is indicative of a very low amount of graphitic phase in the films.

A simple variant of photoelectron spectroscopies, such as XPS or UPS, is the total-photoelectron-yield spectroscopy (sometimes called external photoemission). Instead of measuring the energy distribution of electrons excited by monochromatic light, in total photoyield spectroscopy the energy of the incident light is tuned and all electrons are collected regardless of their kinetic energy. The resulting total yield spectrum is proportional to the convolution of the initial and final states' densities, and the onset of photoemission yields information about the position of the vacuum level relative to the Fermi level [77,78]. It should be noted, however, that interpretation of the threshold energy of photoemission in photoyield spectra is different for films of positive and of negative electron affinity: for films of positive electron affinity the threshold energy gives the work function whereas in the case of negative electron affinity it gives a measure for the transport gap.

As an example, the use of the total-photoelectron-yield spectroscopy to determine the temperature dependence of the conduction and valence band tails for a-Si:H and a-Si_XGe_Y:H films can be reported. It has been found that the films possess purely exponential conduction- and valence-band-tail densities of states. It has also been shown that the conduction band tail is considerably more susceptible to thermal disorder than to structural disorder, whereas the reverse holds for the valence band tail [79].

The total-photoelectron-yield spectroscopy also has been utilized to determine the work function of semiconducting a-Ge_XC_Y:H films prepared from tetraethylgermanium in an r.f. discharge [50]. The total photoelectron yield was measured for films (about 800 nm thick) deposited on a glass plate with evaporated Au strip as the bottom electrode. The top electrode, made of Pt plate, was placed 3.0 mm in front of the sample surface. The system was illuminated by monochromatic light (tuned in the range of 2 – 6 eV) through a hole in the center of the Pt electrode and the current flowing between the bottom electrode and the Pt plate was measured. Two procedures were used to determine the threshold energy of photoemission E_{th}: (*1*) – studies of the relation between the

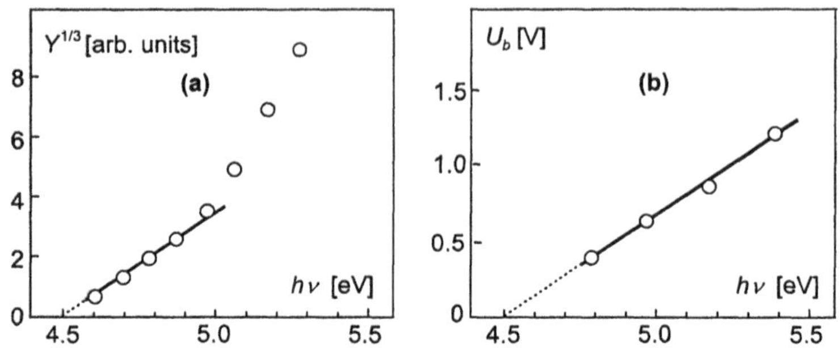

Figure 5.9. Total yield photoemission for semiconducting a-Ge_XC_Y:H: (a) – relation of the quantum yield Y vs. photon energy $h\nu$ plotted according to Eq. (5.8), (b) – the braking potential plotted according to Eq. (5.9) [50].

quantum yield of the current Y and photon energy $h\nu$, which near the threshold energy can be described by:

$$Y \propto (h\nu - E_{th})^3, \tag{5.8}$$

and (*2*) – studies of the braking potential U_b according to relation:

$$U_b \mathrm{e} = h\nu - E_{th}. \tag{5.9}$$

In Fig. 5.9 (a) and (b) results of the total yield photoemission for both procedures of measurements are presented. The threshold energy is identical in both cases and amounts to 4.5 eV. This value expresses the work function of the a-Ge_XC_Y:H films under study.

Recently, similar investigations of the work function have been performed for pure and lithium-containing a-C:H films. Its values are 3.1 eV and 2.55 eV (20 at. % lithium in the film), respectively, and speak volumes for evident changes in the a-C:H electronic structure caused by lithium dopant [61].

Finally, one should mention other photoelectron spectroscopies, namely the electron energy loss spectroscopy (EELS) and Auger electron spectroscopy (AES), which in some cases are also helpful for investigations on the electronic structure of plasma polymers [69,80,81].

5.2.2.4. Electrical Transport Methods

Investigations on electrical properties of plasma polymers are a good source of information about their electronic structure. They are widely used in the case of semiconducting films [8], but also for amorphous insulators some of these methods can be very helpful [11].

One major method based on electrical transport, which can enable us to determine the reliable quantitative DOS distribution in the transport gap of a semiconductor, is the field effect measurement. Unfortunately, although this method has been utilized for plasma polymers since 1972 [82], nevertheless it has achieved in practice a success only in the case of a-Si:H films [83-85].

A typical system for measurements of the field effect is depicted schematically in Fig. 5.10. The sample is constructed as a capacitor, in which a thin insulating film is placed between two electrodes: a metal electrode called gate and the second one made of the investigated semiconducting film. The current I_{SD} flowing through the semiconductor between the source and drain contacts, driven by the voltage U_{SD}, is controlled by the gate voltage U_G, applied between the gate and the semiconductor. The gate voltage induces charge in a thin surface layer of the semiconductor, near the insulator interface.

The interpretation of the field effect is at first sight straightforward. The induced charge condenses primarily into states above or below (depending on the charge polarity) the Fermi level E_F, which causes a bending the energy bands near the surface and, in consequence, a change in the semiconductor film conductivity. The higher the density of states at E_F the slower the change of the conductivity with an increase in induced charge. Thus, measuring I_{SD} as a function of U_G one can derive the DOS distribution in the transport gap.

The field effect can be detected only if the density of states near E_F is relatively low and, furthermore, it dominates over the density of surface states. This is probable reason for the fact that the field effect measurements have been failed in many cases, for example, for semiconducting a-Ge_XC_Y:H films [86].

Another possibility that can deliver us with information on DOS distribution in the transport gap of various materials is a space-charge-limited-current (SCLC) measurement. Such measurements are also used for plasma deposited thin films, e.g. a-Si:H [8,87] and a-C:H [88]. In a SCLC experiment a nonequilibrium density of charge is injected at higher voltages, which extended throughout the sample. Most of this extra charge is trapped around E_F and changes locally the quasi-Fermi level position in relation to the DOS distribution function. From the resulting increase in current, measured as a function of the applied

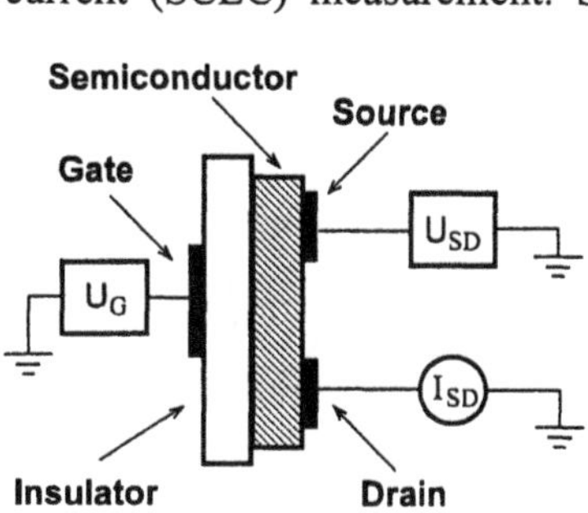

Figure 5.10. Sketch of the setup for measurements of the field effect.

voltage *I*(*U*), DOS can in principle be obtained.

The exact analytical relation between *I*(*U*) and DOS is, however, somewhat complicated because of the spatial dependence of the charge density, the electric field (that is not homogenous), the quasi-Fermi level, as well as the carrier diffusion [11]. The next problem being revealed while SCLC measuring is connected with the fact that the *I*(*U*) and *I*(*d*) functions (*d* is the film thickness), predicted for SCLC conditions, can be observed in a very similar form when other charge carrier transport mechanisms dominate [89]. Thus, if we want to use the SCLC method to obtain information about the transport gap DOS, we have to be sure that the conductivity is just governed by the space charge limited current.

An important and convenient method for determining trapping levels in amorphous semiconductors and insulators, which are closely connected with the transport gap DOS, is thermally stimulated detrapping [90]. This method is generally referred to the thermally stimulated current (TSC) and thermally stimulated depolarization (TSD) measurements, which have been carried out for a long time, very often to characterize the conventional polymers [91]. Such measurements have been also performed for plasma polymers, both semiconducting [92-94] and insulating films [95].

A new experimental technique that has been used recently to study the electronic structure of plasma polymers is the voltage-pulse induced time-domain depolarization (VPITDD). This method, especially useful in the case of amorphous insulators, consists of recording of a long-time isothermal relaxation (for hours) after a short voltage pulse (for milliseconds) [96] or an electron-beam pulse (for microseconds) [97]. Transient currents recorded by means of the VPITDD measurements are related to bulk properties of material and yield information on the generation centers and the transport states.

Figure 5.11 shows results of the DOS distributions for plasma polymers deposited from hexamethyldisilazane (HMDSN) and hexamethyldisiloxane (HMDSO), calculated from VPITDD measurements [98]. Two well defined DOS peaks, at 0.55 eV and 0.7 eV, were found for pp-HMDSO whereas in the case of pp-HMDSN the single broad peak (about 0.65 eV) is visible. Moreover, the results indicate that the density of states lying close to the Fermi level is two orders of magnitude higher for pp-HMDSN than for pp-HMDSO. Very likely this difference is responsible for the occurring low-temperature-conductivity (which is connected with the transport through localized states at the Fermi level) in the former films [99].

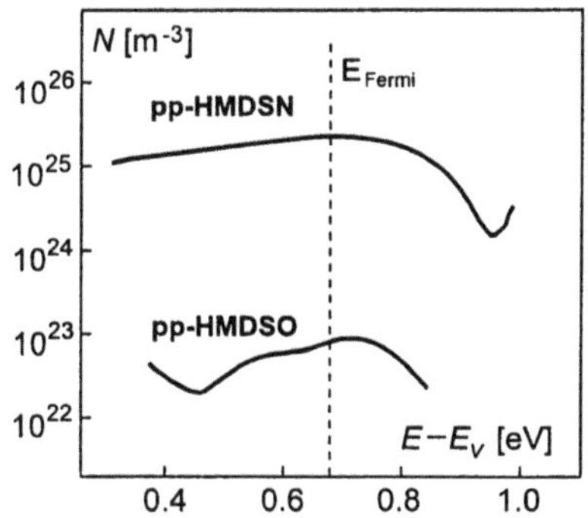

Figure 5.11. DOS distribution calculated from VPITDD for plasma polymerized HMDSN and HMDSO films [98].

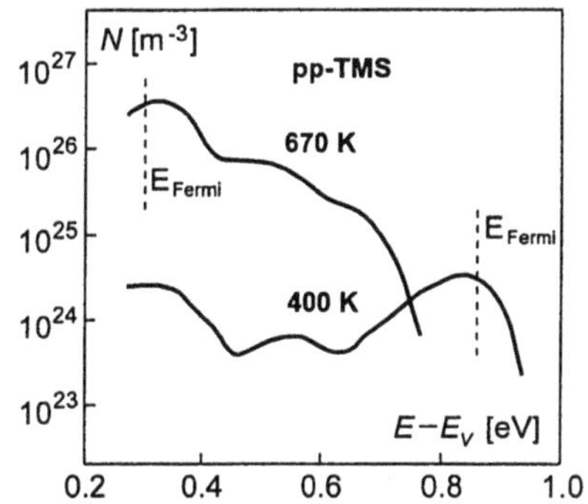

Figure 5.12. DOS distribution calculated from VPITDD for plasma polymerized TMS, deposited at different temperatures [98].

Another example of the use of VPITDD to study plasma polymers is presented in Fig. 5.12. Measurements were performed for films of plasma polymerized tetramethylsilane (TMS) deposited at different temperatures, namely 400 K and 670 K. As one can see, the substrate temperature drastically affects the DOS distribution in the films and, in consequence, evidently shifts the Fermi level position, which influences many important electronic properties of this material [98].

For regularity one should mention here the junction capacitance measurements, which are also connected with trapping and detrapping processes. These methods include, for instance, deep level transient spectroscopy (DLTS) [100], isothermal capacitance transient spectroscopy (ICTS) [101], and that already mentioned, namely the transient photocapacitance spectroscopy [38]. They generally consist of the transient capacitance detection (as a function of time) under thermal, isothermal or illumination conditions depending on the measurement variant.

Although these methods have turned out to be very attractive to study the gap states in semiconductors, they are quite useless for materials of lower conductivity. Furthermore, they need the compliance with some requirements, e.g. appropriate construction of the junctions. As far as plasma deposited films are concerned, the transient junction capacitance measurements are used mainly for a-Si:H films [38,100-102], as well as sometimes for other typical amorphous semiconductors [103,104].

5.3. Amorphous Semiconductors

The plasma polymerized amorphous semiconductor, which has so far been studied in greatest detail and also appears to be of greatest applied interest, is hydrogenated silicon (a-Si:H). Initial studies of this material were done in the sixties by Sterling and co-workers [105,106], who demonstrated the radio-frequency (r.f.) glow discharge deposition of a-Si:H from silane (SiH_4). This was followed by the very important works of Spear and LeComber [107,108], who reported on successful *n*- and *p*-doping a-Si:H. In the same time, the first thin-film photovoltaic solar cell using a-Si:H was fabricated [109]. In 1975 and 1976, research confirmed that plasma-deposited films from silane contained hydrogen [110,111]. The first commercial product based on a-Si:H – photovoltaic batteries for calculators – was introduced in 1980. Since then a-Si:H has found increased applications as large-area, thin-film amorphous semiconductor for photovoltaic solar energy modules [112] and thin-film transistors (TFT) that are most often used in controlling active-matrix liquid crystals displays (LCD) [113].

5.3.1. Band Model

Although the electronic structure as well as electrical and optical properties of a-Si:H films have been widely characterized and extensively reviewed [8,114-116], its DOS distribution function has been the subject of much controversy in the last 25 years. Nevertheless, one can acknowledge undoubtedly this material to be a typical representative of the hydrogenated amorphous semiconductors. A band model of a-Si:H commonly used nowadays, resulting from a variety of experimental findings, which are summarized in the previous section, is shown in Fig. 5.13. Of course, the exact shape and local densities of states depend strongly on the plasma deposition conditions and the further history of the sample, but a general character of the DOS function is

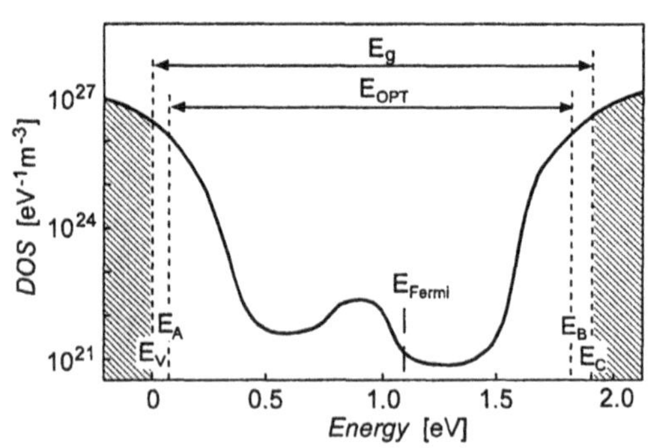

Figure 5.13. Density of states (*DOS*) distribution for a typical a-Si:H. The regions of extended states are shaded.

maintained and is similar to that shown. The most common a-Si:H films have an optical gap of 1.7 to 1.8 eV with a hydrogen content of about 10%. The minimum value of the DOS function can be lower than 10^{21} $eV^{-1}m^{-3}$, whereas the density of states at the E_V and E_C regions attains 10^{27} $eV^{-1}m^{-3}$ [114,116].

According to the Mott-CFO model (Fig. 5.2(b)), the DOS function in Fig. 5.13 can be divided into extended states (below E_V and above E_C), localized band-tail states (regions between E_g and E_{OPT}), and optical gap states (region E_{OPT}). For a typical amorphous semiconductor it should be expected that the difference between E_g and E_{OPT} ought to be only a few tenths of an electron volt (*Sec.* 5.2.1). In fact, this was excellently shown by Mimura and Hatanaka [117]. From their internal photoemission measurements E_g was estimated to be 1.92 eV, whereas E_{OPT} was about 1.8 eV.

Distorted or deformed bonds between Si atoms in a-Si:H, where the bond angles deviate from the normal cubic tetrahedral angle 109°28', can cause the band-tail states. In turn, threefold-coordinated Si atoms with dangling (broken) bonds give rise to localized defect states in the optical gap (mid-gap). The dangling bonds can be positively charged when unoccupied D^+, neutral when singly occupied D^0, and negatively charged when doubly occupied D^-. Although the microscopic details of the conversion between these states are still debated, it is now well established that the density of these states is determined by chemical equilibrium reactions [118,119]:

$$\text{Si–Si} \leftrightarrows 2\text{Si}^\bullet \equiv 2D^0, \tag{5.10}$$

$$D^0 + e \leftrightarrows D^- \quad \text{and} \quad D^0 + h \leftrightarrows D^+, \tag{5.11}$$

where *e* and *h* are electron and hole, respectively.

In recent years the model suggested by Street [115,120], interpreting the DOS function in the mid-gap region (schematically shown in Fig. 5.14 (a)), has become the basic model generally used for the explanation of many results

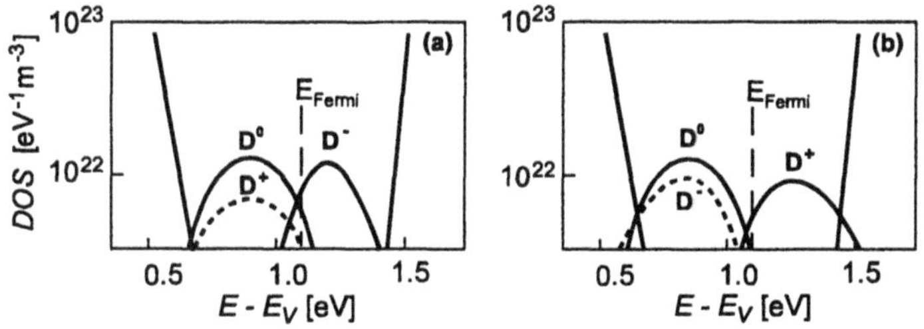

Figure 5.14. Schematic diagrams of the density of states in the mid-gap region of a-Si:H films: (a) – the Street model, (b) – the defect pool model.

concerning electrical and optical properties of a-Si:H. This model has also been adopted for other plasma deposited semiconductors, e.g. a-Si_XN_Y:H [121] or a-Ge:H [122].
There are, however, some experimental results that cannot be explained by this model [37,123]. That is the reason why a more general defect pool model has been introduced [124]. Its main assumption is that the energy of the dangling-bond states can take a range of values, due to the inherent disorder of the amorphous network. The calculation of the density of states, according to this model, produced the key result that D^- defects can be lower in energy than D^+ defects, and these latter are predicted to lie between E_F and E_C [119,125] (Fig. 5.14 (b)). Although the defect pool model has been successfully used to explain many results concerning a-Si:H as well as other plasma deposited amorphous semiconductors, e.g. a-Ge:H (polymerized from GeH_4) [126] or a-Si_XGe_Y:H (polymerized from mixture of SiH_4 and GeH_4) [127], some points of controversy still exist. For example, the model predicts for undoped a-Si:H films that there are approximately four times as many charged defects as D^0 defects [125]. However, from experiments, it has been found that the number of D^0 defects exceeds that of charged defects by a factor of the order of ten [128].

5.3.2. Doping

Apart from the models describing the DOS function in plasma-deposited semiconducting films, it has been recognized that the prime reason for their superior electronic properties is hydrogen incorporated into the structure of the materials during plasma processes. Hydrogen serves primarily to passivate the dangling-bond defects and thus to decrease the density of localized states in the mid-gap. It is important to note that, for instance, evaporated a-Si has about 10^{26} $eV^{-1}m^{-3}$ states in the mid-gap, i.e. about five orders of magnitude higher than for a typical plasma deposited a-Si:H. This just explains why one can control the electronic properties of a-Si:H by doping with donor and acceptor centers, contrary to a-Si [129].

Much of the current interest in plasma deposited amorphous semiconductors stems exactly from the fact that they can be doped. Films of a-Si:H doped with donors and acceptors are at present the most extensively studied and the best understood amorphous semiconducting system. For a few years, however, huge progress in investigations on doping of a-Ge:H with diverse elements of group III, IV, V, and VI has been also achieved [126,130].

Furthermore, some attention is simultaneously devoted to other doped hydrogenated amorphous semiconductors fabricated by plasma deposition, such as a-Si_XC_Y:H [131-134], a-Si_XN_Y:H [135,136], a-C:H [1,137,138], and a-Ge_XC_Y:H [139].

The most frequent method used to prepare doped films is plasma codeposition (copolimerization) from a mixture of the film precursor and the dopant agent. Recently, liquid compounds instead of gases (e.g. triethylboron $B(C_2H_5)_3$ and trimethylphosphine $P(CH_3)_3$ in the place of diborane B_2H_6 and phosphine PH_3) have become more and more popular dopant agents [133,140]. They are less toxic, more stable and their low vapor pressure offers facilities for precise controlling of the doping process.

Although the electronic properties of doped films are presented very often as a function of dopant gas (or vapor) partial pressure, one can realize that generally the concentration of dopant atoms in the solid phase C_{sol} for a given gas-phase concentration C_{gas} can vary by up to 2 orders of magnitude, depending on the gases involved and the nature of the deposition process. It has been found that the relation between C_{sol} and C_{gas} is the following: $C_{sol} \propto (C_{gas})^N$, where the exponent N takes values between 0.5 and 1 [122].

The electrical activity of the dopant atoms is related to a complex reaction with both dangling bonds and hydrogen and this process is basically different from the crystalline case. In a doped crystal, the density of localized states introduced into the energy gap equals C_{sol} and they are a primary source of the increase in the band state electron or hole density. In the meanwhile, changes in electrical properties of amorphous materials are brought about primarily by changes in the gap-state occupation.

If we consider, for example, a donor atom like P (phosphorus) in an amorphous network of a-Si:H, two extreme cases can occur. Firstly, an arrangement of the threefold-coordinated dopant atom can be created

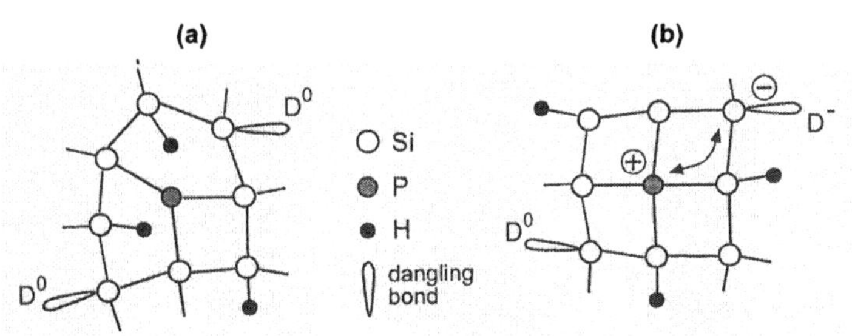

Figure 5.15. Schematic models of dopant configurations (phosphorous atom) in hydrogenated amorphous silicon: (a) – threefold-coordinated dopant (alloying), (b) – fourfold-coordinated dopant (active doping).

(Fig. 5.15 (a)). In this case, the P atom does not play a role of the doping state. Secondly, the dopant atom can be fourfold-coordinated (Fig. 5.15 (b)), which is connected with the generation of an additional dangling bond state, according to the process:

$$Si_4^0 + P_3^0 \; \varsigma \; Si_3^- + P_4^+ \;\; ; \;\; Si_3^- \equiv D^-, \tag{5.12}$$

where subscripts determine the coordination *n*-fold. The changes in the gap-state density cause the shift of the Fermi level and, in consequence, a change in the electrical conductivity of the doped film. Of course, a similar process (in which D^+ are created) occurs after doping with acceptor atoms (e.g. boron).

Figure 5.16 summarizes the results of the conductivity that have been obtained for a-Si:H films doped with acceptors (boron) and donors (phosphorus). The room-temperature conductivity σ of the specimens is plotted against the ratio of the number of dopant agent molecules to the number of silane molecules in the gaseous mixture. In the center of the graph, the conductivity around 10^{-6} S/m is representative of undoped a-Si:H films, which typically are *n*-type material (the Fermi level position is usually slightly closer to E_C than to E_V (Fig. 5.13)). Thus, even a minute quantity of P atoms (donors) increases σ rapidly. In the case of B atoms (acceptors), however, we see that initially σ decreases to about 10^{-10} S/m. This is connected with the transition from *n*-type to *p*-type material [129,141].

Similar changes in σ resulting from the doping have been also observed for a-C:H plasma deposited from acetylene [137], and a-Si_XN_Y:H plasma polymerized from a mixture of SiH_4 and N_2 [136]. An initial drop of σ after doping with B atoms is visible for these materials. However, such behavior is not general. Investigations performed very recently on a-Ge_XC_Y:H films doped with very low concentration (lower than 0.05 %) of Al (acceptor) and P (donor) do not reveal any initial decrease in σ (Fig. 5.17). Based on more thorough examinations of

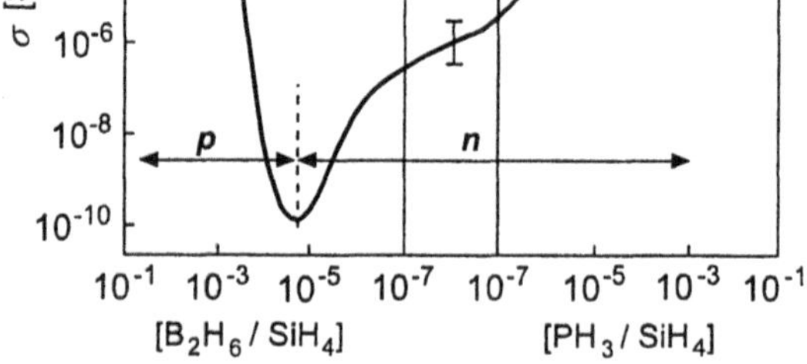

Figure 5.16. Room-temperature conductivity σ of *n*- and *p*-type a-Si:H, plotted as a function of the gaseous impurity ratio [129,141].

a-Ge_XC_Y:H films doped with boron and phosphorous, it has been suggested that both acceptor and donor dopants modify only *p*-type conductivity of the undoped films and any transition to *n*-type is not observed. Boron creates an additional band of acceptors whereas phosphorous eliminates the oxygen donor centers reducing thereby the compensation of the native acceptors [139].

5.3.3. Alloying

As far as extra atoms in a "primary" plasma polymer structure are concerned, two cases should be taken into account: the doping that was presented above (see also: Fig. 5.15 (b)) and another effect called alloying (Fig. 5.15 (a)). Based on this concept we can treat, for example, plasma deposited a-Si_XC_Y:H film as a-Si:H film alloyed with carbon or as a-C:H film alloyed with silicon. We have to realize, however, that the basic electronic structure and, in consequence, the electrical and optical properties of the discussed a-Si_XC_Y:H film can be quantitatively different from those of a-Si:H and a-C:H films fabricated at the same conditions.

Generally, all plasma polymers are alloys. One of the most numerous impurities that occur in such materials is of course hydrogen. Apart from the reduction of defect states in the mid-gap (which has been discussed already), hydrogen can also modify the band structure. It has been found that hydrogen in some a-Si:H films (those which are a silicon–hydrogen alloys with only monohydride form, i.e. Si–H groups), causes the top of the valence band to move downward, and hence the transport gap increases [142].

The role of hydrogen depends not only on its concentration but also on forms in which it exists in the film. If a-Si:H is deposited from silane at substrate temperature below 200° C or from disilane (Si_2H_6), polysilane chains (–SiH_2–) occur in the structure. In this case the transport gap (expressed by E_{OPT}) increases up to 2.5 eV. The polysilane

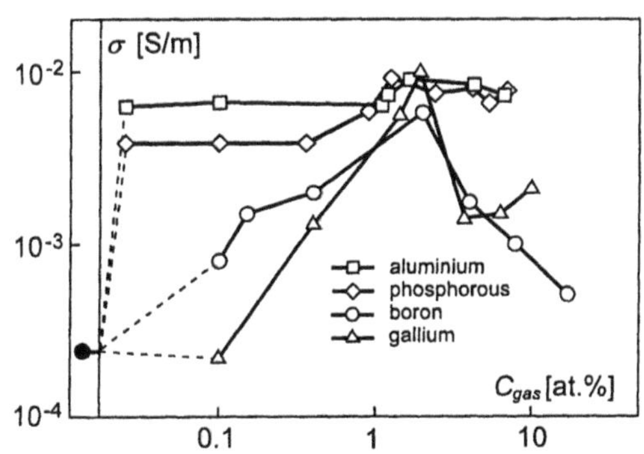

Figure 5.17. Room-temperature conductivity σ of a-Ge_XC_Y:H films doped with various impurity atoms, as a function of the dopant agent concentration in the gaseous mixture. (Black circle presents the conductivity of the undoped film.) [139].

structure obtained by plasma polymerization is an attractive model element for studying condensed matter because it bridges the gap between tetrahedrally coordinated three-dimensional amorphous silicon and typical chainlike organic polymers [143-145].

Among the plasma deposited alloys the amorphous carbon-based films are especially interesting. It has been found that alloying with carbon decreases the conductivity σ both of a-Si:H and a-Ge:H. For a definite carbon content x a minimum value of σ is attained and then, with increasing x, σ grows. This minimum in conductivity occurs at about $x = 30$ at.% for a-Si:H [146] and at $x \approx 12$ at.% for a-Ge:H [147]. The origin of this carbonation effect is attributed to two processes: a significant shift of the Fermi level towards the valence band from its initial position nearing the conduction band, and widening the transport gap. The change in E_G is a result of a typical alloying whereas the Fermi level shift can be rationalized by the explanation that carbon introduction neutralizes the unintentionally generated donors in the films. This might be either to a direct passivation by forming bonds with impurity atoms, changing electric active atoms into inactive ones or to the formation of C-related acceptor states which compensate the unintentionally introduced donors [147]. Although the "doping" aspect of the carbonation effect is not clear yet, one can conclude that it has a certain similarity with the action of the boron acceptor. This effect is evident for a-Ge_XC_Y:H that reveals *p*-type conductivity (the majority carriers are holes) [50] in contradiction to undoped a-Ge:H numbered among *n*-type semiconductors [126]. The same effect is reported for a-Si_XC_Y:H films that transform from *n*-type to *p*-type material when the concentration of CH_4 in the mixture with SiH_4 during the plasma polymerization exceeds $C_{gas} = 0.35$ [131].

A similar "doping" phenomenon resulting from alloying between elements belonging to the same IV group of the periodic system is also observed for hydrogenated silicon-tin films (a-Si_XSn_Y:H). It has been found that the addition of Sn into the a-Si:H plasma polymerized material changes the predominant conduction mechanism from *n* to *p* type [148].

Very recently a new class of hydrogenated amorphous films deposited by plasma polymerization from organic compounds of the carbon family has been obtained and thoroughly investigated [53,149-152]. All these materials may be acknowledged as alloys composed of carbon and other elements of IV group, namely silicon, germanium, tin or lead. Using the deposition parameters from the precisely determined range, the amorphous semiconductors are fabricated. (More details on these films are given in *Sec.* 5.5).

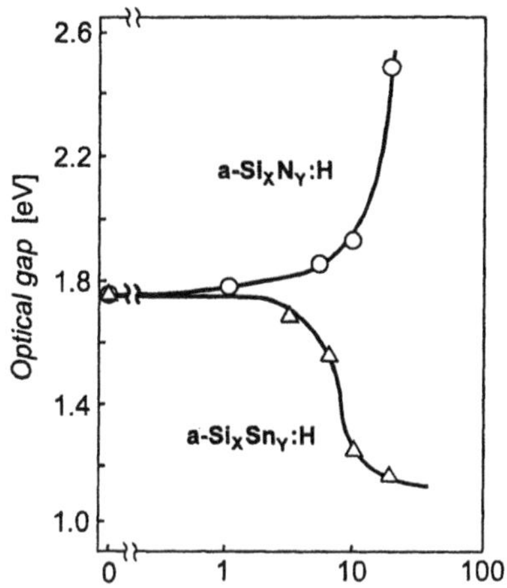

Figure 5.18. Optical gap of a-Si_XN_Y:H and a-Si_XSn_Y:H films as a function of the gaseous mixture ratio [153] (with permission).

The control of the basic parameters of the band structure by alloying process has been investigated widely for a long time and such a procedure is frequently used to prepare a material with desired properties, e.g., with a given E_{OPT}. Fig. 5.18 shows how E_{OPT} changes for films plasma polymerized from SiH_4 and ammonia NH_3 or tetramethyltin $Sn(CH_3)_4$, depending on the mixture composition. The optical gap increases or decreases smoothly from about 1.8 eV to about 2.5 eV and 1.1eV for a-Si_XN_Y:H and a-Si_XSn_Y:H, respectively [153]. Apart from the alloying, the doping effect is also observed in the both cases. For Sn, as has been already mentioned above, *p*-type material occurs for $E_{OPT} < 1.7$ eV. The nitrogen contamination instead acts as typical donor centers and causes a shift of the Fermi level E_F towards the conduction band. The activation energy of the conductivity (the distance between E_F and E_C) of about 0.5–0.6 eV is reached for nitrogen concentrations in the a-Si_XN_Y:H film between X/Y = 0.1 and 0.2 [154].

A subsequent increase in the molar fraction NH_3 to SiH_4 causes a further drastic broadening of the transport gap E_G (expressed by E_{OPT}) that can reach about 5.5 eV when the composition amounts to Y/X = 1.3 – 1.4 [71,153]. A comprehensive study on the electronic structure of a-Si_XN_Y:H films has shown that the position of the conduction band (E_C) exhibits a sharp, upward shift at Y/X ≈ 1.25. It is suggested that we observe here the percolationlike transition from an amorphous network dominated by the connectivity of the Si–Si bonds to that of the Si–N bonds [154]. In this moment, however, an important question emerges: whether it is, in fact, a rapid transition from the amorphous semiconductor to another semiconducting structure with a very wide E_G but with the extended-state-transport, as Tessler and Alvarez propose [155], or it is perhaps a transition from the amorphous semiconductor to an amorphous insulator (see: *Sec.*5.5). The latter option is supported at least by electrical conductivity measurements and testing of the pre-exponential factor σ_0 in the basic relation [6]:

$$\sigma = \sigma_0 \exp(E_A/kT), \qquad (5.13)$$

where E_A is the activation energy of the conductivity, k is the Boltzmann constant and T is the temperature. A rapid decrease of σ_0 (approx. 3 orders of magnitude) when Y/X exceeds a definite value could be indicative of a change in the nature of the transport processes: from the extended state transport to the hopping transport. A significant drop of the electron drift mobility starts at the same Y/X value, from about $10^{-4}\,cm^2V^{-1}s^{-1}$ (typical for amorphous semiconductors) to non-measurable values, which is also typical for such a transition [156].

5.3.4. Charge Carrier Transport

The Mott-CFO model, presented in *Sec.* 5.2.1, predicts two fundamental mechanisms of charge carrier transport through the amorphous material: transport in extended states (band transport) and transport in localized states (hopping transport). Both these mechanisms occur in a-semiconductors.

Most of the transport results concerning plasma deposited materials have been obtained from the time-of-flight (TOF) experiment. The principle of the experimental approach is illustrated in Fig. 5.19. The sample is in a sandwich configuration with non-ohmic contacts. The excess carriers of both polarities are created as a thin sheet below the transparent electrode using a laser light or electron beam pulse. The applied voltage U_D moves carriers of one polarity into the transparent electrode while the carriers of the opposite polarity are dragged through the sample, which induces a current signal in the external circuit.

In the case of the ideal coherent transport we expect a constant current in the time until the electrode at the substrate collects the carriers. Then the current rapidly drops to zero. This time interval is called the transit time t_T, and enables us to calculate the drift carrier mobility μ_{dryf}, according to the following relation:

$$\mu_{dryf} = \frac{d^2}{t_T U_D}, \qquad (5.14)$$

where d is the film thickness.

In practice, however, the current-time relation is more complicated and there are some problems to determine t_T, and hence μ_{dryf}. During transit

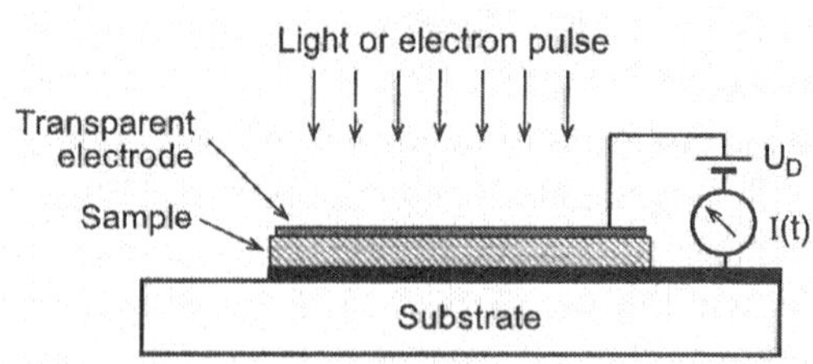

Figure 5.19. Sketch of the setup for a time-of-flight experiment.

through the sample the carriers experience a lot of events such as trapping or hopping, or both simultaneously, which causes statistical fluctuations of their motion. Three general possibilities can be distinguished in this case: (*1*) the sheet of carriers arrives as a coherent packet which is somewhat broadened by a Gaussian distribution (conventionally dispersive transport); (*2*) the packet is so diffuse that the current transient is no longer coherent (anomalously dispersive transport); (*3*) the carriers are trapped by deep traps which have release times long compared to the timescale of the experiment, thus the packet does not arrive to the opposite electrode.

The conventionally dispersive transport was observed for a-Si:H films above 250 K and 410 K for electrons and holes, respectively. Below these temperatures the anomalous dispersion occurs. Typically current signals for both these types of transport are shown in Fig. 5.20 [157]. The signals are also drawn in the bilogarithmic coordinate system, which enable us to determine t_T. Such a procedure is especially helpful for analysis of the anomalously dispersive transport. According to the theoretical approach to the problem (both

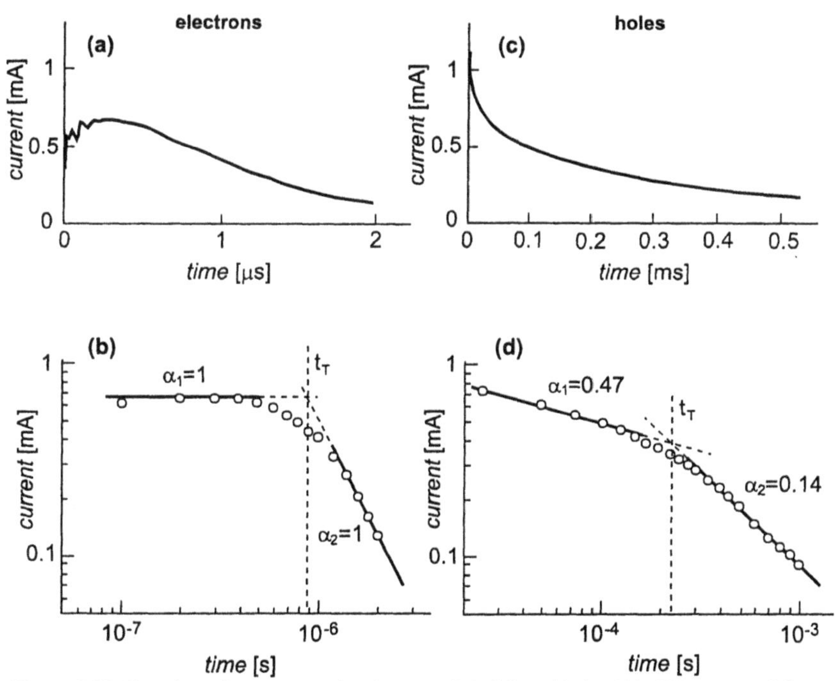

Figure 5.20. Transient photocurrents for electrons ((a), (b)) and holes ((c), (d)) measured for a typical a-Si:H film at room temperature: (a) and (c) – linear scale, (b) and (d) – bilogarithmic plots [157] (with permission).

for the transport via extended states with multiple trapping and the hopping transport [158-160] the current signal decreases following a power law:

$$I(t) = t^{-(1-\alpha_1)} \tag{5.15}$$

up to the break point, attributed to t_T, after which the current is given by:

$$I(t) = t^{-(1+\alpha_2)} \tag{5.16}$$

Parameters α_1 and α_2 are related to the dispersion and it has been established that generally $\alpha_1 \neq \alpha_2$. As one can see, in the case of the coherent transport, $\alpha_1 \approx 1$.

In the a-Si:H films discussed above, both the conventionally and anomalously dispersive transports of electrons and holes are attributed to the process in which the generated carriers move in a band of extended states and are multiply trapped in the tail states. The trapping phenomenon manifests itself by an activation character of the drift mobility. The activation energies near room temperature are about 0.15 and 0.3 eV for electrons and holes, respectively. If the temperature dependence and the field dependence of the current transients are evaluated properly, the mobility of carriers in the extended states can be determined. The electron mobility in extended states (microscopic mobility) thus obtained is with $2 \cdot 10^{-3}$ $m^2V^{-1}s^{-1}$ by a factor of 2 smaller than the respective value for holes. Of course, the drift mobility, which results in this case from the microscopic mobility and the trapping, depends on the DOS distribution in the tails and generally the broader band tail the lower values of μ_{dryf} and the larger activation energy of the mobility are. Typical room-temperature drift mobilities for a-Si:H of good quality are about 10^{-4} $m^2V^{-1}s^{-1}$ for electrons and 10^{-7}–10^{-6} $m^2V^{-1}s^{-1}$ for holes [161].

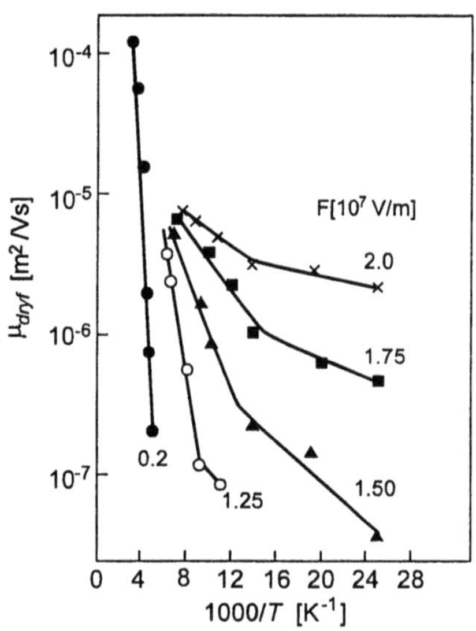

Figure 5.21. Temperature- and field-dependent electron drift mobilities measured on 2-μm a-Si:H film [162] (with permission).

The multiple trapping description is limited to medium and high temperatures, and low fields. The freezing in of vibrational energy at low temperature increasingly favors tunneling propagation (hopping transport) because the time to perform a thermal reemission from traps becomes long compared to that for a tunneling transition. Hopping transport at low temperatures is also favored by the preferential population of deep tail states. Generally, with decreasing temperature the transport comes from the multiple trapping to the hopping, and at the same time the drift mobilities continuously decrease over many orders of magnitude. It has been found, however, that the mobilities at low temperatures, both of electrons and holes, can be greatly enhanced by high electric fields. In Fig. 5.21 such an effect is demonstrated. The results show that the field enhancements are predominantly below 100 K and at applied fields above $1\text{–}2\cdot 10^7$ V/m. It is also worthy of notice that the activation energy of mobility decreases continuously below 100 K approaching zero at 40 K for the highest applied fields. To interpret the data a model of field-enhanced hopping through localized states in the band tails has been developed. It has been shown that the electric field creates a quasimobility edge in the tail region at a certain energy level that depends on the field strength and the tail state distribution [161-163].

The transport processes have a strong influence on the conductivity of a-Si:H films. The field enhancement of carrier mobility is also manifested as a field-enhanced DC dark conductivity σ. Using high electric fields, greater than $F = 10^7$ V/m, we can change a-Si:H films from poorly to very conductive at low temperatures. The field dependence is described empirically by a power law $\sigma \propto F^{\gamma}$ with γ in the range $10 \leq \gamma \leq 17$. The enhancement and the γ parameter depend among others on the doping level of the material, with significant differences between electron and hole currents [163,164].

The existence of the extended states in plasma deposited materials has been confirmed, also besides a-Si:H, for other a-semiconductors. For both a-Si_XN_Y:H films prepared by the r.f. glow-discharge from ammonia–silane mixtures [156] and a-Si_XC_Y:H films deposited from methane–silane mixtures or disilymethane [165], it was suggested from drift mobility measurements that the extended-state conduction with multiple trapping predominates. It should be emphasized, however, that these films contain very low amounts of the alloyed atoms (N and C) in relation to Si.

The transport via extended states has been also proposed for plasma-polymerized acrylonitrile films subjected to pyrolysis. A dramatic change in the drift mobility, which is observed when the pyrolysed film attains the conductivity of about 10^{-9} S/m, has been attributed to a transition from the hopping to extended state transport [166].

5.4. Amorphous insulators

Although amorphous covalent insulators are a natural consequence of the theory of amorphous state, an interest in these systems has been generally limited. The term of amorphous insulator has appeared occasionally in literature for a long time, e.g. [167], but it rather expresses materials with a very low conductivity and has no concern with the present conception of their band structure. The amorphous insulator, to our mind, according to the model presented in *Sec.*5.2.1, is an amorphous material without the extended states in the conduction and the valence bands. Thus, only localized states exist there. Of course, we will confine our discussion to materials fabricated by plasma polymerization. It should be, however, pointed out at once that most of the plasma deposited films seem to be rather a-insulators than a-semiconductors, though in general, sometimes uncritically indeed, their electronic structure is described just as for an a-semiconductor.

5.4.1. Band Model

The model for the density of states of amorphous insulator was proposed in 1984 [22]. Since that time, investigations performed on various thin plasma films have given more experimental evidence for the existence of the a-insulator state [9,19,52,166,168]. Figure 5.22 shows a sketch of the band model for the best understood at present amorphous insulator, namely a-Si_XC_Y:H material plasma polymerized from organosilicon precursors [19,22]. Main features of this model, however, should be recognized as characteristic for the whole group of a-insulators.

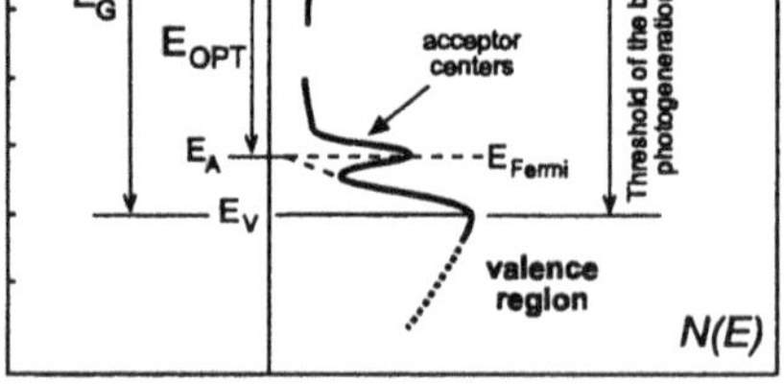

Figure 5.22. Basic model of the electronic structure for insulating a-Si_XC_Y:H films [19,22].

The first striking property of the a-insulator band structure is a very large difference between E_G and E_{OPT}. The transport gap, determined from photoinjection measurements (*Sec.* 5.2.2.2), is about 6.3 eV while the optical gap (*Sec.* 5.2.2.1) is much more narrow and amounts to about 3.2 eV. A similar discrepancy between E_G and E_{OPT} has been also recorded by x-ray spectroscopy (*Sec.* 5.2.2.3) and optical absorption for a-Si_XC_Y:H films with a high concentration of carbon [169]. Although the absolute density of states is unknown as yet (excluding a narrow region around the Fermi level (Figs. 5.11 and 5.12)), the presence of long linear tails of localized states in the mid-gap (between E_V and E_C) can be predicted from investigations on optical absorption. Transport of holes and electrons takes place there below E_V and above E_C, respectively. However, it should be emphasized that in such a structure the transport of charge carriers can proceed by hopping only, and E_V and E_C are not the mobility edges in the sense in which it is used in a-semiconductors. In the case of a-insulators the E_V and E_C values indicate energy levels where the effective hopping transport occurs [10].

As one can see in Fig. 5.22, a narrow band of native acceptors determines the position of the Fermi level. The fact that the native acceptors govern the conductivity process in the films under discussion (holes are the majority carriers in this case) has resulted from the measurements of the anomalous transient photocurrent effect [170].

Some attention should be paid to the physical meaning of the Fermi level in the case of a-insulators. The lack of extended states, which is attributed to these materials, can result in the fact that not all localized states participate in the electronic equilibrium. The concept of isolated states in the films is supported by photoluminescence investigations [19]. In such a case one should bear in mind that the Fermi level is the equilibrium level only for some of the states, e.g. those involved in the electrical conductivity. The presence of isolated states that do not take part in the electronic equilibrium is the second important property of the a-insulators, making a fundamental distinction between these materials and a-semiconductors.

The lack of carrier exchange between the isolated states and the other states in the material makes it impossible to adjust the position of E_G (determined from measurements involved states in equilibrium) and E_{OPT} (including all states in the film) in relation one to another. Thus, it should not be surprising when E_A is positioned above E_F. The next important consequence of this model is that the charge carrier hopping, responsible for DC (direct current) conductivity, proceeds non-randomly in the whole volume of the material, but only throughout certain one-dimensional paths of localized states [171]. In that case one can predict the presence of two independent paths: for electrons and holes. This approach does not ignore the attaining of the equilibrium, which is secured by path-to-path "shortings" (e.g. delivered by metal electrodes), and is useful to explain, for instance, the fact that the electron and hole transport in the

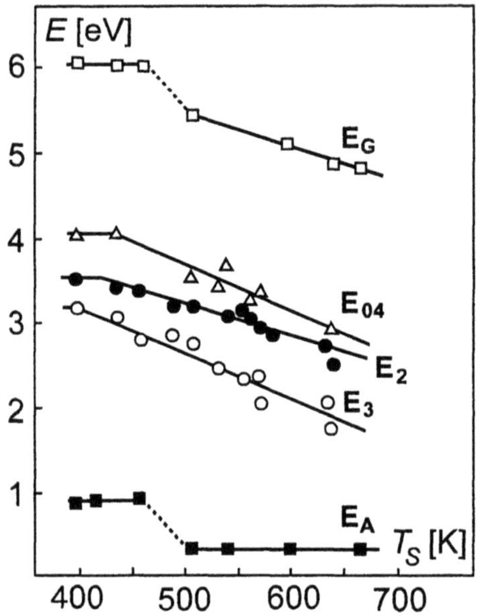

Figure 5.23. Transport gap E_G, optical gaps E_{04}, E_2, E_3, and activation energy E_A of conductivity for a-Si_XC_Y:H films prepared from tetramethylsilane at various substrate temperatures T_s [19].

photoinjection process can be observed simultaneously without effective carrier recombination [19,46].

It is obvious that the character of plasma polymerized materials depends strongly on the deposition conditions. They control the chemical composition and the molecular constitution, both determining the electronic structure of the deposited film. Thus, it should not surprise anybody if seemingly the same material, for example a-Si_XC_Y:H, is treated by some investigators as a-semiconductor [16,165] and others as a-insulator [19,172]. (Of course, we assume the correct interpretation of the results.) Keeping this in mind, it was attempted to find "critical" conditions of plasma deposition process, in which a transition from a-semiconductor to a-insulator takes place.

Fig. 5.23 presents the basic parameters of the electronic structure for a-Si_XC_Y:H films prepared from tetramethylsilane at various substrate temperature T_s, from 400 to 670 K. The transport gap E_G has been obtained from photoinjection measurements, E_2, E_3 and E_{04} denote E_{OPT} determined for $n = 2$ and $n = 3$ in Eq. (5.1), and $\alpha = 10^6\ m^{-1}$, respectively (*Sec.* 5.2.2.1), and E_A is the activation energy of the DC conductivity. As one can see, the insulating character of the films is maintained regardless of T_s used, and no transition to semiconductor structure is observed. In the entire T_s range the E_G values are much higher than the widths of the optical gaps and long linear tails of localized states have been found in every case from the shape of the optical absorption edges. The step decrease in E_G and E_A (at about 480 K) is attributed to a rapid shift (about 0.6 eV) of the valence band towards the conduction band, but the insulator structure still remains [19].

The efforts towards finding the transition in a-C:H films plasma-deposited from pentane under various flow rate and pressure of the precursor vapor as well as various energy of ions bombarding the growing film, have also misfired. In the whole range of the used deposition parameters the fabricated films are classified as a-insulators [173]. This is not, however, a rule. As will be shown in *Sec.* 5.5, there are cases when two types of material, semiconductor and insulator, for the same species of plasma polymer can be obtained changing the precursor composition, energetic parameters during deposition or conditions of the post-deposition treatment.

Finally, it should be noted that although the band model for a-insulator requires further detailed studies, it already gives us a good tool to explain various properties of amorphous insulators. The energy of bulk photogeneration, for example, which can be predicted from the model (Fig. 5.22), is compatible with experimental results [19,22,52].

The band model of a-insulator has been also used to describe such electrical properties as the conductivity under influence of water and oxygen molecules [174] and low-temperature transport in plasma deposited organosilicon films [99,175]. The most important, however, is that the model enables us to understand details of electrical conductivity mechanisms in a-insulators, which reveal considerable dissimilarity to those characteristic for a-semiconductors.

5.4.2. Electrical Conductivity

Generally, a-insulators are characterized by a very low DC conductivity ($< 10^{-10}$ S/m). Investigations of this property, however, aside from the verification of their applicable feature, i.e. the insulating character of the film, are very useful to obtain some information on the electronic structure. It has appeared to be especially instrumental in the case of insulating plasma polymers that are very complex and inconvenient for study systems.

The phenomenon of electrical conductivity in all materials consists of two basic processes: charge carrier generation and transport of generated carriers. When the carriers can be transported through the bulk of material faster than they are generated, generation processes play the dominant role in the conduction phenomenon. In the opposite case, transport processes decide about the electrical conductivity. The essential problem, however, that should be solved before the basic processes are discussed, is the type of charge carriers responsible for the conductivity.

5.4.2.1. Charge Carriers

For a large majority of plasma polymerized a-insulators, as for the whole class of a-semiconductors, the charge carriers responsible for conductivity processes are electrons and holes. Unfortunately, methods of the carrier type determination commonly used for a-semiconductors, e.g. investigations on the field effect, the thermoelectric power, the doping process and sometimes the Hall effect, fail in the case of a-insulators due to their very low conductivity. Thus, the differentiation of electrons and holes in these materials creates a considerable problem and there is not any standard method in this sphere.

In the first attempts to determine the charge type, measurements of the voltage–capacitance characteristics were used. It was suggested from such investigations performed for plasma polymerized hexamethyldisiloxane (pp-HMDSO) films, that electrons are the majority carriers in this material [176]. On the other hand, studies on the band model of the pp-HMDSO films indicated rather holes as carriers responsible for conductivity [22]. The hole conductivity in the pp-HMDSO films was confirmed by measurements of the anomalous transient photocurrent effect [170]. The same type of conductivity was also proposed for plasma-polymerized styrene films [43]. Studies of the conductivity changes under the influence of oxygen in plasma polymerized malononitrile films suggested, in turn, that electrons are the majority carriers. As one can see, the determination of the charge carrier sign in a-insulators is rather speculative.

This is a yet more difficult task when ionic carriers occur in films. It has been found that sometimes, especially in plasma films deposited at low temperatures and under low-power conditions, ions can participate themselves in the conductivity as well as can modify the electronic carrier current. Such effects were observed, for example, for plasma polymerized organosilicon films. The ion mobility in pp-HMDSO films was estimated at about 10^{-20} $m^2V^{-1}s^{-1}$. It should be emphasized, however, that for most of plasma deposited a-insulators fabricated at more drastic conditions or subjected thermal treatment after deposition, the ionic component of the conductivity may be quite neglected [172,176-179].

In some cases ions are introduced to plasma films with intent. An example is polymeric solid electrolyte membranes prepared by hybridization of plasma polymer formed from octamethylcyclotriasiloxane, poly(propylene oxide), and lithium perchlorate. The films, in which lithium ions play a role of charge carriers, show high ionic conductivity of the order of 10^{-4} S/m at 60°C [180]. Such membranes can be used in thin solid-state lithium polymer batteries.

5.4.2.2. Generation Mechanisms

For a-semiconductors, where E_G is narrow, the generation process consists in thermal excitation of carriers: electrons and holes from the Fermi level to the conduction and valence band of extended states, respectively. In the case of a-insulators, the problem is more complicated due to the wide E_G. If the Fermi level E_F is in the middle of the gap, the distances between E_F and the transport edges amount to a few electron volts, thus any thermal generation should not be observed. However, if a band of acceptors or donors lying in the vicinity of one of the transport edges, with the density of centers sufficient to shift E_F to the position inside the band or between the band and the transport edge, occurs in the insulator, we can observe the bulk generation. It is a result of thermal ionization of donor or acceptor centers. This process is called the Poole-Frenkel (P-F) generation [11].

The density of generated carriers in the P-F process depends at a given temperature on the electrical field applied to the sample. Hole or electron released from the center has to overcome a Coulombic potential barrier created between the carrier and the oppositely charged center. If the external electric field F is applied, the barrier height Φ_{PF} is lowered according to the relation:

$$\Delta\Phi_{\mathrm{PF}} = \beta_{\mathrm{PF}}\, F^{1/2}. \tag{5.17}$$

$\Delta\Phi_{\mathrm{PF}}$ is the decrease (lowering) of the Coulombic barrier and β_{PF} is the Poole-Frenkel coefficient given by:

$$\beta_{\mathrm{PF}} = \left(\frac{e^3}{\pi\varepsilon\varepsilon_0}\right)^{1/2}\frac{1}{r}, \tag{5.18}$$

where e is the elementary charge, $\varepsilon\varepsilon_0$ is the permittivity of the material, and r is a parameter connected with the compensation process. For very high and very low compensation, $r = 1$ and $r = 2$, respectively, and generally $1 \le r \le 2$. If we assume, for example, that $\varepsilon = 3.5$, which is a typical value for plasma-polymerized a-insulators [181], we obtain, depending on r, $\beta_{\mathrm{PF}} = (1\text{–}2)\cdot 10^{-5}$ eV(m/V)$^{1/2}$. Thus, it is obvious that the lowering of the barrier can be observable only for very high electric fields ($F > 10^6$ V/m).

The lower barrier the higher density of generated carriers and the higher density of the conduction current. A dependence of this current on the external electric field (expressed by the applied voltage U and the film thickness d) is shown in Table 5.1.

Carriers of the electrical conductivity in plasma-deposited a-insulators can be generated not only by the P-F process but also in another way, namely by injection from electrodes. Three main processes of the injection (in dark) can be distinguished: thermionic emission, thermionic field (thermally assisted field) emission, and field emission. For intermediate field and temperature most often the thermionic emission is the dominant process of the charge carrier injection from contacts. This is the Schottky generation [11].

In this process carriers have to overcome the contact barrier, and this barrier can be modified, as for the P-F mechanism, by the external electric field F. The field lowering of the contact barrier Φ is expressed by:

$$\Delta\Phi = \beta_{\text{Sch}} F^{1/2}, \tag{5.19}$$

where the Schottky coefficient β_{Sch} is:

$$\beta_{\text{Sch}} = \left(\frac{e^3}{4\pi\varepsilon\varepsilon_0} \right)^{1/2} \tag{5.20}$$

In Table 5.1 a relation of the conduction current versus the applied voltage for the case when the Schottky process dominates, is presented. As one can see, both the P-F generation and the Schottky generation give in practice the same current-voltage dependence: $\ln j$ is proportional to $U^{1/2}$.

The generation by injection processes is associated with electrodes that control the conduction current. However, if an electrode emits more electrons or holes than the bulk can accept, the remainder will form a space charge with a distribution that is dependent on the DOS distribution. The charge creates a field that reduces the rate of carrier emission from the electrode. Thus the current is now controlled not by the charge-injection electrode, but by the bulk of the material. In such a case the conductivity is described by the theory of space charge limited current (SCLC) [11]. The general relation describing the conduction current according to the theory is shown in Table 5.1.

The parameter Θ, connected with the space charge distribution, is a function of the DOS and temperature. A bilogarithmic plot of the j-U characteristic should give in this case a straight line with the slope of n.

The basic processes of conductivity in a-insulators, relations describing their conduction current (as a function of U, d and T) and information that we can obtain if the given mechanism is dominant, are compared in Table 5.1. Unfortunately, although the mechanisms of these processes differ essentially, it is a great problem to establish explicitly which process is dominant in the material studied. Investigations on the conductivity of plasma-polymerized hexamethyldisilazane films will be discussed for example.

Table 5.1. The basic processes of conduction in amorphous insulators (relations described in the text).

PROCESS	RELATION	INFORMATION
Poole-Frenkel Mechanism	$\ln j \propto a U^{1/2}$ (5.21) $a = \frac{\beta_{PF}}{kT} d^{-1/2}$	Energy and density of generation centers Position of the Fermi level
Schottky Mechanism	$\ln j \propto b U^{1/2}$ (5.22) $b = \frac{\beta_{Sch}}{kT} d^{-1/2}$	Contact barriers
SCLC	$j = \Theta U^n d^{-m}$ (5.23) $n \geq 2 \quad m = 2n - 1$	Distributions of density-of-states

Typical $\ln j$ versus $\ln U$ plots for two different thicknesses of the films are presented in Fig. 5.24 (a). The characteristics show three branches with different slopes. Both the second and the third branch suggest, according to Eq. (5.23), the occurrence of SCLC ($n \geq 2$). However, plotting these characteristics in the $\ln j$ versus $U^{1/2}$ coordinate system (Fig. 5.24 (b)) we obtain almost straight lines for field strengths exceeding about $7 \cdot 10^6$ V/m, which may point to one of generation processes (Poole-Frenkel or Schottky). From other measurements (temperature dependences, photoinjection) it has been found that the Poole-

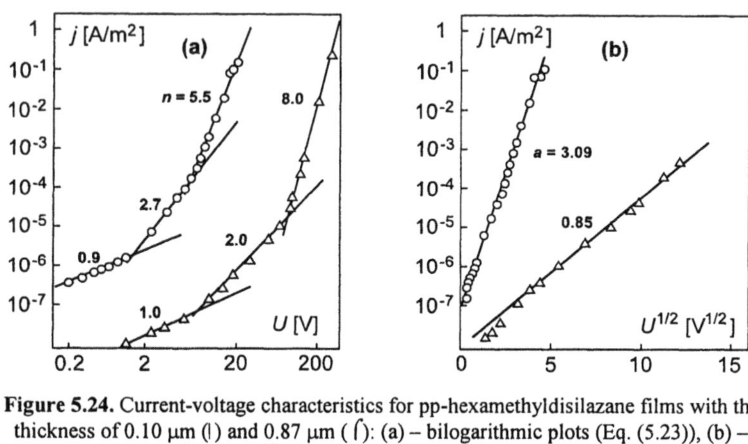

Figure 5.24. Current-voltage characteristics for pp-hexamethyldisilazane films with the thickness of 0.10 μm (|) and 0.87 μm (ſ): (a) – bilogarithmic plots (Eq. (5.23)), (b) – $\ln j$ vs. $U^{1/2}$ plots (Eqs. (5.21) and (5.22)) [89].

Frenkel generation is the dominant process in this case. Thus, the branches in Fig. 5.24 (a) whose slopes indicate the occurrence of SCLC would only be the approximation of the exponential curve described by Eq. (5.21) [89].

5.4.2.3. Transport Mechanisms

In contrast with a-semiconductors, we can predict for a-insulators only one type of charge carrier transport mechanism, namely hopping transport. The lack of extended states excludes the band transport. In that case, investigations on the transport of charge carriers should provide useful information on the electronic nature of the material, and among others, to confirm or exclude the existence of the extended states.

A lot of attempts were undertaken in the eighties to investigate the transport phenomena in plasma polymerized insulating films [166,182-184]. It was, however, a difficult task due to the very low conductivity of the films and generally the lack of photosensitivity, thus the only way to generate a charge packet was an electron beam pulse. A relatively great deal of attention has been focused on plasma polymerized organosilicon films. Figure 5.25 presents a typical transient current for holes in a pp-HMDSO film at room temperature, plotted according to Eqs. 5.15 and 5.16. On this basis the transit time t_T can be determined, and hence the drift mobility is estimated. Its value amounts to approx. $3.5 \cdot 10^{-10}\ m^2V^{-1}s^{-1}$ [182] that is, for example, one order of magnitude higher than that calculated for conventional amorphous polyethylene [185].

Very weak temperature dependence of the mobility (its activation energy is between 0.07 and 0.02 eV [183]) confirms the hopping character of the hole transport in pp-HMDSO films. A similar conclusion has been also drawn from measurements of transient currents for electrons. Thus, for both holes and electrons and in the whole range of the used temperature and applied

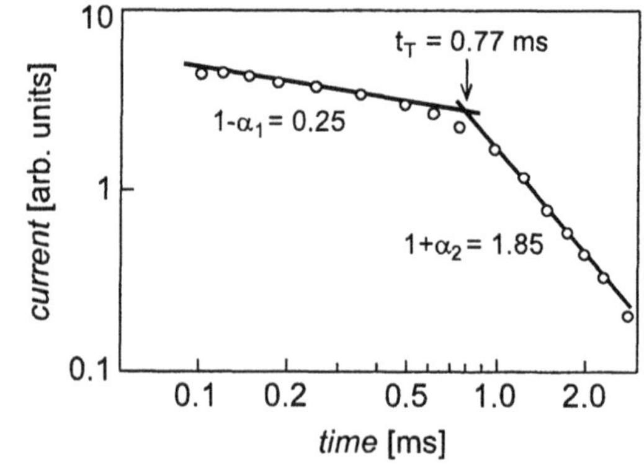

Figure 5.25. Transient hole current for pp-hexamethyldisiloxane (pp-HMDSO) film, plotted in the bilogarithmic coordinate systém. (Excitation by an electron beam pulse.) [182].

field it has not been successful in obtaining any evidence of the band transport.

Results presented above should be treated, however, very carefully. Recently, it has been shown for plasma polymerized tetramethylsilane films that carriers generated by an electron-beam pulse do not reach the collecting electrode but are deeply trapped after a few jumps, and after all the transit current can be described by the standard formalism used for calculation of the carrier mobility. Of course, the estimation of the carrier mobility has no physical meaning in this case [97].

As far as the drift mobility in a-insulators is concerned, a very interesting result was obtained for pp-acrylonitrile films subjected to pyrolysis. As it was mentioned in *Sec. 5.3.4*, a drastic change in the charge carrier transport process is observed when the pyrolysed films attain the conductivity of approx. 10^{-9} S/m. For the films with lower conductivity, $\mu_{dryf} = 10^{-10} - 10^{-8}\ m^2V^{-1}s^{-1}$, and these films are numbered among the a-insulators. The films with $\sigma > 10^{-9}$ S/m reveal, in turn, much higher μ_{dryf}, approximately $10^{-4}\ m^2V^{-1}s^{-1}$. It is suggested that a transition from a-insulator to a-semiconductor state occurs in this case [166].

5.5. Amorphous Insulator – Amorphous Semiconductor Transition

The critical strength of disorder, which determines the type of electronic structure, is closely connected with the molecular structure of the material. Going along a theoretical way, for instance, from frozen mixture of silane (SiH_4) and methane (CH_4) (which is a molecularly amorphous material) via a mixture of polymer-like chains ([–C–C–], [–Si–Si–] and [–C–Si–]), then more and more cross-linked chains (covalent amorphous material), hydrogenated carbon-silicon glasses (a-Si_XC_Y:H), up to a-Si_XC_Y:H alloys, one should agree to the fact that completely different models describe the electronic structure of the materials lying at both ends of the way. Thus, we have to expect that at some point of the way a rapid transition in the electronic structure from one to another type must occur.

In fact, it has been found that contrary to a-Si_XC_Y:H alloys, which are generally known as typical semiconducting materials (*Sec.* 5.3.3), a-Si_XC_Y:H glasses (pp-organosilicon glasses) rank rather among a-insulators (*Sec.* 5.4). If these considerations are correct, the monotonic change of the molecular structure from glass to alloy, which results in a change of the strength of disorder, should cause a transition from a-insulator to a-semiconductor, i.e. the transition from a material with no extended states to a material with extended states. In fact, such an effect has been observed in films plasma-deposited from methylosilanes with varying methyl contents (from 0 to 4 methyl groups). The films produced from compounds with low silicon content, $[Si]/([Si]+[C]) \leq 0.33$, are characterized by wide E_G (much wider that E_{OPT}) and long tails of localized states. On the contrary, the films obtained when $[Si]/([Si]+[C]) \geq 0.5$ reveal narrower E_G close to E_{OPT}. It has been suggested that in this case the a-insulator–a-semiconductor (a-I–a-S) transition takes place [9].

A similar conclusion has been drawn from results concerning pp-acrylonitrile films. During pyrolysis a drastic change in the charge carrier transport was recorded (*Sec.* 5.3.4 *and* 5.4.2.3). This effect was attributed to the a-I–a-S transition [166]. One of the consequences of such a transition is the occurrence of high-conducting filaments in the semiconducting form of the films and their absolute absence in the insulator. They are responsible for a bistable switching effect in the semiconducting pp-acrylonitryle films [186].

An especially pronounced instance of the a-I–a-S transition in plasma polymers is the effect observed for a-Ge_XC_Y:H films [187,188], deposited in the three-electrode reactor with an audio frequency (20 kHz) glow discharge [152]. It is particularly interesting that in this case both electronic structures (a-semiconductor and a-insulator) are fabricated from the same precursor (organogermanium compound) and at the same deposition process, in which only one of the deposition parameters is changed, namely a coupling capacitance C responsible for the energy of ions bombarding the growing film. The most striking feature of these films is a dramatic change in their electronic properties with small variations of C in a defined range of its values. The transition in the properties is shown very well by electrical conductivity measurements. In Fig. 5.26 (a) the dark conductivity σ of the films measured at room temperature is presented as a function of C. For $C < 40$ pF, σ is extremely low and amounts to 10^{-18} S/m. From $C = 40$ pF a rapid increase of σ occurs attaining more than 10^{-3} S/m. This is 15 orders of magnitude higher than that of the films prepared at $C < 40$ pF. Similar abrupt changes have been also found in the activation energy E_A (Fig. 5.26 (b)) and pre-exponential factor σ_0 calculated according to Eq. (5.13) (Fig. 5.26 (c)). As one can see, E_A drops from approx. 0.9 eV to 0.26 eV, and σ_0 rapidly increases from $3 \cdot 10^{-3}$ S/m to 20 S/m, i.e. almost 4 orders of magnitude. Such a rise in σ_0 has to result from a change in the nature of transport processes and the transition from the hopping transport to the extended state transport is the most probable.

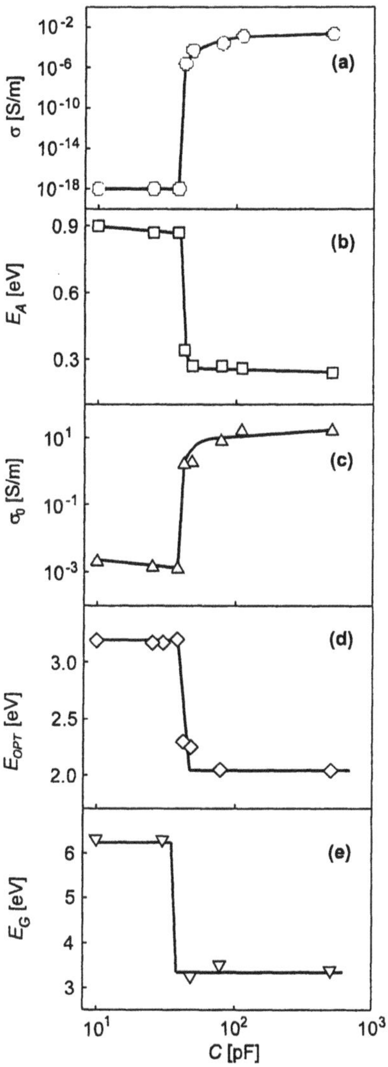

Figure 5.26. Electronic properties of a-Ge_XC_Y:H films as a function of the coupling capacitance C: (a)–the conductivity at room temperature σ, (b)–the activation energy of the conductivity E_A, (c)–the preexponential factor σ_0 calculated from Eq. (5.13), (d)–the optical gap E_{OPT}, and (e)–the transport gap E_G [188,189].

A dramatic change in the electronic structure of the a-Ge_XC_Y:H films, when the critical value of C is surpassed, is also manifested by the optical gap (Fig. 5.26 (d)) and the transport gap (Fig. 5.26 (e)) determined from internal photoemission measurements (*Section* 5.2.2.2). A very large difference between E_{OPT} and E_G (approx. 3 eV) for the films deposited at $C < 40$ pF, is considerably reduced to 1 eV when $C > 40$ pF.

What is, however, really surprising is that both the elemental composition and the basic molecular constitution of the a-Ge_XC_Y:H films are practically unchanged when the electronic structure reveals the step transition at $C = 40$ pF. Exemplary results concerning the elemental composition determined by XPS, Auger spectroscopy, and microanalysis method, are presented in Fig. 5.27 (a) [189]. These results show that a reason of the a-I–a-S transition lies in some more refined changes in the molecular structure. Indeed, it has been found that with growing C the atom packing of films monotonically increases (Fig. 5.27 (b)) [189]. It seems to be obvious that an increase of packing results in a decrease in the strength of disorder of the material and such a process can cause the creation of extended states in amorphous insulator.

Similar effects consisting in step changes in the electronic properties of a film, when some critical value of C is attained during deposition from the same single-source precursor in the three-electrode reactor, have been also observed for other types of plasma polymers, namely: a-Si_XC_Y:H [53], a-Sn_XC_Y:H [149], and a-Pb_XC_Y:H [151]. The lack of such a phenomenon in a-C:H films fabricated from pentane under the analogous plasma deposition conditions (all these films have revealed only insulating properties (*Sec.* 5.4.1)) is indicative of the fact that atoms of the other elements than carbon, belonging to the carbon family, are necessary for the a-I–a-S transition to occur [173]. The dark conductivities of all investigated films presented as a function of the coupling capacitance *C*, are compared in Fig. 5.28.

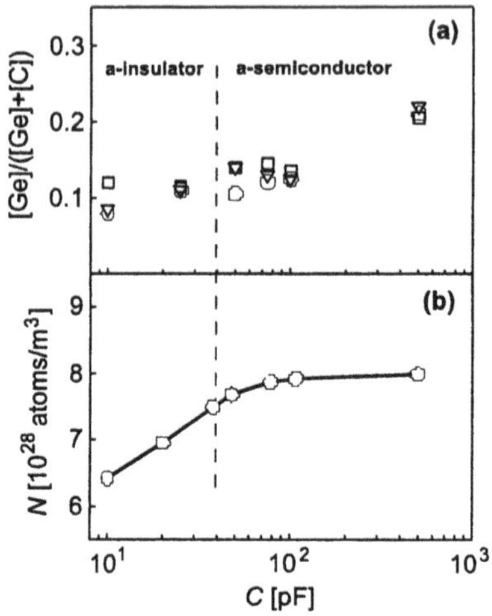

Figure 5.27. a-Ge_XC_Y:H films deposited at various values of the coupling capacitance: (a) – molar fraction of Ge measured by XPS ((), Auger spectroscopy (J), and chemical micro-analysis (|); (b) – packing of atoms (normalized to a pure carbon structure [189].

The results presented above do not imply, however, that we cannot prepare semiconducting a-C:H films in general, and moreover, we cannot obtain the a-I–a-S transition in these films. The a-C:H semiconductors have been widely discussed, for example, by Robertson [1] who predicted the transition effect in such materials. Interesting results in this domain were obtained by Weissmantel et al. [190] for a-C:H prepared from benzene. A drastic increase in σ (approx. 8 orders of magnitude) was observed if one of the deposition parameters, the bias voltage, had been changed from 100 to 300 V. Although the results were interpreted in another way, they are strongly indicative of the a-I–a-S transition. Recently, Helmbold et al. [191] based on the wide confrontation of electrical conductivity measurements performed on a-C:H films by various authors, have suggested that structural changes in the medium-range order of the films can be responsible for the dramatic changes in the conductivity. However, they have also found, what is a striking result, that irrespective of the conductivity values (σ from 10^{-1} to

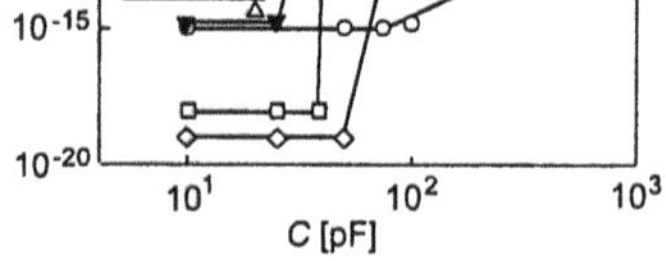

Figure 5.28. Electrical conductivity σ of various plasma-polymer films deposited in the three-electrode reactor (20 kHz), as a function of the coupling capacitance *C* [152,173].

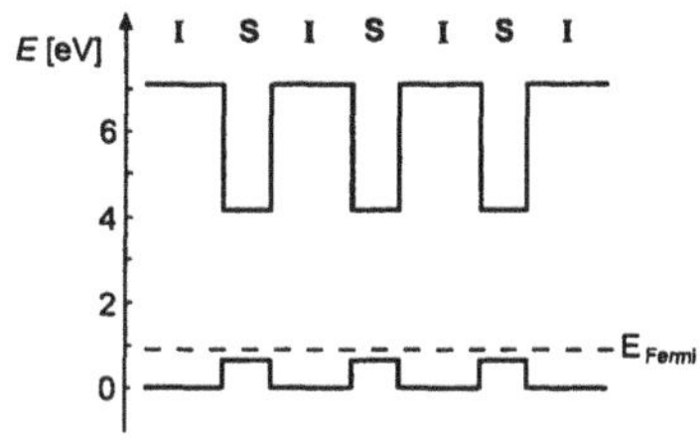

Figure 5.29. A sketch of the predicted band model for a twin amorphous superlattice system of a-Ge_XC_Y:H.

10^{-12} S/m at room temp.) the films show a uniform behavior which can be described by the multiphonon hopping transport model and the transport via extended states is excluded in all the cases.

The existence of two qualitatively different types of plasma polymers, which has been already evidently shown in some cases, is a very exciting phenomenon from the basic research point of view. Moreover, it also offers new possibilities for thin film technology consisting in preparation of a novel class of layer structures (superlattices) formed from twin films of insulator and semiconductor deposited from the same single-source precursor and in the same deposition process. Attempts towards obtaining such systems have been already undertaken, using a-C:H [192-194] and a-Ge_XC_Y:H [195]. In Fig. 5.29 a sketch of the predicted band model for a twin amorphous superlattice system of a-Ge_XC_Y:H is presented. The system is plasma deposited from tetramethylgermanium in the three-electrode reactor, in which the *C* parameter is changed periodically from 10 pF do 500 pF.

5.6. Electronic Properties and Potential Applications

5.6.1. Electrical Properties

The potential applications of plasma polymers in electronics, as it follows from the previous parts of the chapter, are wide and very promising. The main advantage of these materials is that they can be deposited easily in the form of thin films, at low temperature and low costs, on inexpensive substrates of almost any size or shape. In some cases the plasma polymer films have been already brought into practice, in others – considerable effort is focused on finding their technological application.

In the beginning when plasma polymers emerged, i.e. in the sixties, a great deal of attention was paid first of all to dielectric and insulating properties of the films. The attempts to apply these materials in the electronics field, as thin-film capacitors as well as insulating and corrosion-protecting coatings of electronic devices, were undertaken. The works of Goodman [196], Bradley and Hammes

[197,198], Stuart [199,200], Ozawa [201], Bashara and Doty [202], Gregor [203], Mearns [204], and Babcock and Christy [205] constituted the most significant basis to develop this new class of amorphous insulators.

At the same time, in 1965, the first semiconducting films (a-Si:H and a-Ge:H) were fabricated as well [105]. However, they were not "discovered" for electronics before the mid-1970s, when the doping effect was found out [107-109]. A huge progress in the technological application of semiconducting plasma polymers has been achieved ever since.

The special interest, which has existed since the earliest days of plasma polymerization, is also focused on films deposited from organometallics [206]. Among the earliest reported electrical conductivity measurements performed on plasma deposited organometallic films are those of Bradley and Hammes [197] for diphenyl mercury, ferrocene, and hexa-n-butylditin, which were found to be relatively good insulators (10^{-11}–10^{-13} S/m). In the following years, however, a number of papers have appeared which show that, depending on the deposition conditions, the electrical properties of organometallic plasma polymers may vary from insulating ($\sigma < 10^{-10}$ S/m) through semiconducting ($10^{-10} < \sigma < 10$ S/m) to even conducting ($\sigma > 10$ S/m) [181,206]. It should be emphasized, however, that the introduced classification is only tentative and has no connection with the real electronic structure of the materials (see: *Sec.* 5.3 and 5.4).

In Fig. 5.30 a diagram of the conductivity typical for various plasma polymer types is presented. As one can see, the whole range from 10^{-18} S/m to 10^{6} S/m is covered

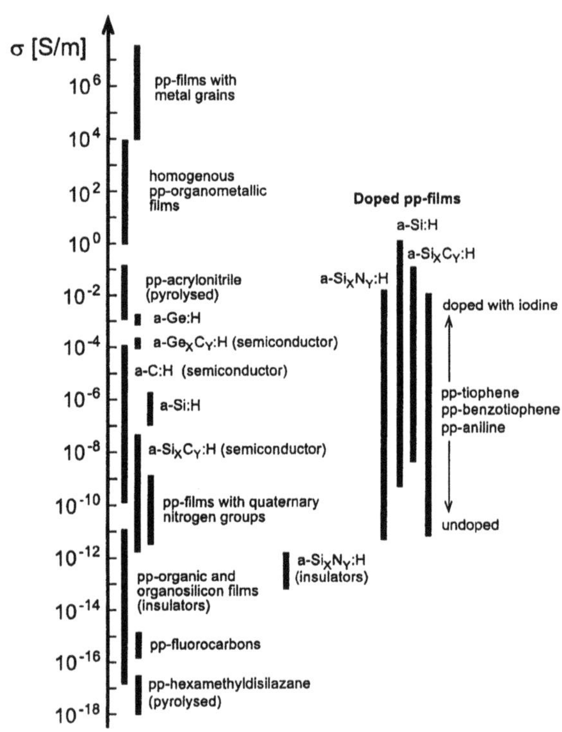

Figure 5.30. Room-temperature conductivity of the main groups of plasma polymers.

by plasma polymers. In other words – if we need a thin film with a given conductivity, we can always do it by plasma polymerization.

5.6.1.1. Low Conducting Films

The films that are most insulating and therefore the most interesting from the application point of view are those deposited from fluorinated organic [207,208] and organosilicon compounds [206]. To improve their insulating properties sometimes they are subjected to pyrolysis. For example, the thermal treatment of pp-hexamethyldisilazane films carried out under nitrogen atmosphere at 623 K gives a material of extremely low conductivity, of the order of 10^{-18} S/m [174]. Furthermore, the pyrolysis of pp-hexamethylcyclotrisilazane films in air and at the temperature linearly increasing up to 1500 K produces high-thermal-resistant insulating coatings that can work at 1400 K for a very long time without destruction. These coatings have been used as electrical insulator of kanthal wires in heaters for sealing TV-kinescope tubes [209].

From the chemical composition of such films the presence of many polar groups in the structure was revealed. These materials were expected to be good dielectrics in the film form and many investigations on dielectric properties (the dielectric constant and dielectric loss, dielectric relaxations, dielectric breakdowns, charge carrier storage etc.) were carried out. The history of the progress of these investigations is described in detail, for example, by Yasuda [181]. The main conclusion that emerges from these studies is that plasma polymers, in general, are poor dielectric materials for the commercial application. This is a result of the instability of the dielectric properties against aging and humidity. Capacitors made of these films change their capacitances during a long time and these values are dependent on humidity [201,210,211].

Although most investigators lost interest in plasma polymers as dielectric materials, some works dating from the mid-eighties began to be promising. For example, Ramu at al. [212,213] found that pp-hexamethyldisiloxane films deposited by microwave (2.45 GHz) discharge and at elevated substrate temperature T_S (400°C) show low losses ($\tan\delta \approx 7 \cdot 10^{-4}$), excellent resistance to aging, and low moisture absorption. Figure 5.31 illustrates the relative resistance to "aging" of sample materials produced at different substrate temperatures. Samples were kept in air at 300°C for periods of up to 10 days and their $\tan\delta$ values were monitored during this period. In the case of all four T_S values, samples are seen to undergo "aging", that is, an increase in $\tan\delta$ with

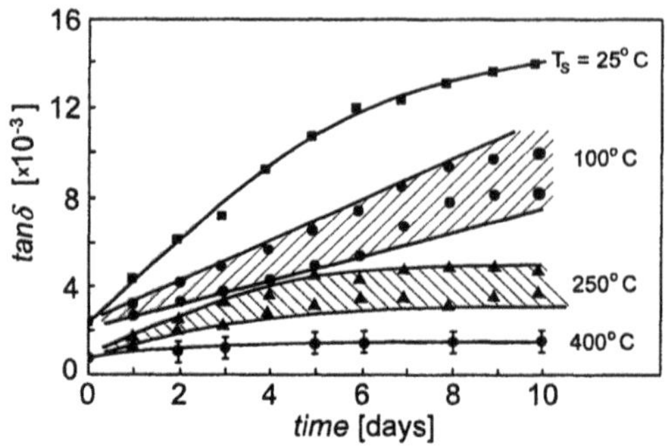

Figure 5.31. Effect of aging in air at 300°C of pp-hexamethyldisiloxane films deposited at different substrate temperatures T_S [212] (with permission).

time. For $T_S = 400°C$, however, this effect is least pronounced, and tan δ reaches an asymptote after two or three days. It was also pointed out that the dielectric breakdown strength of the films deposited at $T_S = 400°C$ is very high and amounts to $12.5 \cdot 10^8$ V/m, close to the value for thermally-grown SiO_2. These results have evidently suggested that the plasma polymers, provided that the films are properly prepared, show considerable promise as dielectrics for various applications, for instance, in capacitor technology.

Lately, research in the field of plasma dielectrics has come into renaissance. Both fluorocarbon and organosilicon films are the center of attention [51,214-220]. Most of the authors predict that the properties of the films including permittivity less than 3, low stress, low leakage current density, very good adhesion to substrates and high thermal and oxidative stability can meet the requirements imposed on new materials intended for state-of-the-art integrated circuit manufacturing technology.

When discussing plasma deposited insulators, it must be at least mentioned about a-Si_XN_Y:H films. They are mostly used as the gate dielectric in thin-film transistors and as passivation layer coatings over integrated circuits made on crystalline semiconductors [221-224]. Besides very high resistivity, the a-Si_XN_Y:H films exhibit especially useful properties of the charge trapping and transport when the interfaces between them and a-Si:H are prepared. By controlling deposition conditions, very smooth a-Si_XN_Y:H/a-Si:H interfaces with high stability important to microelectronics applications can be fabricated [225].

5.6.1.2. High Conducting Films

At the top end of the diagram shown in Fig. 5. 30, we can find plasma polymerized conductive or even typical metallic films. The films are obtained both as composites (by simultaneous plasma polymerization of an organic precursor and sputtering, etching or thermal evaporation of a metal) and as homogenous polymers made by plasma polymerization of organometallic compounds. In the latter case, however, the films are homogenous and amorphous only when they are deposited under certain polymerization conditions (e.g. low substrate temperature, low discharge power density). For other conditions the deposits very often consist of metallic clusters and become typical heterogeneous metal/plasma polymer systems [206,226,227].

From the point of view of the electrical properties, the films fabricated by plasma polymerization of organometallic precursors reveal a very complex character. First of all, we have to distinguish two structural regimes: (*1*) amorphous and homogenous films, in which the molecularly dispersed metal atoms can play a role of electrically active centers (*Sec.* 5.3.2 and 5.3.3) as well as the transition from a-insulator to a-semiconductor states can be observed (*Sec.* 5.5), and (*2*) heterogeneous films with metal grains (see: Chapter 9). Now, the next three regimes should be distinguished: (*a*) the grains can be so dispersed in the polymer matrix that they have no influence in practice on the DC electrical conductivity mechanisms (dielectric regime) [226,228], (*b*) the grains are "in contact" and they control the conduction process (involving the matrix itself or by charge carrier tunneling directly across the insulating barriers between them) (transition regime) [229,230], and finally, (*c*) the grains form a metallic continuum with polymer inclusions (metallic regime). The metallic films fabricated by plasma polymerization from organometallic compounds can reveal that the conductivities approximate those for the bulk of metals. For example, the conductivity for some gold films deposited from dimethyl(2,4-pentane-dionato)gold(III) is only two times lower than σ for bulk Au ($4.1 \cdot 10^7$ S/m) [231]. Similarly, high conductive films were also obtained from other organometallic compounds, containing e.g. Mo, Ag, Rh, Cu, Sn [232-236]. The metallic nature of the films was confirmed by a decrease of the conductivity with increasing temperature [237].

Plasma deposited films with metallic conductivity have been recently used as a gate metal for high temperature self-aligned GaAs metal-semiconductor field-effect transistor devices. Schottky contacts to GaAs were produced by deposition of tungsten or tungsten nitride films [238]. Conducting metal tracings have been also realized on gold-containing hydrocarbon matrix films by laser irradiation. The metal tracings were formed by the generation of localized heating that decomposed the hydrocarbon matrix and agglomerated the gold particles leaving conducting patterns [239].

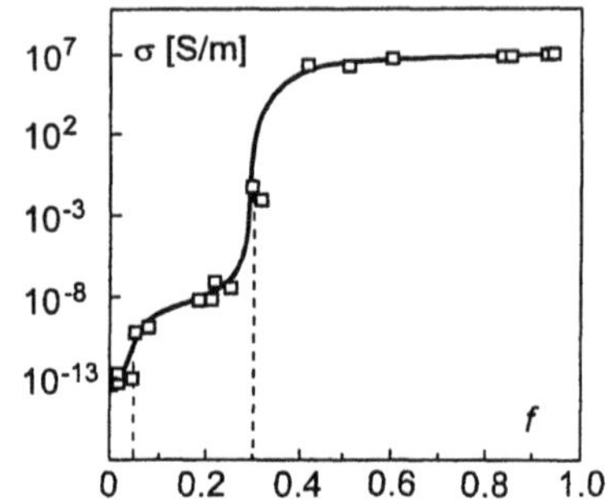

Figure 5.32. Conductivity σ of gold-hydrocarbon composite films as a function of gold volume fraction f [230] (with permission).

To describe the electrical conductivity of metal/polymer plasma thin films a percolation theory may be used [226]. Figure 5.32 shows an example of the electrical conductivity as a function of metal volume fraction f for gold-hydrocarbon composite films. The main percolation threshold, which occurs for $f \approx 0.3$, is connected with the formation of the metallic continuum (transition between (*b*) and (*c*) regimes). In this case, however, another transition characterized by a rapid increase in σ is also seen at $f \approx 0.05$. The conductivity changed from 10^{-13} S/m to approx. 10^{-8} S/m and its activation energy drops from 0.85 to 0.1 eV, respectively. This critical variation is associated, according to Flouttard et al. [230], to the occurrence of well-conjugated graphitic structures in the presence of metallic inclusions. It is not unlikely, however, that this phenomenon could be reflecting the a-I–a-S transitions (Sec. 5.5).

5.6.1.3. Intermediately Conducting Films

Among all plasma polymerized films, taking into account their applications, amorphous semiconductors, in which the electrical properties can be controlled by doping and alloying, have risen to dizzy heights. The interest in these materials is still progressing at an undiminished pace.

Plasma deposited amorphous semiconductors are increasingly being used in applications that require large-area, thin-film semiconducting materials. One of these applications is the fabrication of thin-film transistors (TFT) that are most often used in active-matrix liquid crystal displays (LCD). Flat panel displays have the potential capability to reproduce large-capacity, high-resolution pictures in full color on a large screen that takes the place of conventional cathode ray tube. TFT/LCDs are now the leading candidates for these applications and a-Si:H thin film transistors are currently the leading technologies for this purpose [113,116].

Many excellent TFT/LCDs have been demonstrated to be useful in various applications such as video monitors, computer terminals, projection television, etc. Pocket TV sets with 5-inch full-color screens were placed on sale in 1984

[240]. Recently, the Sharp Corp. has announced that 40-inch full-color display based on TFT technology have been developed [241].

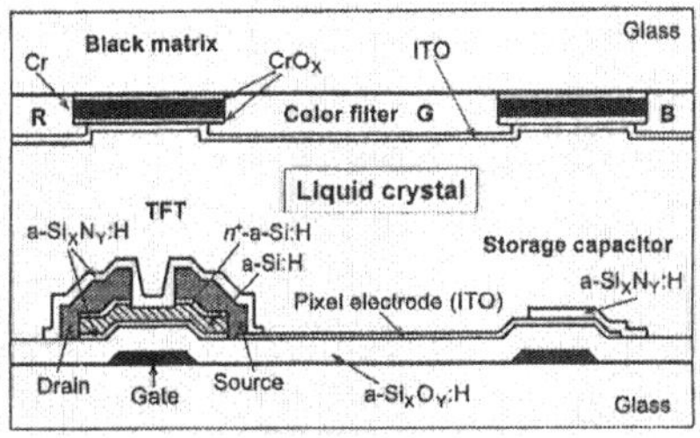

Figure 5.33. Cross-sectional view of a thin-film-transistor (TFT) / liquid crystal display (LCD) for color system [113] (with permission).

A cross-sectional view of such a device is shown in Fig. 5.33. Key components are a TFT, a storage capacitor (which is necessary for good gray scale reproduction), a color filter and a black matrix. The current flowing through the a-Si:H thin film (100 to 200 nm-thick) between the source and drain contacts is modulated by the gate electrode. Current modulation ratios of more than 10^7 can be achieved. If an appropriate voltage occurs at the pixel electrode, a liquid crystal is oriented. By controlling the voltage of pixel electrodes, we can switch pixels between ON and OFF states and thereby generate a picture. In Fig. 5.34 an arrangement of the color pixels is shown. They are located in horizontal RGB (red, green, blue) triplets. The pitches of dots are 184×220 μm [113].

With the development of TFT/LCDs technology, investigations on electrical properties of a-Si:H thin film transistors fabricated at various deposition conditions, with various post-deposition treatments and in various configurations have been carried out simultaneously, giving newer and newer proposals [242-246].

Although hydrogenated amorphous silicon films and hydrogenated amorphous silicon-based alloys are undeniable leaders among plasma deposited films having applications in electronics, quite a large number of other plasma-polymerized films revealing semiconducting character have drawn the attention of investigators. However, it should be noted at once that such films, in spite of the intermediate conductivity. are not necessarily amorphous

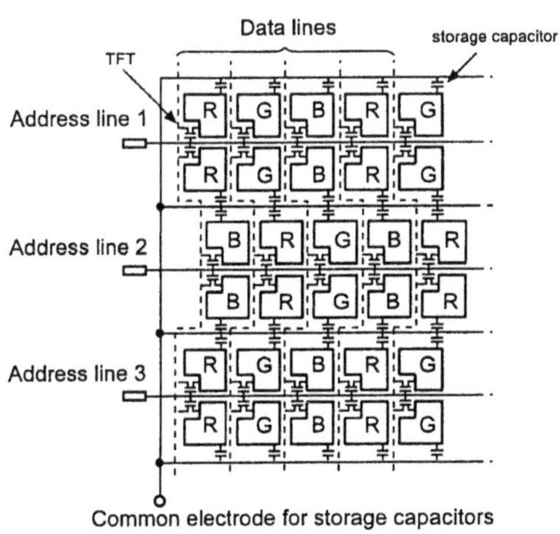

Figure 5.34. Arrangement of the color pixels in the TFT/LDC device [113] (with permission).

semiconductors in the meaning of the electronic structure; some of them can be only amorphous insulators with a relatively high conductivity.

For some time past an interest has been focused on plasma-deposited films that are counterparts of conventional organic materials (crystals or polymers) which are expected to be candidates for microelectronic organic systems. The films plasma-polymerized from such precursors as copper phtalocyanine [247], TCNQ (tetracyanoquinodimethane) and TCNE (tetracyanoethylene) [248], thiophene [249], 1-benzothiofene [250], and aniline [251,252], have been recently thoroughly investigated. Although these films are generally different in their molecular and electronic structures from the related conventional materials, many exciting results concerning their electrical properties have been achieved. It has been found, for example, that doping with iodine can result in some cases in an increase in the conductivity by many orders of magnitude (e.g. from 10^{-11} S/m to 10^{-2} S/m for pp-benzotiophene [250]). The pp-copper phtalocyanine films interposed between aluminum and ITO (indium tin oxide) electrodes have exhibited good rectification, photovoltaic and electrochromic characteristics [247]. A similar behavior of the Al/plasma polymer/ITO sandwich cell has been found when a pp-TCNQ or pp-TCNE film was used as the plasma polymer [248]. The conductivity of some of the films under discussion also reveals the high susceptibility to water vapor and other low-molecular gases. This effect is especially useful to apply it to sensor devices.

The undoped pp-aniline films exhibit the conductivity between 10^{-10} S/m and 10^{-12} S/m for low values of relative humidity *RH*, but, as *RH* increases from 40 % to 90 %, σ increases by around 2 orders of magnitude. When aniline is combined with iodine during deposition, σ changed from 10^{-10}–10^{-11} S/m up to 10^{-4} S/m (this is by a factor of aprox. 10^7) with *RH* growing from 40 % to 90 % [251]. The pp-aniline films can be also used to design CO_2 sensors [252].

The field of sensor devices, in which various plasma polymerized films are (or can be) used, is already widespread. A review of such sensors with their characteristics and the basis of sensing mechanisms is given by Inagaki [253]. The sensors have been prepared using planar comb-type evaporated electrodes of roughly 1 cm^2 surface area on glass or ceramic substrates. Schematic representations of typical sensor devices are shown in Fig. 5.35. For humidity sensors, various nitrogen-containing plasma-polymerized organosilicones were deposited (a mixture of tetramethylsilane (TMS) and ammonia, trimethylsilyldimethylamine (TMSDMA), bis(dimethylamino)-methyl-vinylsilane (BDMAMVS)). To improve the sensitivity the films containing amino groups (e.g. deposited from TMSDMA) were quaternized with methyl bromide.

A typical relationship between the electrical resistance and the relative humidity *RH* (for quaternized and unquaternized pp-TMSDMA films) is shown in Fig. 5.36. The electrical resistance of the quaternized film is found to drop by 4 orders of magnitude between 30 % and 90 % *RH*. The response time of this sensor is, however, relatively long (from several to tens of seconds).

A very rapid response to humidity changes was observed in pp-hexamethyldisilazane (pp-HMDSN) films. Unfortunately, the conductivity of these films was too low to use them in simple sensor devices [174]. Meanwhile, promising results have been shown by Sadhir and Sanjana [254] who deposited films from a mixture of hexamethyldisiloxane (HMDSO) and NH_3 directly onto a microdielectrometer chip. Microdielectrometry was chosen as the measuring technique because of its fast response and real time detection capability.

The humidity-sensitive plasma polymers show great promise as commercial sensor devices but have not been commercialized yet because of the instability of their electrical properties. Very recently it has been shown, however, that these difficulties can be overcome [255].

Plasma polymers, apart from humidity sensors, have been found as good candidates for other low-molecular gas detectors. For example, pp-tetramethyltin films show promise as hydrocarbon (propane C_3H_8) sensors

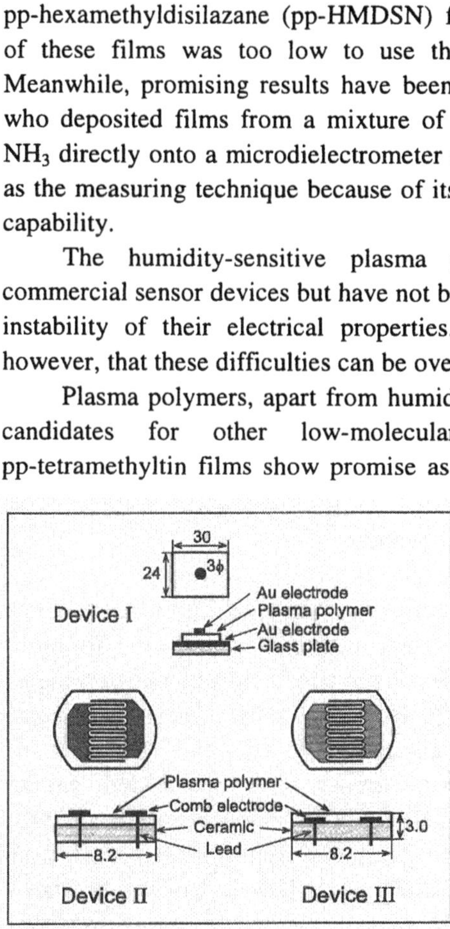

Figure 5.35. Schematic representations of typical sensor devices [253] (with permission).

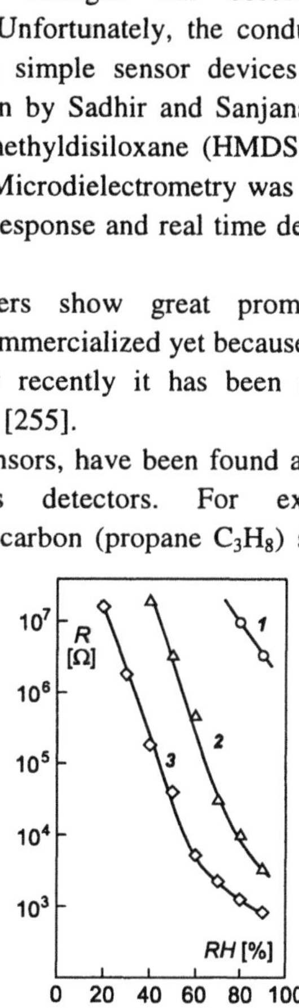

Figure 5.36. Electrical resistance *R* vs. relative humidity *RH* for pp-TMSDMA: *1* – unquaternized (0.45 μm thick), *2* – quaternized (0.45 μm thick), *3* – quaternized (0.82 μm thick) [253] (with permission).

[253], films deposited from metal acetyloacetonates are proposed as CO sensors [253], pp-aniline, as it is mentioned above, can be use to detect CO_2 [252], and pp-styrene, in turn, is presented as a film to formation a quick response NO_2 sensing device [256]. In all these cases, changes in electrical properties of the films under the influence of a given gas-agent are the essence of the sensor operation [174,253].

5.6.2. Optical and Optoelectronic Properties

Many properties of plasma deposited films, sometimes unconventional indeed, render these materials attractive for optical and optoelectronic applications. A good transparency, high hardness and scratch resistance, intimate contact with substrates, and also ability to tailor optical parameters, such as refractive index *n* and extinction coefficient *k*, are important characteristics recommending the materials to be utilized as optical coatings, from protective layers on optical elements to planar wave guides in integrated optical circuitry. On the other hand, the electronic properties of plasma polymers, discussed above in this chapter, turned out to be very useful to design active photoelectronic elements, e.g. solar cells as well as luminescence devices and microdevices.

5.6.2.1. Coatings and Films for Optical Devices

Plasma polymers that can be used as scratch resistant wear protective coatings for various optical elements, water vapor resistant coatings for IR optics, antireflection coatings, and thin-film filters, have been already widely presented in several reviews [206,226,257]. It has been shown that hydrogenated amorphous silicon alloys are especially useful to this end. The films are deposited both directly from organosilicones, e.g. hexamethyldisiloxane (HMDSO) [258], tetramethylsilane (TMS) [259], tetraethoxysilane (TEOS) [260], vinyltrimethoxysilane (VTMOS) [261], and from mixtures of SiH_4 and, e.g. N_2O [262,263], NH_3 [264], CH_4 [265], C_2H_4 [266]. Optical properties of these films can be varied over a wide range depending on the plasma conditions, the reactive gas composition and the post-plasma treatment. For example, in the case of the hydrogenated amorphous silicon oxynitride (a-SiO_XN_Y:H) films prepared by r.f. glow discharge from a SiH_4, N_2O and H_2 mixture, *n* decreases monotonically from 3.5 to 1.5 as the N_2O content in the reactive mixture increases. Figure 5.37 shows the change in n for low and high frequencies as a function of the reactive mixture composition [263]. Correlation of refractive index and the film composition enables us to fabricate films with graded refractive index [258].

Recently, great attention has been attracted by hydrogenated amorphous carbon (a-C:H) films obtained in the diamond-like form (DLC) by plasma polymerization of hydrocarbons. These films as a coating material offer the unique combination of extreme hardness, chemical inertness, and optical transparency over a wide spectral range (in the visible and IR spectral regions). Although there is a variety of hard carbon films produced by different methods, only DLC appears to have found practical applications so far. It is used, for example, as antireflective and scratch resistant wear protective coatings for IR optics (at wavelengths of 8–13 μm) made of ZnS and ZnSe. The low deposition temperature of DLC also allows its use as a wear-protective layer on sunglass lenses made of polycarbonate [267]. Excellent surveys concerning optical properties of the DLC films and their applications have been given very recently by Lettington [268] as well as by Grill [269].

Plasma polymers can be used as optical filters, both one-sided and two-sided units. As early as the 1980s, it was found that some of such films, e.g. pp-tetramethyltin, are good candidates for one-sided UV cut filters [270]. Some metal/plasma polymer composite films are, it turn, very attractive materials for two-sided filters working at the visible range. Optical properties of these films have been widely investigated and discussed [226,227]. In general, the effective medium theory model (with various refinements) has been developed to describe their optical response. The type of doped metal, the shape and the size of its clusters, as well as the filling factor affected the optical transmission, thus resulting in different colors of the films. Using, for example, gold-containing pp-fluorocarbon films we can fabricate filters of different colors, from red to blue in the transmitted light [271].

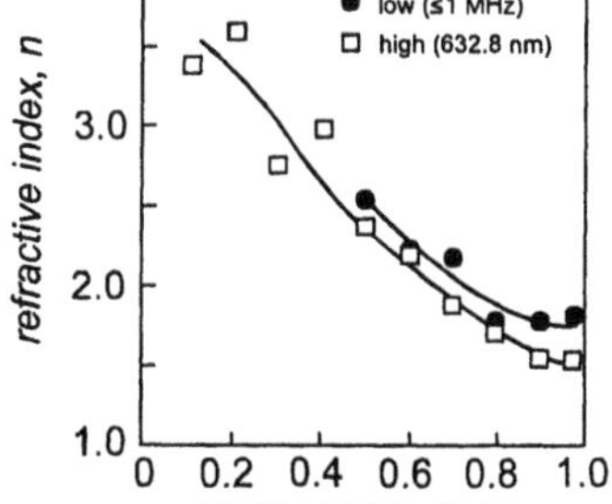

Figure 5.37. Low frequency and high frequency refractive indexes of a-$Si_XO_YN_Z$:H films as a function of the reactive mixture composition [263] (with permission).

The plasma polymerization procedure enables us to prepare more complex filters, with precisely determined optical transmission window. Curtis et al. [272] presented mosaic color-encoding filters for color displays, fabricated as a multilayer structure by the alternating plasma deposition from gas mixtures of $TiCl_4 + O_2$ and $SiCl_4 + O_2$. The *n* index of these films was 2.25–2.38 and 1.44–1.49, respectively. The filter with the thickness of 1.3 μm (composed from 19 layers) transmitted green light (between 487 and 567 nm). By dry etching such a green filter, a high-resolution green/cyan/white mosaic filter pattern was obtained. Recently, multilayered optical filters have been fabricated from amorphous hydrogenated silicon nitride (a-Si_XN_Y:H), oxide (a-Si_XO_Y:H) and oxynidride (a-SiO_XN_Y:H) films. The filters are prepared by varying the film composition either abruptly (discrete, homogeneous multilayer structure) or continuously (graded-index, inhomogeneous structure) [273].

5.6.2.2. Integrated Optics

In the seventies, Tien and coworkers at Bell Telephone Laboratories published several papers pertaining to pp-organosilicon films for integrated optics [274-277]. A key feature of this research was the ability to prepare perfectly homogeneous films exhibiting very low attenuation for light-wave propagation. The films were deposited mainly from vinyltrimethylsilane (VTMS) and hexamethyldisiloxane (HMDSO), as well as their mixtures. The loss of the best films was <0.04 dB/cm at 632.8 nm, which is a tolerable value for utilization of the coating in integrated optical circuitry as a wave guide. In addition to optical waveguides, various other types of devices were prepared with plasma polymers, namely interconnections [275], thin film prism, lenses, and reflectors [276].

The field of plasma polymer applications in the integrated optics is currently experiencing renewed growth. Amorphous pp-fluorocarbon films have been proposed, for example, to use as passive and active waveguide in integrated optical devices [278]. The existence of an electro-optic response in a-Si:H suggests potential use of these films as the core of a waveguide with dielectric claddings [279]. It has also succeeded in preparing plasma polymerized a-SiO_XN_Y:H films that offer a low-loss optical waveguide with well-controlled refractive index and a low-internal stress. The films have been utilized as the core layer of a strip-loaded waveguide [280]. Finally, it should be noted that a simple configuration of an amorphous optoelectronic integrated circuit has been fabricated by a conventional r.f. plasma polymerization as well. The circuit consists of an a-Si_XC_Y:H thin film light emitting diode, an optical waveguide and an a-Si:H thin film photodiode [281] (see also the next Section).

5.6.2.3. Optoelectronic Devices

The thin-film solar cells are without doubt the most spectacular application of semiconducting materials obtained by plasma polymerization processes. Since the demonstration of the first a-Si:H photovoltaic devices at RCA laboratories by Carlson and Wronski [109], there has been remarkable progress in the development of a-Si:H solar cells, spurred, first of all, by the wide demand for a low cost, clean and safe energy [282-284].

Basically, photovoltaic conversion occurs through three separate processes: (*1*) the absorption of light to create electron–hole pairs in an appropriate semiconductor; (*2*) collection and separation of these carriers by an internal electric field; and (*3*) distribution to an external load. The most extensively studied a-Si:H solar cells, due to their highest conversion efficiencies, are those fabricated in the form of *p-i-n* devices. Typically, the *p*-a-Si:H film is less than 10 nm, the undoped *i*-a-Si:H is between 200 and 700 nm and *n*-a-Si:H layer is approx. 30–50 nm. In Fig. 5.38 the energy band diagram for such a cell is sketched. Photons with energy greater than E_G of the *i*-a-Si:H layer are absorbed via interband transitions and free electrons and holes are transported by drift and/or diffusion creating, in consequence, a voltage across the *p-i-n* cell.

The output characteristics for a typical, simply a-Si:H solar cell is shown in Fig. 5.39. The current at zero voltage is the short-circuit current I_{SC} and the voltage at zero current is the open-circuit voltage U_{SC}. The ratio of the product of current and voltage at the maximum power output (black point in the curve) to the product of I_{SC} and U_{SC} is called the fill factor *FF*. The conversion efficiency *Eff*, in turn, is the ratio of the maximum power output to the solar energy input. As one can see, *Eff* = 5.71 % in this case [285].

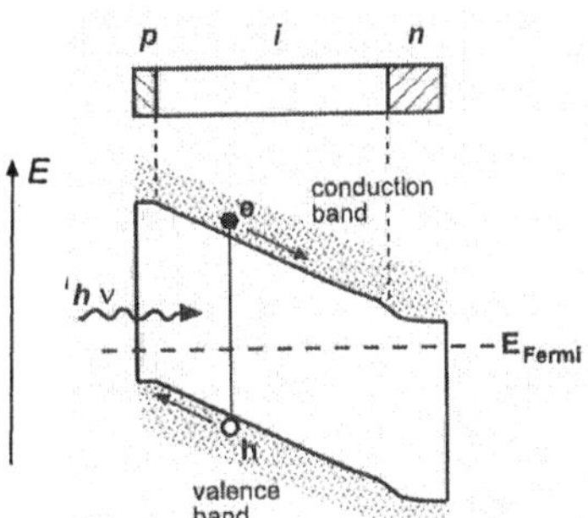

Figure 5.38. Energy band diagram for a *p-i-n* a-Si:H solar cell.

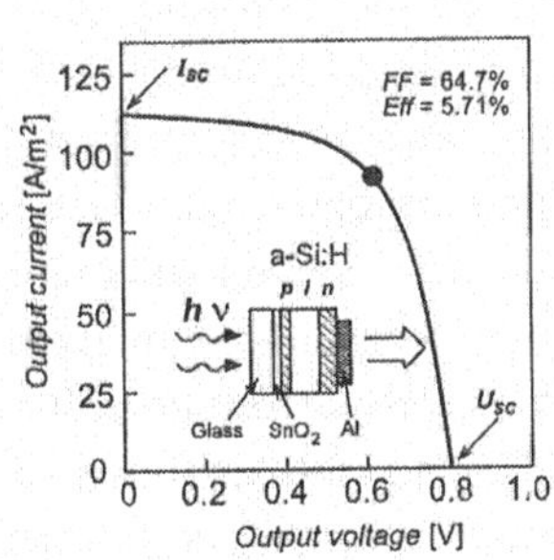

Figure 5.39. Current – voltage characteristics for a simply a-Si:H solar cell [285] (with permission).

To improve the efficiency and stability of a-Si:H solar cells a lot of various experimental investigations coupled with theoretical device modelling and design analysis have been carried out over the past dozen years [112,282,286,287]. This has considerably clarified the physical process occurring in the solar cell and has been instrumental in identifying the material characteristics governing cell performance and losses limiting conversion efficiencies. It should be emphasized, however, that although many changes in the a-Si:H solar cell construction have been proposed, the main principle of operation, based on the *p-i-n* or *n-i-p* junctions, is generally maintained.

Among the most important ideas are multiple-junction solar cell structures deposited from different amorphous semiconducting films, such as a-Si:H, a-Si_XC_Y:H and a-Si_XGe_Y:H. An example of double *n-i-p* junction and the related energy band diagram are presented in Fig. 5.40. This structure utilizes two materials with different E_G to increase optical absorption efficiency by limiting heat loss of photon energies above the transport gap. It also allows the use of thinner *i*-layers, which improves cell stability. For example, a double-junction cell obtained with a-Si:H having $E_G = 1.75$ eV and a-Si_XGe_Y:H, whose $E_G = 1.41$ eV, reveals in laboratory $Eff = 11.16\%$. The other parameters of the cell are: $I_{SC} = 106.1$ A/m^2, $U_{SC} = 1.61$ V, and $FF = 0.655$ [288].

Today the commercial solar cell modules based on a-Si:H are fabricated with a stabilized efficiency greater than 10%. It is believed that the efficiency around 15% could be reached for single-junction a-Si:H cells in the foreseeable future. The maximum conversion efficiency for triple-junction a-Si:H cells is estimated to be about 24 % [112,116,287]. There are, however, much more optimistic prognoses [284].

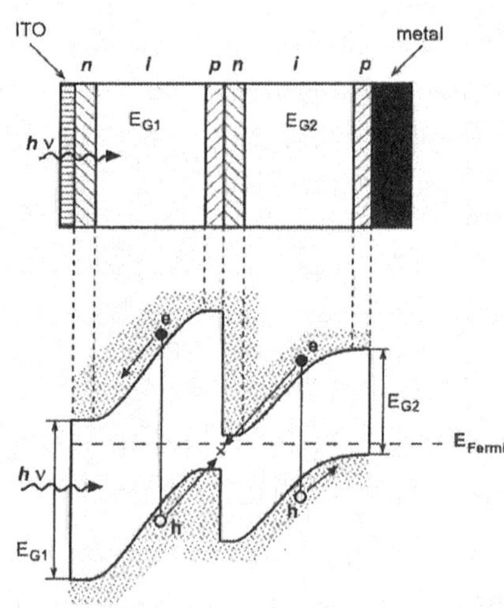

Figure 5.40. A double junction solar cell structure and the related energy band diagram.

Besides the thin-film solar cells, plasma-deposited amorphous semiconductors are proposed as promising materials to prepare photo- and electroluminescent devices. The photoluminescence properties of plasma-polymerized films have been widely investigated for a long time [289]. The main attention has been focused, first of all, on a-Si:H and silicon-based alloy films. Only recently, investigators have begun to take more interest in other plasma polymers. One of the objects of such studies has been pp-methyl methacrylate (pp-MMA). It has been found that when pp-MMA films are excited with blue, violet or UV light, they emit light over a wide range of wavelengths with sharp peaks around 525 and 560 nm, i.e. in green and yellow. This photoluminescence characteristic is ideal for windows for solar cells and therefore the pp-MMA films are proposed as wavelength transformers to this end [290]. Other investigations have been performed on plasma polymerized aromatic hydrocarbons. The spectral responses (at an excitation wavelength of 350 nm) depend on the plasma conditions and the gas mixture used. The luminescence of pp-benzene films is orange, the polymer of toluene yellow and that of styrene yellow-green. The addition of hydrogen or ethylene changes the luminescence color and intensity [291].

A fascinating group of plasma polymerized photoluminescence materials are films doped with lanthanide atoms (more precisely speaking – their cations). In particular, doping with erbium (Er^{3+}) is of great interest due to its optical transition at wavelengths around 1.54 μm (infrared region), one of the standard wavelengths in optical telecommunications. For the first time the observation of a sharp peak in the broad luminescence spectrum of a-Si:H was achieved in 1990, using Er as a dopant [292]. The first Er^{3+}-doped waveguide amplifier fabricated by plasma deposition, using SiH_4, N_2O and P alkoxide to prepare an amorphous phosphosilicate matrix, and a volatile Er chalete as an erbium source, was described in 1993 [293]. Recently, the a-Si:H and a-Si_XC_Y:H plasma deposited films have been mainly taken into consideration as hosts for Er^{3+} [294,295].

No less important range of light for modern optoelectronics is the ultraviolet region. Unfortunately, very little attention has been paid as yet to plasma-deposited thin films doped with optically active centers emitting UV light. Some efforts towards plasma deposition of such films have been only just undertaken. As a matrix, plasma polymerized HMDSO has been proposed. This material is known as an excellent medium for guiding light waves, also in the near-UV region (*Sec.* 5.6.2.2). The films have been doped with gadolinium, which turns out to be a good candidate to create active centers (Gd^{3+}) with an optical transition at short wavelengths. As a precursor of Gd^{3+} for plasma deposition process, the Gd-organic compound, tris(2,2,6,6-tetramethyl-3-5-heptanedionato)gadolinium(III) [Gd(thd)$_3$], has been chosen. Fig. 5.41 presents the emission spectra excited by 200 nm for pp-HMDSO doped with Gd (a), thermally evaporated Gd(thd)$_3$, in which Gd does not exist in the form of Gd^{3+} ions (b), and pure pp-HMDSO matrix (c). The spectrum (a) reveals a broad

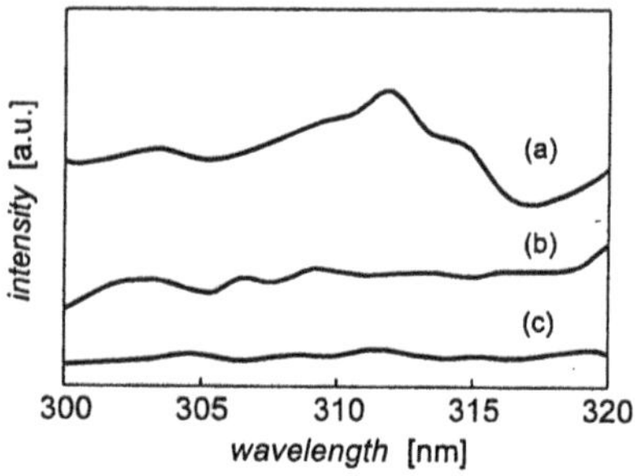

Figure 5.41. Photoluminescence emission spectra excited by 200 nm: (a) – Gd^{3+} doped pp-HMDSO film (4 μm thick), (b) – thermally evaporated $Gd(thd)_3$ film (0,7 μm thick), (c) – pp-HMDSO matrix (4 μm thick) [296].

emission peak with a maximum at 312 nm, which is attributed to the $^6P_{7/2} \rightarrow {}^8S_{7/2}$ transition in Gd^{3+} [296].

The electroluminescent devices based on plasma deposited materials were prepared for the first time 25 years ago [297]. These were *p-i-n* diodes and Schottky barrier diodes fabricated from a-Si:H films, which revealed at 77 K infrared electroluminescence with the maximum emission at 1.3 eV. Since then many new devices have been proposed, focusing the main interest on visible light-emitting diodes operating at room temperature. These systems have been based on a-Si_XC_Y:H, a-Si_XN_Y:H, a-Si_XO_Y:H, and a-C:H films [298-301]. One of the unique innovations is the utilization, as the luminescent active layer, of plasma polymers in which the optical gap E_{OPT} can be changed over a wide range. In consequence, it makes it possible to prepare diodes emitting required color light. It has been demonstrated that the emitting color can be changed from red to blue by choosing E_{OPT} (from 2.4 to 4.0 eV) of a-Si_XN_Y:H layer in the structure shown in Fig. 5.42. As it was discussed in *Sec.* 5.3.3, E_{OPT} of a-Si_XN_Y:H can be easily controlled by deposition conditions. The *p* and *n* doped a-Si_XC_Y:H layers in the device act as the injectors of holes and electrons into the luminescent active *i*-a-Si_XN_Y:H layer.

Although the plasma deposited films are assessed as one of the promising candidates for preparation of luminescent devices, they have not yet reached commercialization due to their low brightness and low total electroluminescence efficiency. In general, the efficiency at room temperature is of the order of $10^{-6} - 10^{-5}$ [298,301], whereas this parameter for conventional organic electroluminescent diodes amounts to approximately 10^{-2}. Taking into account, however, other advantages of these systems, for example, the possibil-

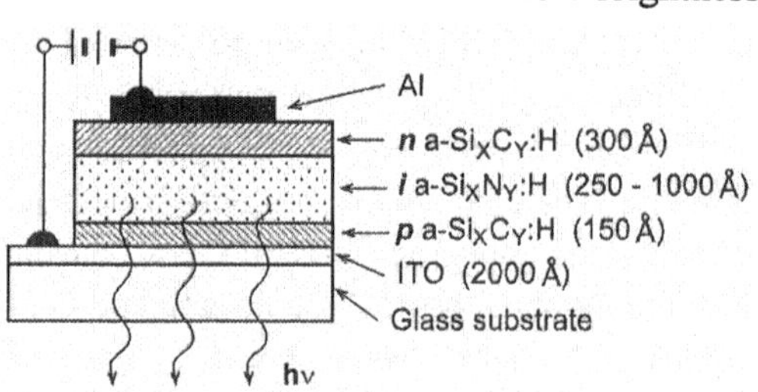

Figure 5.42. Structure of an a-Si_XN_Y:H / a-Si_XC_Y:H thin-film light emitting diode. Changing E_{OPT} of the a-Si_XN_Y:H layer, the emitting color can be changed from red to blue [300] (with permission).

ity of fabrication of optoelectronic integrated circuits [281], further investigations and progress in this field is expected.

5.6.3. Current Problems

Among current problems concerning plasma polymers and their electrical properties that could be used in electronics, a-C:H films exhibiting a low or even negative electron affinity χ come to the forefront. Owing to comparably simple deposition techniques and low process costs, the films may offer a promising way to small-scale electron field emitters, for example, for display applications [302]. In an attempt to lower the work function W of the a-C:H films, lithium-containing films were deposited by lithium evaporation from dispensers during methane polymerization. A reduction of W from 3.4 eV in pure a-C:H to 2.45 eV in samples containing 35 at.% lithium is observed [61]. It has been also found that beyond a chemical composition of 14 at.% nitrogen in a-C:H films, the films become self-texturing, and multiple "domelike" cathodes of nanometer scale occur. When the electric field emission of these textured films (N content 15 at.%) is measured, an enhancement in the emitted current density of approx. 2 orders of magnitude at an electric field of $2 \cdot 10^7$ V/m is observed, in comparison to the untextured films with a nitrogen content of 11 at.% [303].

As far as intermediately conducting plasma polymers are concerned, there has been recently a growing interest in such films as active materials for electrochemical processes. Several laboratories have already reported research on the use of plasma polymerized coatings to modify electrodes for electrocatalysis. The coatings prepared, among other precursors, from vinylferrocene and 4-vinylpyridine have revealed electrocatalytic activity and thus deposited on an inactive conducting support, e.g. carbon, may render inexpensive active electrode materials [304-306].

Lately, plasma-polymerized carbon disulfide thin films have been proposed as the cathode material for solid-state lithium rechargeable batteries. The deposited films have shown better electrochemical activity than poly(carbon disulfide) prepared by chemical means [307]. Plasma-polymerized films are also considered as a candidate for the solid electrolyte to the lithium batteries. Thin films have been formed by complexation of plasma-polymerized tetrasiloxane or vinylsilane derivatives and lithium perchlorate [308-310].

Finally, the heterogeneous systems fabricated by plasma polymerization, both multilayers and nanodisperse structures should be reported. The amorphous multilayer structures deposited by plasma processes have been known since the early eighties. In the beginning the structures composed of the alternating layers such as a-Si:H/a-Si_XN_Y:H, a-Si:H/a-Si_XC_Y:H, a-Si:H/a-Si_XO_Y:H, a-Si:H/a-Ge:H were investigated [311-315]. The layered structures were formed by a rapid periodical switching of the gas composition in the plasma reactor. The thickness of the alternating layers was fixed (between several and several hundred angstroms). Recently, a novel class of layer structures has been found. They are formed from twin films of a-insulator and a-semiconductor deposited from the same single-source precursor and in the same deposition process, by changing only one of the deposition parameters [192,193,195] (*Sec.* 5.5). An interesting system was also proposed by Nitta [316,317], who fabricated multilayers in which the thickness of the alternating films was not fixed but changed randomly. Apart from their practical application (e.g. for solar cells, and $\lambda/4$ reflection mirrors), the random amorphous multilayers are a good model for work on the properties of randomness in condensed matter.

Unfortunately, the electronic properties of the plasma deposited multilayer systems are not quite clear as yet. In practice all earlier papers on amorphous multilayers explained their optical and electrical properties based on the quantum size effect. However, because of the widely accepted electron mean free path of at most 1 nm in a-Si:H the existence of quantized energy levels seems to be a surprising result from an intuitive point of view as well as from a more sophisticated theoretical investigation [318]. In fact, it has been shown that changes in the electrical conductivity and the optical gap measured as a function of the layer thicknesses for a few different systems (e.g. a-Si:H/a-Si_XC_Y:H, twin a-Ge_XC_Y:H multilayer) can be satisfyingly explained without taking any quantum size effect into account [195,319]. As one can see, the problem concerning a role that is played by the quantum effects in plasma-deposited amorphous multilayers is more general and require a further investigation.

Another thin film system, to which quite a lot of attention has been paid recently, is based on nanodisperse structures. Such systems produced by the metal dispersion in plasma-polymerized films have been already mentioned in *Sec.* 5.6.1.2 (see also: Chapter 9). An interesting hybrid is a combination of the multilayer and metal cluster ideas. Such structures are fabricated by alternate deposition: an insulating or semiconducting layer by plasma deposition and a metal layer by evaporation. For example, Drüsedau et al. [320] investigated systems prepared from a-Si:H films and molybdenum, titanium or vanadium, with maximum number of layers equal to 21.

The metal/a-semiconductor/metal systems that exhibit fast electrically programmable digital and analogue memory switching can be also numbered among this group of structures. Owen et al. [321-323] have studied metal/a-Si:H/metal systems using various metals. Particular attention has been, however, given to the chromium/p^+-a-Si:H/vanadium system that shows promise as non-volatile memory devices.

A curious group of nanodisperse structures is that composed of either a-insulators with a-semiconductor nano-grains or a-semiconductor matrixes "doped" with a-insulator grains. The former systems have appeared to be especially interesting as a medium for reversible charge storage. It has been found by AFM (atomic force microscope) measurements that the a-Ge_XC_Y:H films deposited from tetramethylgermanium in the three-electrode reactor, due to their domain structure reveal reversible charge storage effect. In data storage application the information density of the order of 10^{14} bit/m^2 can be achieved [324,325]. Such systems can also reveal switching effects. For example, this phenomenon was observed in pp-acrylonitrile films and it was attributed to semiconducting filaments occurring in the film [186].

As far as the applications of plasma polymers in optoelectronics are concerned, the current challenge is in terms of integrated circuits consisting of a waveguide matrix in which active elements are created by local doping with appropriate optical centers. Such a system would be, in the full sense of the word, integrated. In particular, doping with the lanthanide ions is of great interest (*Sec.* 5.6.2.3). The next way in this field is focused at present on all-silicon based optoelectronic devices, in which plasma deposited hydrogenated amorphous silicon and silicon-based films play a significant role.

References

1. Robertson, J., *Adv. Phys.*, **35** (1986) 317.
2. Mott, N. F. and Davis, E. A., *Electronic Processes in Non-Crystalline Materials* (Clarendon Press, Oxford, 1979).
3. Dong, J. and Drabold, D. A., *Phys. Rev. B*, **54** (1996) 10284.
4. Anderson, P. W., *Phys. Rev.*, **109** (1958) 1492.

5. Fritzsche, H., *Electronic Properties of Amorphous Semiconductors*, in *Amorphous and Liquid Semiconductors*, ed. Tauc, J. (Plenum Press, London, 1974) Chap.5.
6. Nagels, P., *Electronic Transport in Amorphous Semiconductors*, in *Amorphous Semiconductors*, ed. Brodsky, M. H. (Springer Verlag, Berlin, 1979) Chap.5.
7. Mott, N. F., *J. Non-Cryst. Solids*, **90** (1987) 1.
8. Overhof, H. and Thomas, P., *Electronic Transport in Hydrogenated Amorphous Semiconductors* (Springer Verlag, Berlin, 1989).
9. Tyczkowski, J., *Thin Solid Films*, **168** (1989) 175.
10. Adler, D., *Sol. Energy Mater.*, **8** (1982) 53.
11. Kao, K. C. and Hwang, W., *Electrical Transport in Solids* (Pergamon Press, Oxford, 1981).
12. Cohen, M. L. and Chelikowsky, J. R., *Electronic Structure and Optical Properties of Semiconductors* (Springer-Verlag, Berlin, 1989).
13. Robertson, J., *Phil. Mag. B*, **66** (1992) 615.
14. Drabold, D. A. and Dong, J., *J. Non-Cryst. Solids*, **227/230** (1998) 153.
15. Chen, C. W. and Robertson, J., *J. Non-Cryst. Sölids*, **227/230** (1998) 602.
16. Sussmann, S. and Ogden, R., *Phil. Mag. B*, **44** (1981) 137.
17. Bullot, J., *et al.*, *Phil. Mag. B*, **49** (1984) 489.
18. Siebert, W., *et al.*, *phys. stat. sol. (b)*, **140** (1987) 311.
19. Tyczkowski, J., *et al.*, *Thin Solid Films*, **209** (1992) 250.
20. Tauc, J., *Optical Properties of Amorphous Semiconductors*, in *Amorphous and Liquid Semiconductors*, ed. Tauc, J. (Plenum Press, London, 1974) Chap.4.
21. Saraie, J., *et al.*, *Thin Solid Films*, **117** (1984) 59.
22. Tyczkowski, J. and Kryszewski, M., *J. Phys. D: Appl. Phys.*, **17** (1984) 2053.
23. Klazes, R. H., *et al.*, *Phil. Mag. B*, **45** (1982) 377.
24. Nitta, S., *et al.*, *Solar Energy Mater.*, **8** (1982) 249.
25. Curtins, H. and Favre, M., *Surface and Bulk States Determined by Photothermal Deflection Spectroscopy*, in *Amorphous Silicon and Related Materials*, ed. Fritzsche, H. (World Scientific, Singapore, 1989) 329.
26. Kondilis, A., *Phys. Rev. B*, **62** (2000) 10526.
27. Conde, J. P., *et al.*, *J. Appl. Phys.*, **85** (1999) 3327.
28. Giorgis, F., *et al.*, *J. Non-Cryst. Solids*, **266/269** (2000) 588.
29. Leung, T. Y., *et al.*, *J. Non-Cryst. Solids*, **254** (1999) 151.
30. Vandevelde, T., *et al.*, *Diamond & Related Mater.*, 7 (1998) 152.

31. Mpawenayo, P., *et al.*, *Chemotronics*, **3** (1988) 111.
32. Al-Dallal, S., *et al.*, *J. Non-Cryst. Solids*, **196** (1996) 168.
33. Zedlitz, R., *et al.*, *J. Non-Cryst. Solids*, **198/200** (1996) 403.
34. Payson, J. S. and Ross, R. C., *J. Non-Cryst. Solids*, **77/78** (1985) 579.
35. Comedi, D., *et al.*, *Phys. Rev. B*, **52** (1995) 4974.
36. Sala, D. D., *et al.*, *J. Appl. Phys.*, **67** (1990) 814.
37. Kočka, J., *et al.*, *Energy and Density of Gap States in a-Si:H*, in *Amorphous Silicon and Related Materials*, ed. Fritzsche, H. (World Scientific, Singapore, 1989) 297.
38. Cohen, J. D. and Gelatos, A. V., *Transient Photocapacitance Studies of Deep Defect Transitions in Hydrogenated Amorphous Silicon*, in *Amorphous Silicon and Related Materials*, ed. Fritzsche, H. (World Scientific, Singapore, 1989) 475.
39. Naito, H., *et al.*, *Meas. Sci. Technol.*, **2** (1991) 912.
40. Hattori, K., *et al.*, *J. Non-Cryst. Solids*, **198/200** (1996) 288.
41. Main, C., *et al.*, *Sol. State Commun.*, **83** (1992) 401.
42. Demichelis, F., *et al.*, *Phil. Mag. B*, **60** (1989) 713.
43. Morita, S., *et al.*, *J. Polym. Sci: Polym. Phys. Ed.*, **15** (1977) 981.
44. Wronski, C. R., *et al.*, *Appl. Phys. Lett.*, **37** (1980) 96.
45. Yamamoto, T., *et al.*, *Jpn. J. Appl. Phys.*, **20 Suppl. 20-2** (1981) 185.
46. Tyczkowski, J. and Kryszewski, M., *J. Phys. D: Appl. Phys.*, **14** (1981) 1877.
47. Hirose, M. and Miyazaki, S., *J. Non-Cryst. Solids*, **66** (1984) 327.
48. Takai, Y., *et al.*, *J. Phys. D: Appl. Phys.*, **17** (1984) 399.
49. Mimura, H. and Hatanaka, Y., *Appl. Phys. Lett.*, **50** (1987) 326.
50. Tyczkowski, J., *et al.*, *J. Non-Cryst. Solids*, **137/138** (1991) 875.
51. Ishii, K., *et al.*, *J. Photopolymer Sci. Technol.*, 7 (1994) 353.
52. Tyczkowski, J. and Ledzion, R., *J. Appl. Phys.*, **86** (1999) 4412.
53. Tyczkowski, J., *et al.*, *Surf. Coat. Technol.*, **142/144** (2001) 843.
54. Caywood, J. M., *Mol. Crystals and Liquid Crystals*, **12** (1970) 1.
55. Mizutani, T., *et al.*, *J. Phys. D: Appl. Phys.*, **9** (1976) 2253.
56. Klöpffer, W., *Introduction to Polymer Spectroscopy* (Springer-Verlag, Berlin, 1984) Part B.
57. Christmann, K., *Introduction to Surface Physical Chemistry* (Springer-Verlag, New York, 1991) Chap. 4.
58. Oelhafen, P., *et al.*, *Thin Solid Films*, **120** (1984) 231.
59. Tsai, H. and Bogy, D. B., *J. Vac. Sci. Technol. A*, **5** (1987) 3287.
60. Schäfer, J., *et al.*, *Phys. Rev. B*, **53** (1996) 7762.

61. Thiele, J. U., *et al.*, *J. Vac. Sci. Technol. A*, **15** (1997) 1739.
62. Sénémaud, C., *et al.*, *J. Non-Cryst. Solids*, **227/230** (1998) 447.
63. Robertson, J., *J. Non-Cryst. Solids*, **137/138** (1991) 825.
64. Fourches, N. and Turban, G., *Thin Solid Films*, **240** (1994) 28.
65. Ley, L., *Photoemission and Optical Properties*, in *Physics of Hydrogenated Amorphous Silicon II. Electronic and Vibrational Properties*, eds. Joannopoulos, J. D. and Lucovsky, G. (Springer-Verlag, Berlin, 1984) Chap.3.
66. Jackson, W. B., *et al.*, *Phys. Rev. B*, **31** (1985) 5187.
67. Gertkemper, T., *et al.*, *J. Non-Cryst. Solids*, **164/166** (1993) 123.
68. Seta, M. D., *et al.*, *J. Non-Cryst. Solids*, **137/138** (1991) 851.
69. Evangelisti, F., *J. Non-Cryst. Solids*, **164/166** (1993) 1009.
70. Hammer, P., *et al.*, *J. Non-Cryst. Solids*, **227/230** (1998) 645.
71. Iqbal, A., *et al.*, *J. Appl. Phys.*, **61** (1987) 2947.
72. Sénémaud, C., *et al.*, *J. Non-Cryst. Solids*, **164/166** (1993) 1073.
73. Hayes, T. M., *J. Non-Cryst. Solids*, **31** (1978) 57.
74. Fink, J., *et al.*, *Solid St. Commun.*, **47** (1983) 687.
75. Ziethen, C., *et al.*, *J. Electron Spectrosc. Relat. Phenom.*, **107** (2000) 261.
76. Krieg, J., *et al.*, *Solid St. Commun.*, **42** (1982) 831.
77. Winer, K. and Ley, L., *Phys. Rev. B*, **36** (1987) 6072.
78. Winer, K. and Ley, L., *Near-Surface Electronic Properties of a-Si:H*, in *Amorphous Silicon and Related Materials*, ed. Fritzsche, H. (World Scientific, Singapore, 1989) 365.
79. Aljishi, S., *et al.*, *Phys. Rev. Lett.*, **64** (1990) 2811.
80. Shinar, R., *et al.*, *j. Vac. Sci. Technol. A*, **5** (1987) 2804.
81. McKenzie, D. R., *et al.*, *Solid. St. Commun.*, **59** (1986) 325.
82. Spear, W. E. and LeComber, P. G., *J. Non-Cryst. Solids*, **8/10** (1972) 727.
83. Guha, S., *Solar Energy Mater.*, **8** (1982) 269.
84. Shumacher, R., *et al.*, *Phil. Mag. B*, **58** (1988) 389.
85. Deane, S. C. and Powell, M. J., *J. Non-Cryst. Solids*, **164/166** (1993) 323.
86. Tyczkowski, J. and Kazimierski, P., *unpublished data.*
87. Fritzsche, H., *J. Non-Cryst. Solids*, **77/78** (1985) 273.
88. Silva, S. R. P. and Amaratunga, G. A. J., *Thin Solid Films*, **253** (1994) 146.
89. Tyczkowski, J., *et al.*, *phys. stat. sol. (a)* **72** (1982) 751.
90. Arkhipov, V. I. and Adriaenssens, G. J., *J. Non-Cryst. Solids*, **181** (1995) 274.
91. vanTurnhout, J., *Thermally Stimulated Discharge of Polymer Electrets* (Elsevier, Amsterdam, 1975).

92. Ibaraki, N. and Fritzsche, H., *J. Non-Cryst. Solids*, **66** (1984) 231.
93. Landweer, G. and Bezemer, J., *Thermally Stimulated Current Method for Determining Density of States*, in *Amorphous Silicon and Related Materials*, ed. Fritzsche, H. (World Scientific, Singapore, 1989) 525.
94. Lee, K. B. and Lee, K. H., *J. Non-Cryst. Solids*, **181** (1995) 116.
95. Kazimierski, P., *et al.*, *J. Macromol. Sci.-Phys.*, **B29** (1990) 63.
96. Kazimierski, P. and Kryszewski, M., *J. Phys. D: Appl. Phys.*, **24** (1991) 2115.
97. Kazimierski, P. and Tyczkowski, J., in *5th Inter. Conf. on Conduction and Breakdown in Solid Dielectrics* (Leicester (England) 1995) IEEE, 706.
98. Kazimierski, P., *et al.*, *Folia Sci. Univ. Tech. Resoviensis*, **102** (1992) 41.
99. Tyczkowski, J. and Kazimierski, P., *Vacuum*, **39** (1989) 11.
100. Lang, D. V., *et al.*, *Phys. Rev. B*, **25** (1982) 5285.
101. Okushi, H., *et al.*, *Thermally and Light-Induced Changes in the Gap States of p-Doped a-Si:H*, in *Amorphous Silicon and Related Materials*, ed. Fritzsche, H. (World Scientific, Singapore, 1989) 657.
102. Cohen, J. D., *et al.*, *J. Non-Cryst. Solids*, **164/166** (1993) 327.
103. Matsuura, H., *et al.*, *Phil. Mag. Lett.*, **59** (1989) 109.
104. Crandall, R. S., *et al.*, *J. Non-Cryst. Solids*, **137/138** (1991) 311.
105. Sterling, H. F. and Swann, R. C. G., *Solid-State Electron.*, **8** (1965) 653.
106. Chittick, R. C., *et al.*, *J. Electrochem. Soc.*, **116** (1969) 77.
107. Spear, W. E. and LeComber, P. G., *Sol. State Commun.*, **17** (1975) 1193.
108. Spear, W. E. and LeComber, P. G., *Phil. Mag.*, **33** (1976) 935.
109. Carlson, D. E. and Wronski, C. R., *Appl. Phys. Lett.*, **28** (1976) 671.
110. Triska, A., *et al.*, *Bull. Am. Phys. Soc.*, **20** (1975) 392.
111. Knights, J. C., *AIP Conf. Proc.*, **31** (1976) 296.
112. Wronski, C. R., *Solar Energy Mater. Solar Cells*, **41/42** (1996) 427.
113. Ibaraki, N. and Oana, Y., *Optoelectronics - Devices and Technologies*, **4** (1989) 209.
114. Morigaki, K. and Nitta, S., *Amorphous Silicon*, in *Nonctystalline Semiconductors*, ed. Pollak, M. (CRC, Boca Raton, 1987) 53.
115. Street, R. A., *Hydrogenated Amorphous Silicon* (Cambridge University Press, New York, 1991).
116. Luft, W. and Tsuo, Y. S., *Hydrogenated Amorphous Silicon Alloy Deposition Processes* (Marcel Dekker, New York, 1993).
117. Mimura, H. and Hatanaka, Y., *Appl. Phys. Lett.*, **50** (1987) 326.
118. Schumm, G., *J. Non-Cryst. Solids*, **164/166** (1993) 317.
119. Schumm, G., *Phys. Rev. B*, **49** (1994) 2427.

120. Street, R. A., *J. Non-Cryst. Solids*, **77/78** (1985) 1.
121. Meaudre, R., *et al.*, *J. Non-Cryst. Solids*, **87** (1986) 321.
122. Stutzmann, M., *et al.*, *Phys. Rev. B*, **35** (1987) 5666.
123. Chen, L., *et al.*, *Phys. Rev. B*, **46** (1992) 2050.
124. Smith, Z. E. and Wagner, S., *Implication of the "Defect Pool" Concept for "Metastable" and "Stable" Defects in Amorphous Silicon*, in *Amorphous Silicon and Related Materials*, ed. Fritzsche, H. (World Scientific, Singapore, 1989) 409.
125. Powell, M. J. and Deane, S. C., *Phys. Rev. B*, **48** (1993) 10815.
126. Drüsedau, T. P. and Schröder, B., *Solid State Phenomena*, **44/46** (1995) 425.
127. Carius, R., *et al.*, *J. Non-Cryst. Solids*, **227/230** (1998) 432.
128. Hattori, K., *J. Appl. Phys.*, 77 (1995) 2989.
129. LeComber, P. G. and Spear, W. E., *Doped Amorphous Semiconductors*, in *Amorphous Semiconductors*, ed. Brodsky, M. H. (Springer Verlag, Berlin, 1979) Chap. 9.
130. Chambouleyron, I. and Comedi, D., *J. Non-Cryst. Solids*, **227/230** (1998) 411.
131. Chaudhuri, P., *et al.*, *Thin Solid Films*, **121** (1984) 233.
132. Ray, S., *et al.*, *J. Appl. Phys.*, **62** (1987) 3917.
133. Suchaneck, G., *et al.*, *J. Non-Cryst. Solids*, **137/138** (1991) 701.
134. Carreno, M. N. P., *et al.*, *J. Non-Cryst. Solids*, **227/230** (1998) 483.
135. Kurata, H., *et al.*, *Jpn. J. Appl. Phys.*, **20** (1981) L811.
136. Alvarez, F. and Chambouleyron, I., *Solar Energy Mater.*, **10** (1984) 151.
137. Meyerson, B. and Smith, F. W., *Sol. State Commun.*, **34** (1980) 531.
138. Matsushita, M. and Harum, M., *J. Photopolymer Sci. Technol.*, **12** (1999) 11.
139. Tyczkowski, J., *et al.*, *Surf. Coat. Technol.*, **142/144** (2001) 792.
140. Schmal, J., *et al.*, *J. Non-Cryst. Solids*, **164/166** (1993) 415.
141. Spear, W. E., *et al.*, *Physica*, **117B/118B** (1983) 908.
142. Carlson, D. E. and Wronski, C. R., *Doped Amorphous Semiconductors*, in *Amorphous Semiconductors*, ed. Brodsky, M. H. (Springer Verlag, Berlin, 1979) Chap.10.
143. Nitta, S., *et al.*, *On the properties of quasi one dimensional hydrogenated amorphous silicon films*, in *Tetrahedrally-Bonded Amorphous Semiconductors*, eds. Adler, D. and Fritzsche, H. (Plenum Press, New York, 1985) 501.
144. Tanaka, K. and Nitta, S., *Phys. Rev. B*, **39** (1989) 3258.

145. Nonomura, S., *et al.*, *J. Non-Cryst. Solids*, **114** (1989) 729.
146. Anderson, D. A. and Spear, W. E., *Phil. Mag.*, **35** (1976) 1.
147. Drüsedau, T. P., *et al.*, *Phil. Mag. B*, **69** (1994) 1.
148. Mahan, A. H., *et al.*, *Appl. Phys. Lett.*, **44** (1984) 220.
149. Tyczkowski, J., *et al.*, *Appl. Surf. Sci.*, **113/114** (1997) 534.
150. Tyczkowski, J. and Pietrzyk, B., *Mater. Sci. Forum*, **287/288** (1998) 431.
151. Tyczkowski, J., *et al.*, *Chem. Mater.*, **10** (1998) 3879.
152. Tyczkowski, J., *J. Vac. Sci. Technol. A*, **17** (1999) 470.
153. Kuwano, Y., *et al.*, in *16th IEEE Photovoltaic Specialists' Conf.* (San Diego, 1982) IEEE, 1338.
154. Kärcher, R., *et al.*, *Phys. Rev. B*, **30** (1984) 1896.
155. Tessler, L. R. and Alvarez, F., *J. Non-Cryst. Solids*, **83** (1986) 1.
156. Dunnett, B., *et al.*, *Phil. Mag. B*, **53** (1986) 159.
157. Shirafuji, J., *et al.*, *Jpn. J. Appl. Phys.*, **22** (1983) 775.
158. Scher, H. and Montroll, E. W., *Phys. Rev. B*, **12** (1975) 2455.
159. Pfister, G. and Scher, H., *Adv. Phys.*, **27** (1978) 747.
160. Marshall, J. M., *et al.*, *Phil. Mag. B*, **47** (1983) 211.
161. Nebel, C. E. and Street, R. A., *Phil. Mag. B*, **67** (1993) 407.
162. Nebel, C. E., *et al.*, *Phys. Rev. B*, **46** (1992) 6789.
163. Nebel, C. E., *et al.*, *Phys. Rev. B*, **46** (1992) 6803.
164. Nebel, C. E., *J. Non-Cryst. Solids*, **137/138** (1991) 395.
165. Gu, Q., *et al.*, *J. Appl. Phys.*, **76** (1994) 2310.
166. Tyczkowski, J. and Sielski, J., *Thin Solid Films*, **187** (1990) 1.
167. Griscom, D. L., *J. Non-Cryst. Solids*, **31** (1978) 241.
168. Dworschak, W., *et al.*, *Thin Solid Films*, **189** (1990) 257.
169. Solomon, I., *et al.*, *J. Non-Cryst. Solids*, **97/98** (1987) 1091.
170. Tyczkowski, J. and Kryszewski, M., *phys. stat. sol. (a)* **78** (1983) 259.
171. Böttger, H. and Bryskin, V. V., *Hopping Conduction in Solids* (VCH, Berlin, 1985), Chap.III.
172. Szeto, R. and Hess, D. W., *J. Appl. Phys.*, **52** (1981) 903.
173. Tyczkowski, J., *et al.*, *J. Chem. Vap. Dep.*, **4** (1996) 261.
174. Tyczkowski, J. and Kryszewski, M., *J. Appl. Polym. Sci.: Appl. Polym. Symp.*, **38** (1984) 149.
175. Tyczkowski, J. and Kazimierski, P., in *4th Inter. Conf. "Conduction and Breakdown in Solid Dielectrics"* (Sestri Levante (Italy), 1992), IEEE, 77.
176. Maisonneuve, M., *et al.*, *Thin Solid Films*, **44** (1977) 209.
177. Maisonneuve, M., *et al.*, *Thin Solid Films*, **33** (1976) 35.
178. Segui, Y., *et al.*, *Thin Solid Films*, **120** (1984) 37.

179. Szeto, R. and Hess, D. W., *Thin Solid Films*, **78** (1981) 125.
180. Ogumi, Z., *et al.*, *J. Electorchem. Soc.*, **136** (1989) 625.
181. Yasuda, H., *Plasma Polymerization* (Academic Press, Orlando, 1985) Chap.11.
182. Tyczkowski, J., *et al.*, *Polym. Bull.*, **8** (1982) 117.
183. Tyczkowski, J., *et al.*, in *Symp. on Electrets* (Heidelberg (Germany), 1985), IEEE, 340.
184. Tyczkowski, J. and Sielski, J., in *3rd Inter. Conf. on Conduction and Breakdown in Solid Dielectrics* (Trondheim (Norway), 1989), IEEE, 177.
185. Yoshino, K., *et al.*, *J. Appl. Phys.*, **49** (1978) 4849.
186. Tyczkowski, J., *Thin Solid Films*, **199** (1991) 335.
187. Tyczkowski, J., *et al.*, *Thin Solid Films*, **241** (1993) 291.
188. Tyczkowski, J. and Kazimierski, P., *J. Phys. D: Appl. Phys.*, **27** (1994) 179.
189. Tyczkowski, J. and Szymanowski, H., *Mater. Sci. (Medzagotyra)* **2(7)** (1998) 3.
190. Weissmantel, C., *et al.*, *Thin Solid Films*, **72** (1980) 19.
191. Helmbold, A., *et al.*, *Phil. Mag. B*, **72** (1995) 335.
192. Silva, S. R. P., *et al.*, *Thin Solid Films*, **253** (1994) 20.
193. Davis, C. A., *et al.*, *Phys. Rev. Lett.*, **75** (1995) 4258.
194. Silva, S. R. P., *et al.*, *J. Non-Cryst. Solids*, **227/230** (1998) 1137.
195. Kazimierski, P. and Tyczkowski, J., *Surf. Coat. Technol.*, **116/119** (1999) 614.
196. Goodman, J., *J. Polym. Sci., Lett. Ed*, **44** (1960) 551.
197. Bradley, A. and Hammes, J. P., *J. Electrochem. Soc.*, **110** (1963) 15.
198. Bradley, A. and Hammes, J. P., *J. Electrochem. Soc.*, **110** (1963) 543.
199. Stuart, M., *Nature*, **199** (1963) 59.
200. Stuart, M., *Proc. Inst. Electr. Eng.*, **112** (1965) 1614.
201. Ozawa, P. J., *IEEE Trans. Parts, Mater. Packag*, **PMP-5** (1969) 112.
202. Bashara, N. M. and Doty, C. T., *J. Appl. Phys.*, **35** (1964) 3498.
203. Gregor, L. V., *Thin Solid Films*, **2** (1968) 235.
204. Mearns, A. M., *Thin Solid Films*, **3** (1969) 201.
205. Babcock, L. E. and Christy, R. W., *J. Appl. Phys.*, **43** (1972) 1423.
206. Wrobel, A. M. and Wertheimer, M. R., *Plasma-Polymerized Organosilicones and Organometallics*, in *Plasma Deposition, Treatment, and Etching of Polymers*, ed. d'Agostino, R. (Academic Press, Boston, 1990), Chap.3.
207. Hetzler, U. and Kay, E., *J. Appl. Phys.*, **49** (1978) 5617.
208. Vollmann, W. and Poll, H. U., *Thin Solid Films*, **26** (1975) 201.

209.Tyczkowski, J. and Szymanowski, H., *Mater. Sci. Eng. A*, **139** (1991) 120.
210.Tuzov, L. S., *et al.*, *Int. Chem. Eng.*, **11** (1971) 60.
211.Ristov, D., *J. Mater. Sci.*, **12** (1977) 1411.
212.Ramu, T. S., *et al.*, *IEEE Trans. Electr. Insul.*, **EI-21** (1986) 549.
213.Ramu, T. S. and Werheimer, M. R., *IEEE Trans. Electr. Insul.*, **EI-21** (1986) 557.
214.Weber, A., *et al.*, *J. Vac. Sci. Technol. A*, **16** (1998) 2120.
215.Loboda, M. J., *et al.*, in *Proc. of the Symposia on Electrochemical Processing in ULSI Fabrication I and Interconnect and Contact Metallization: Materials, Processes, and Reliability* (San Diego, CA, & Boston, MA, 1998) Electrochem. Soc., 145.
216.Lee, W. W., *et al.*, in *Plasma Deposition and Treatment of Polymers - Symposium* (Boston, MA, 1998) Materials Research Society, 161.
217.Ping, X., *et al.*, in *Proc. of the IEEE 1999 International Interconnected Technology Conference* (San Francisco, CA, 1999) IEEE, 109.
218.Lin, C. T. and Mantei, T. D., *J. Vac. Sci. Technol. A*, **17** (1999) 735.
219.Loboda, M. J., *Microelectronic Eng.* **50** (2000) 15.
220.Silverstein, M. S., *et al.*, *Polymer*, **42** (2001) 4299.
221.Olcaytug, F., *et al.*, *Thin Solid Films*, **67** (1980) 321.
222.Powell, M. J., *et al.*, *Appl. Phys. Lett.*, **38** (1981) 794.
223.Rieder, G. and Olcaytug, F., *Thin Solid Films*, **89** (1982) 95.
224.Adams, A. C., *Silicon Nitride and Other Insulator Films*, in *Plasma Deposited Thin Films*, eds. Mort, J. and Jansen, F. (CRC Press, Boca Raton, 1986), Chap.5.
225.Uchida, H., *et al.*, *Jpn. J. Appl. Phys.*, **30** (1991) 3691.
226.Shi, F. F., *Surf. Coat. Technol.*, **82** (1996) 1.
227.Biederman, H. and Martinu, L., *Plasma Polymer-Metal Composite Films*, in *Plasma Deposition, Treatment, and Etching of Polymers*, ed. d'Agostino, R. (Academic Press, Boston, 1990), Chap.4.
228.Tyczkowski, J., *et al.*, in *9th Inter. Symp. on "Plasma Chemistry"* (Pugnachiuso (Italy), 1989), IUPAC, 1224.
229.Canet, P., *et al.*, *J. Appl. Phys.*, **72** (1992) 2423.
230.Flouttard, J. L., *et al.*, *J. Appl. Phys.*, **70** (1991) 798.
231.Feurer, E. and Suhr, H., *Appl. Phys. A*, **44** (1987) 171.
232.Ianno, N. J. and Plaster, J. A., *Thin Solid Films*, **147** (1987) 193.
233.Oehr, C. and Suhr, H., *Appl. Phys. A*, **49** (1989) 691.
234.Etspüler, A. and Shur, H., *Appl. Phys. A*, **48** (1989) 373.
235.Oehr, C. and Suhr, H., *Appl. Phys. A*, **45** (1988) 151.

236.Oehr, C. and Suhr, H., *Thin Solid Films*, **155** (1987) 65.
237.Sadhir, R. K., *et al.*, *Thin Solid Films*, **97** (1982) 17.
238.Kim, Y. T. and Lee, C. W., *J. Appl. Phys.*, **76** (1994) 542.
239.Guyard, C. and Despax, B., *Surf. Coat. Technol.*, **116/119** (1999) 638.
240.Yamano, M., *et al.*, *IEEE Trans. Consumer Electronics*, **CE-31** (1985) 39.
241.Take, H., in *Proc. International Workshop on Active-Matrix Liquid-Crystal Display in conjuction with IDW'96* (Kobe (Japan), 1996), Japan Soc. Appl. Phys., 369.
242.Rolland, A., *et al.*, *J. Electrochem. Soc.*, **140** (1993) 3679.
243.Morita, T., *et al.*, *Mat. Res. Soc. Symp. Proc.*, **345** (1994) 71.
244.Feng, M. S., *et al.*, *J. Electrochem. Soc.*, **141** (1994) 1040.
245.Uchikoga, S., *et al.*, *J. Appl. Phys.*, **76** (1994) 8150.
246.Kim, J. H., *et al.*, *J. Appl. Phys.*, **76** (1994) 7601.
247.Osada, Y. and Mizumoto, A., *J. Appl. Phys.*, **59** (1986) 1776.
248.Osada, Y., *et al.*, *J. Appl. Phys.*, **64** (1988) 1476.
249.Sadhir, R. K. and K.F. Schoch, J., *Thin Solid Films*, **223** (1993) 154.
250.Tanaka, K., *et al.*, *J. Appl. Phys.*, **70** (1991) 5653.
251.Cruz, G. J., *et al.*, *Synth. Met.*, **88** (1997) 213.
252.Takeda, S., *Thin Solid Films*, **343/344** (1999) 313.
253.Inagaki, N., *Plasma Surface Modification and Plasma Polymerization* (Technomic Publ., Lancaster, 1996).
254.Sadhir, R. K. and Sanjana, Z. N., *J. Mater. Sci.*, **26** (1991) 4261.
255.Takeda, S., *Jpn. J. Appl. Phys.*, **38** (1999) 909.
256.Takeda, S., *Thin Solid Films*, **281/282** (1996) 539.
257.Morita, S. and Hattori, S., *Applications of Plasma Polymers*, in *Plasma Deposition, Treatment, and Etching of Polymers*, ed. d'Agostino, R. (Academic Press, Boston, 1990), Chap.6.
258.Poll, H. U., *et al.*, *Surf. Coat. Technol.*, **59** (1993) 365.
259.Catherine, Y. and Zamouche, A., *Plasma Chem. Plasma Proc.*, **5** (1985) 353.
260.Ishii, K., *et al.*, *J. Appl. Phys.*, **76** (1994) 5418.
261.Wydeven, T., *Appl. Opt.*, **16** (1977) 717.
262.Knolle, W. R., *Thin Solid Films*, **168** (1989) 123.
263.Yen, J. I. and Lee, S. C., *J. Appl. Phys.*, **79** (1996) 656.
264.Demichelis, F., *et al.*, *J. Appl. Phys.*, **79** (1996) 1730.
265.Catherine, Y. and Turban, G., *Thin Solid Films*, **60** (1979) 193.
266.Basa, D. K. and Bose, M., *Thin Solid Films*, **298** (1997) 211.
267.Kimock, F. M. and Knapp, B. J., *Surf. Coat. Technol.*, **56** (1993) 273.

268.Lettington, A. H., *Carbon*, **36** (1998) 555.
269.Grill, A., *Thin Solid Films*, **355/356** (1999) 189.
270.Inagaki, N., *et al.*, *J. Polym. Sci.: Polym. Lett. Ed.*, **18** (1980) 765.
271.Martinu, L. and Biederman, H., *Vacuum*, **35** (1985) 171.
272.Curtis, B. J., *et al.*, *J. Vac. Sci. Technol. A*, **4** (1986) 70.
273.Rats, D., *et al.*, *Surf. Coat. Technol.*, **111** (1999) 220.
274.Tien, P. K., *et al.*, *Appl. Opt.*, **11** (1972) 637.
275.Tien, P. K., *et al.*, *Appl. Opt.*, **12** (1973) 1909.
276.Tien, P. K., *et al.*, *Appl. Phys. Lett.*, **24** (1974) 547.
277.Tien, P. K., *Rev. Mod. Phys.*, **49** (1977) 361.
278.Wirges, W., *et al.*, *Proc. SPIE – the Inter. Soc. for Optical Engineering*, **2213** (1994) 303.
279.Zelikson, M., *et al.*, *Appl. Phys. Lett.*, **61** (1992) 1664.
280.Gorecki, C., *Optics & Lasers in Engineering*, **33** (2000) 15.
281.Kruangam, D., *et al.*, *J. Non-Cryst. Solids*, **227/230** (1998) 1146.
282.Das, S. R., *Thin Solid Films*, **119** (1984) 31.
283.Hamakawa, Y., *phys. stat. sol. (b)*, **194** (1996) 15.
284.Green, M. A., *Progress in Photovoltaics: Research & Applications*, **9** (2001) 123.
285.Hamakawa, Y. and Okamoto, H., *Jpn. Annu. Rev. in Electronics, Computers and Telecommunications*, **6** (1983) 182.
286.Tsuda, S., *et al.*, *Appl. Surf. Sci.*, **113/114** (1997) 734.
287.Rech, B. and Wagner, H., *Applied Physics A*, **69** (1999) 155.
288.Yang, J. and Guha, S., *Appl. Phys. Lett.*, **61** (1992) 2917.
289.Fischer, R., *Luminescence in Amorphous Semiconductors*, in *Amorphous Semiconductors*, ed. Brodsky M. H. (Springer Verlag, Berlin, 1979), Chap.6.
290.Li, G., *et al.*, *Appl. Phys. Lett.*, **62** (1993) 1582.
291.Kaiser, M., *et al.*, *Surf. Coat. Technol.*, **105** (1998) 165.
292.Oestereich, T., *et al.*, *Appl. Phys. Lett.*, **56** (1990) 446.
293.Shuto, K., *et al.*, *Electron. Lett*, **29** (1993) 139.
294.Tessler, L. R. and Zanatta, A. R., *J. Non-Cryst. Solids*, **277/300** (1998) 399.
295.Terukov, E. I., *et al.*, *J. Non-Cryst. Solids*, **227/230** (1998) 488.
296.Tyczkowski, J., *et al.*, *Surf. Coat. Technol.*, **116/119** (1999) 629.
297.Pankove, J. I. and Carlson, D. E., *Appl. Phys. Lett.*, **29** (1976) 620.
298.Munekata, H. and Kukimoto, H., *Appl. Phys. Lett.*, **42** (1983) 432.
299.Shimizu, H., *et al.*, *J. Non-Cryst. Solids*, **137/138** (1991) 1275.
300.Kruangam, D., *et al.*, *J. Non-Cryst. Solids*, **164/166** (1993) 809.

301.Janssen, R., *et al.*, *J. Non-Cryst. Solids*, **227/230** (1998) 1151.
302.Robertson, J. and Milne, W., *J. Non-Cryst. Solids*, **227/230** (1998) 558.
303.Silva, S. R. P., *et al.*, *Appl. Phys. Let.*, **71** (1997) 1477.
304.Doblhofer, K., *Electrochim. Acta*, **25** (1980) 871.
305.Dautartas, M. F. and Evans, J. F., *J. Electroanal. Chem.*, **109** (1980) 301.
306.Nowak, R., *et al.*, *Anal. Chem.*, **52** (1980) 315.
307.Sadhir, R. K. and K.F. Schoch, J., *Chem. Mater.*, **8** (1996) 1281.
308.Ogumi, Z., *et al.*, *J. Electrochem. Soc.*, **135** (1988) 2649.
309.Takehara, Z., *et al.*, in *Proc. "Symp. on High Power, Ambient Temperature Lithium Batteries"* (Phoenix, AZ, 1992), Electrochem. Soc., 179.
310.Ogumi, Z., *et al.*, in *Proc. "Symp. on Lithium Polymer Batteries"* (San Antonio, TX, 1997), Electrochem. Soc. 4-9.
311.Abeles, B. and Tiedje, T., *Phys. Rev. Lett.*, **51** (1983) 2003.
312.Munekata, H., *et al.*, *J. Non-Cryst. Solids*, **59/60** (1983) 1167.
313.Tiedje, T., *et al.*, *J. Non-Cryst. Solids*, **66** (1984) 345.
314.Kakalios, J., *et al.*, *J. Non-Cryst. Solids*, **66** (1984) 339.
315.Tiedje, T., *et al.*, *Phys. Rev. Lett.*, **54** (1985) 2545.
316.Nitta, S., *et al.*, *Optoelectronics - Devices and Technologies*, **9** (1994) 355.
317.Nitta, S., *et al.*, *Superlattices and Microstructures*, **15** (1994) 483.
318.Raikh, M. E., *et al.*, *Phys. Rev. B*, **41** (1990) 7701.
319.Bernhard, N., *et al.*, *J. Non-Cryst. Solids*, **137/138** (1991) 1103.
320.Drüsedau, T. P., *et al.*, *J. Non-Cryst. Solids*, **198/200** (1996) 829.
321.Hajto, J., *et al.*, *J. Non-Cryst. Solids*, **137/138** (1991) 499.
322.Snell, A. J., *et al.*, *J. Non-Cryst. Solids*, **137/138** (1991) 1257/1262.
323.Hu, J., *et al.*, *J. Non-Cryst. Solids*, **227/230** (1998) 1187.
324.Kazimierski, P. and Tyczkowski, J., *Mater. Sci. (Medzagotyra)* **7** (2001) 3.
325.Kazimierski, P. and Tyczkowski, J. Surf. Coat. Technol., 174/175 (2003) 770.

CHAPTER 6

PULSED PLASMA POLYMERIZATIONS

RICHARD B. TIMMONS AND ANDREW J. GRIGGS
Department of Chemistry and Biochemistry
University of Texas at Arlington
Arlington, USA

6.1. Introduction

In this chapter, the use of variable duty cycle pulsed discharges during plasma polymerizations are examined. This review centers on discussion of phenomenological aspects of pulsed plasmas as they relate to film chemistry control during polymer formation. The more physically oriented aspects of pulsed plasma discharges have been treated by others [1,2]. Likewise, the uses of pulsed discharges simply to minimize substrate heating [3,4]; or as a means of effecting reactive ion etch rates [5,6], or to reduce powder formation [7,8] are not included in this review. Instead our emphasis is on the film formation processes which have recently been shown to be strongly influenced by intermittent operation of plasma discharges [9]. As documented below, the pulsed plasma technique provides an especially convenient route to wide ranging film chemistry controllability during plasma polymerizations. Also, the question of whether pulsed plasma polymerizations can provide unique film compositions, unavailable under continuous wave discharge conditions, is examined.

6.1.1. Controlling Film Chemistry during Plasma Polymerizations

It is now well established that plasma polymerization represents an unusually convenient and versatile coatings technology. There is obviously much inherent appeal for a one-step, solventless process, which provides uniform, conformable, pinhole free coatings of controllable thickness. Additional attractive features of this approach are that the plasma technique permits use of an exceptionally wide range of monomers (i.e. virtually any volatile compound) and can provide unique polymeric structures unattainable by any other available technique.

Undoubtedly, the combination of factors cited above accounts for the continued rapid rise in popularity of plasma polymerizations for surface modifications. Nevertheless, a point of continuing discussion and analysis in plasma polymer science is that relating to the issue of compositional controllability of polymeric films generated via this approach. Originally this compositional issue focused predominately on synthesis of unique, highly cross-linked polymers produced under relatively high power conditions. However, in recent times, an increased interest has clearly developed in providing less cross-linked, more highly functionalized films. This development is largely applications driven in that surface molecular tailoring of various materials has become a topic of increased relevancy. The performance of many devices and materials are frequently critically dependent on the composition of the outermost layers of the substrate employed. Examples include biocompatibility of many medical implants; development of improved sensors, catalysts and permselective membranes; biologically non-fouling surfaces; improved optical coatings; etc. These examples, as well as many others, illustrate the need for more exact surface molecular tailoring capability. Other surface structuring techniques such as self-assembled monolayers (SAMs) and Langmuir-Blodgett (L-B) films represent extremely useful techniques to be employed in fundamental studies of molecule-surface interactions. However, these methods are less practical for large-scale applications nor are they useful with respect to long-term applications, particularly those involving deployment in harsh environments or use under abrasive conditions. Clearly a coatings approach, which retains the inherent advantages of plasma polymerizations while providing more exact film chemistry control, will have potential appeal in many applications.

In light of the above considerations, significant effort has been devoted to acquiring improved film chemistry controllability during plasma depositions. The most extensive early work in this area was the pioneering studies of Yasuda and co-workers [10,11]. Yasuda introduced the composite variable W/FM (where W is power input, F and M are the monomer flow rate and molecular mass) as a particularly useful method of correlating plasma polymer film structures with key reaction parameters. The W/FM term provides a quantitative measure of the energy input per unit mass of monomer. In general, lower values of W/FM correlate well with less extensive fragmentation of the monomer and thus formation of less highly cross-linked films. Among the many additional noteworthy reports on compositional control of film chemistry, analyzed in terms of operational parameters, are those of Clark et al [12,13] with fluorocarbons and those of Inagaki et al [14] with various functionalized monomers.

Although W/FM considerations are clearly important in terms of helping rationalize plasma polymer compositions, the range of film chemistry controllability accessible under CW conditions is, in many cases, quite limited. These restrictions are imposed at the high power input extremes by increased domination of ablation processes in lieu of net film formation, whereas a low power input limitation is usually imposed by formation of oily or powdery films. For example, it is frequently not possible to generate films having a high percent retention of monomer reactive functional groups under CW conditions. In light of these well-known limitations, a number of specialized techniques have been utilized under CW conditions to promote increased retention of monomer functionality in the resultant polymeric films. For example, location of substrates upstream or downstream of the plasma discharge zone results in formation of significantly less cross-linked films [15]. In general, substrate location outside the actual plasma discharge zone usually results in much lower film formation rates and, in the case of larger substrates, relatively poor film thickness control. Additionally, film adhesion to the substrate can also be a problem for samples located only in the vicinity of the discharge zone. Another interesting approach to film chemistry control under CW conditions is to utilize lower substrate temperatures, as achieved by controlled substrate cooling during deposition processes [16]. Again, lower substrate temperatures reduce monomer fragmentation and can provide increased incorporation of monomer reactive groups in the film [16,17,18]. However the overall controllability appears to be fairly limited in that the lower temperature extremes are frequently restricted by concerns of simple monomer condensation on the targeted substrates.

The above brief review of film chemistry control issues has been cited to illustrate the continuing interest in achieving improved film chemistry compositional controllability during polymer film formation under plasma conditions.

6.1.2. The Pulsed Plasma Approach

Conceptually, it is easy to visualize that the use of pulsed plasma discharges may provide a more selective chemistry during plasma polymerization. Under pulsed conditions, reactive species created during the brief plasma on periods are subsequently consumed during the plasma off periods. It is believed that a variety of reactive species (e.g. ion and neutral radicals, ionized and neutral atoms, etc.) are produced during the plasma on periods. The more reactive of these species, and thus the more indiscriminate in terms of chemical processes, should be consumed more rapidly during the initial moments of the off periods. In the case of ionic species an important step during the plasma off period may be simply neutralization via electron capture. This would then leave the major role of chain propagation processes during the plasma off times to relatively more selective neutral radical and atomic species. Additionally, the overall free radical concentrations during the plasma off periods will be less than those present under plasma on conditions. Thus, competitive processes involving radical-radical recombination versus radical-monomer reactions will be tilted increasingly towards radical-monomer as the plasma off times are increased.

In addition to the above, admittedly simplistic, view of plasma dynamics, there are several additional plasmas parameters that should be considered in contrasting CW and pulsed operational modes. For example, it is well established that plasma discharges are accompanied by the production of relatively intense photon fluxes involving light emission from electronically excited states of the various gaseous species present in the reactor. Photon wavelengths can extend from the visible down to the vacuum UV region. Some of these energetic UV photons will be absorbed by both the gas phase species as well as the polymeric film formed by the discharge. The photon energies involved are sufficiently energetic to dissociate even relatively strong bonds and have been shown to play an important role in the overall film chemistry composition occurring under typical CW plasma deposition conditions [19,20]. These photochemical induced reactions can thus add further to the relatively indiscriminate array of chemical reactions taking place under CW operation, in many cases contributing to ultimately more highly cross-linked polymer films. In contrast, such added photochemically driven processes are minimized under pulsed conditions. Electronically excited species have lifetimes of fractions of microseconds, whereas the plasma off times typically employed are in range of many milliseconds. Thus any film formation during plasma off times will take place in the absence of perturbing additional photochemically driven reactions.

Other plasma dynamic factors that should be noted in contrasting CW and pulsed plasma operational modes include substrate heating and substrate self-bias. In general, substrate heating occurs to a lesser extent under pulsed operation. This simply reflects the fact that the average power input under pulsed conditions is frequently quite low compared to that involved during CW operations (see below). Additionally, the relatively long plasma off times permit thermal energy transfer from the substrate to the surroundings. As noted earlier, low substrate temperatures promote less monomer fragmentation in the growing films. If the film formation processes involve competitive reactions having different activation energies, the lower substrate temperatures will favor reactions with lower activation energies, thus introducing increased reaction selectivity.

In terms of self-bias considerations, it is well known that substrates acquire a negative-bias potential immediately upon initiation of the discharge. This simply reflects the greater mobility of electrons compared with the more massive positive ions. Subsequently, the positive ions are accelerated, via attractive potential interactions, to impact the substrate with relatively large momenta. These high-energy collisions are believed to account, at least in part, for the occurrence of ablation processes as well as relatively non-selective bond breaking-bond formation reactions. Under pulsed conditions one anticipates rapid decay of self-bias potentials when the discharge is extinguished. Thus film formation occurring during plasma off times will do so in the absence of high-energy impacts involving electrostatic attraction of positively charged particles to a negatively charged substrate. An interesting aspect of films generated under pulsed conditions is, as noted later, the generally progressively smoother films that are obtained with decreasing plasma duty cycles. It is assumed that the decreasing mean film surface roughness with decreasing plasma duty cycles reflect, at least in part, the decreasing relative importance of high energy surface collisions.

Although to date it has not been possible to quantitate the relative importance of each of the individual dynamic factors identified above, it is important to note that all of these parameters operate in a fashion to provide increased reaction selectivity under pulsed versus CW conditions. As detailed below, this increased reaction selectivity under pulsed conditions has now been confirmed experimentally in a number of laboratories, involving a wide range of monomers.

6.2. Experimental Considerations

In general, operating plasma discharges under pulsed conditions is easily achieved within the context of typical or existent plasma discharge systems. At the present time, the overwhelming majority of studies involving plasma polymerizations center on the use of radio frequency (RF) discharges, and to a lesser extent microwave driven systems. In both cases, a pulsed operational mode is achieved by simple insertion of a pulse generator into the electrical system [21]. The pulse operation can typically be operated to provide separately controlled plasma on and off times. An oscilloscope can be employed to tune the electrical circuit so as to minimize reflected power under RF pulsed operation.

In principle a wide range of plasma on and off times, extending from about 0.1 μs to many seconds, are available with relatively inexpensive components. Square wave pulsed modes are easily obtained in that typical voltage rise and decay times are less than 0.1 μs. In practice, it has been observed that the most effective plasma on and off times, in terms of affecting plasma film formation dynamics, lie in the time domains of a few microseconds to tens of milliseconds. Too short of a plasma on time results in inefficient production of reactive species and thus low deposition rates whereas longer plasma on times (regardless of plasma off time) usually result in formation of films more characteristic of those achieved under CW conditions. Based on publications to date, it appears that plasma on and off times ranging from 10 μs to perhaps 500 ms cover the most interesting time domain, although the exact values to be utilized in a given situation are clearly very much monomer dependent. The actual on and off times employed are usually reported in terms of the duty cycle, which is defined as the ratio of plasma on to plasma on + plasma off times. However, it is simpler and, in a way, more informative to provide simply the ratio of on and off times in terms of the pulse width (i.e. ms or μs). For example, studies have shown that films produced at the same duty cycle but employing different pulse widths result in different film compositions and deposition rates [22].

6.2.1. Power Input Considerations

An important consideration in pulsed plasma discharges is that of equivalent average power ($\langle P \rangle$), defined as:

$$\langle P \rangle = \text{Duty Cycle} \times \text{Peak Power} \tag{6.1}$$

Here peak power represents power input during the plasma on portion of the on/off cycle. In effect, it has been observed that although relatively high peak powers may frequently be employed under pulsed conditions (i.e. relative to those typically used under CW conditions), the equivalent powers are usually low, especially when averaged over relatively long plasma off times. Of more significance, in terms of power input considerations, is the fact that the pulsed operational mode permits high quality film formation under extremely low power average inputs. Frequently this involves equivalent power inputs that are simply unavailable under comparable CW conditions. This reflects the fact that CW discharges under the same monomer pressure and flow conditions cannot be maintained for the same low equivalent power input. In contrast, the pulsed mode can be readily maintained since the brief plasma on times can be operated at higher power inputs. The importance of extending effective operation to these low average power regimes has now been underscored by synthesis of films containing relatively high surface densities of extremely reactive groups [23,24] as well as by syntheses of highly linear polymers (by plasma considerations) [25]. These films were achieved under ultra low power plasma power input conditions, which were simply unavailable under CW conditions in the same reactor for the given monomer. The pulsed plasma technique extends considerably the range of the W/FM parameter described earlier at the low value end of the scale. This low value extension introduces some unexpected film compositional changes as documented later in this chapter.

6.2.2. Film Formation Rates

By definition, film deposition times under pulsed condition include significant periods wherein the active electrically conductive plasma discharge is not present. Thus, the usual film formation rates involving film thickness per unit time, as quoted for CW studies, is not particularly informative for pulsed mode operations. Instead, it is more meaningful to present film formation rates in terms of energy considerations (i.e. thickness per joule). This representation in effect provides a useful measure of the overall energy efficiency of film formation processes. For example, interesting results are obtained for a series of runs carried out at a fixed plasma on time and peak power input when film thickness per joule are computed for various plasma off times. The exact shapes of the resulting plots have been observed to be strongly monomer dependent. However, a common feature observed with virtually all monomers studied to date under pulsed plasma conditions is the presence of an off time regime in which dramatic increases in film formation efficiency are observed with increasing plasma off times (see below). In general, this increasing film formation energy efficiency with increasing plasma off times has been interpreted as evidence for significant film formation during the plasma off times [21]. In light of considerations previously discussed, vis-a-vis differences in key plasma dynamic factors in contrasting pulsed versus CW operations, the occurrence of film formation under relatively quiescent plasma off periods is obviously important in analyzing film chemistry controllability issues.

6.2.3. Mean Roughness of Polymer Films Deposited under Pulsed Operation

An interesting physical feature of polymeric films deposited under pulsed conditions is that relating to average surface roughness. In general, it has been observed that for a given monomer, the average surface roughness of films decreases with decreasing plasma duty cycle. For example, the mean roughness of films obtained from pulsed plasma polymerization of acryloyl chloride decreased progressively from 6.7 nm to 0.24 nm with decreasing plasma duty cycles [24]. Comparable changes in average surface roughness have been obtained with many other monomers. Thus, whereas plasma polymer films obtained under CW conditions have usually been characterized as having relatively smooth surface morphology, the low duty cycle pulsed plasmas can be typically an order of magnitude smoother than those obtained under CW conditions. As noted earlier, the increased smoothness of the pulsed plasma films most probably reflects the gentler deposition conditions involved and, in particular, absence of high-energy molecule-surface impact collisions, thus minimizing ablation processes. In this connection, one should note recent studies of Martinu et al in which significantly lower ion energies are observed under pulsed plasma conditions with the mean ion energies increasing with increasing duty cycle [26].

Although it is tempting to generalize that film roughness always decreases with decreasing duty cycles, there are at least two recent reports in which dramatically increased film roughness are observed under pulsed conditions [27,28,29]. These studies involve generation of super hydrophobic films from fluorocarbon monomers, in which the ultra hydrophobicity is clearly associated with the formation of unusual textured surface morphologies (see below). It is important to keep these more recent findings in mind when considering surface roughness of films produced under pulsed conditions.

6.2.4. Film Adhesion and Film Stability

As noted above, ultra low average power input can be employed under pulsed plasma polymerization conditions. However, the use of low power input raises rather obvious questions of film adhesion and overall stability for polymers deposited under such conditions. Although high quality films are readily achieved under pulsed conditions, the overall film adhesion to a given substrate is decreased as the average power employed during film deposition is decreased. This is not surprising given the fact that the low power input conditions represent an overall more gentle reaction condition. Grafting of the plasma-generated polymer to the substrate would be less extensive under low duty cycle conditions. In some cases, one may be forced to compromise between maximizing a given polymeric chemical composition versus strength of film adhesion [30]. The actual adhesion strength of these films is a function of both monomer identity and substrate composition. For example, the adhesion strengths of plasma deposited films on polymeric substrates are generally significantly higher than those observed on silicon or metal substrates. A particularly useful technique in improving film adhesion strength of pulsed plasma synthesized polymers is use of a gradient layering technique [31]. In this approach, the initial deposition is carried out under CW or high duty cycle conditions, with the duty cycle being subsequently decreased to provide a final exterior layer of desired composition. This layering technique helps transition between the dissimilar chemistry of the substrate and the polymeric films via initial strong grafting of films of intermediate compositions. Alternatively, the use of a sub-layer film from a different monomer can be employed to bridge the compositional differences between substrate and outermost film composition.

An additional aspect of pulsed plasma film properties is the relatively facile swelling of these films when immersed in liquids. Subsequent removal of the films from the solvent results in retention of basically the same chemical composition as before the initial immersion but the remaining film is decidedly thinner than the original as deposited structure [32]. Continued solvent immersion and drying cycles of these films produced no further changes in film thickness. The lower the plasma duty cycle employed during film formation, the larger the degree of film shrinkage [32]. These observations have been interpreted in terms of the presence of low molecular weight oligomers (or even possibly individual monomer molecules) being trapped in the film and which are not covalently bonded to the overall polymeric matrix. Immersion of the film in solvent results in extraction of these oligomers, or weakly bound components, with a resultant decrease in film thickness upon drying. Direct evidence for this extraction process has been reported [33].

Further support for the presence of loosely bound species in the polymeric matrix under low duty cycle pulsed conditions has been obtained from thermal analysis studies. For example, TGA analysis of several films revealed significant weight loss at relatively low temperatures [34]. However, if these films are initially subjected to a thermal annealing treatment then the thermal gravimetric analysis of these films reveals a dramatically improved thermal stability [34]. The thermal annealing step is accompanied by a decrease in film thickness, with the extent of film thickness decrease being more pronounced for films produced at lower duty cycles.

Overall, characteristics such as solvent induced film swelling and subsequent thickness loss, as well as thermal stabilities of the pulsed plasma films deposited at low duty cycles appear to differ significantly from films produced at higher duty cycles or CW conditions. Whether these swelling or heat loss characteristics represent significant disadvantages for the pulsed plasma generated films is obviously dependent upon the particular application involved. For example, it is conceivable that the film swelling properties may be utilized to advantage in some applications.

6.3. Results

In light of the various dynamic distinctions offered by pulsed plasmas, as noted above, it is surprising to observe that it is only during relatively recent times that this technique has been commonly employed to control film chemistry during plasma polymerizations. Although the use of pulsed plasmas have been known for a long time, the first mention of this technique with respect to questions concerning plasma polymer compositions appears to have been in 1977 [35]. In this ESR study, the relative densities of free radicals trapped in CW generated films were contrasted with the spin densities obtained from the same monomers under pulsed conditions. All pulsed experiments were carried out at the same duty cycle and the CW and pulsed runs were carried out at the same peak powers. Interestingly, the relative density of free radicals trapped in the films under the two different operational modes was found to be strongly dependent on the nature of the monomer. Some monomers provided higher radical concentrations under pulsed conditions whereas other monomers revealed higher radical concentrations under CW condition [35]. Two years later, a paper by Shen et al reported studies comparing film compositions of pulsed and CW plasma polymerization of fluoroethylene monomers [36]. Perhaps the authors stated conclusion that essentially the same film compositions were obtained under pulsed and CW conditions accounts for the fact that there was no further study of pulsed plasma polymerization during the '80s. It was not until the study of Savage et al in 1991 that large-scale variations in film composition with changes in plasma duty cycles were clearly demonstrated [9]. In this paper, involving the monomer perfluoroprpylene oxide (C_3F_6O), wide ranging progressive changes in plasma film compositions with sequential changes in plasma duty cycles (all other plasma variables being held constant) were reported for the first time. The results from this study, along with selected examples of film chemistry controllability from a number of other monomers are provided below.

6.3.1. Some Examples of Film Chemistry Control under Pulsed Plasma Conditions

The following examples are provided to illustrate the film chemistry variations observed with duty cycle changes, all other plasma variables being held constant. The examples are chosen to encompass a range of monomers and thus illustrate the generality of this approach. Results obtained from the monomer C_3F_6O are shown in Fig. 6.1 [9,21]. Here C (1s) XPS spectra are arranged in order of decreasing duty cycles during deposition (top to bottom). Subsequent studies of C_3F_6O in other laboratories have confirmed the wide-ranging variation in film chemistries as a function of the plasma duty cycle [37]. These latter studies have provided additional valuable information on the detailed structure of these films using NMR [38] and EXAFS [39] techniques. The resolved peak assignments of the C (1s) XPS spectrum have been made in accord with generally accepted binding energies for the various carbon-fluorine functional groups [40]. As shown in Fig. 6.1, the fluorocarbon films exhibit a distinctly less cross-linked structure as the plasma duty cycle employed is decreased. In particular, the films produced at the lowest plasma duty cycles are dominated by the $-CF_2-$ functionality, suggesting formation of a highly linear polymer. This linear-like polymer is further confirmed by FT-IR studies of these films in which the broad C-F stretching-band ~1250 cm^{-1} obtained under high duty cycles is transformed to a much sharper band, incorporating a well-resolved doublet structure, at lower duty cycles (Fig. 6.2). In fact, the IR spectrum of the low-duty cycle produced film is essentially identical in nature that of conventionally polymerized tetrafluoroethylene (Teflon) and can be contrasted with the typical broad C-F stretching band absorptions observed in CW studies of numerous fluorocarbon monomers [11].

An interesting important feature of the C_3F_6O pulsed plasma polymerization is the virtual complete absence of oxygen in the polymeric films, despite the presence of oxygen in the starting monomer. In fact, no measurable oxygen atom content was noted in films deposited at the lowest duty cycles [9]. Furthermore, in contrast with most films produced under CW conditions, essentially no post-plasma oxidation of the low duty cycle films was observed, suggesting the relative absence of trapped free radical species in these films. This observation is consistent with results from Gleason and co-workers of pulsed plasma polymerization of C_3F_6O in which a sharp reduction in free radical films content is observed with decreasing duty cycle during deposition [41].

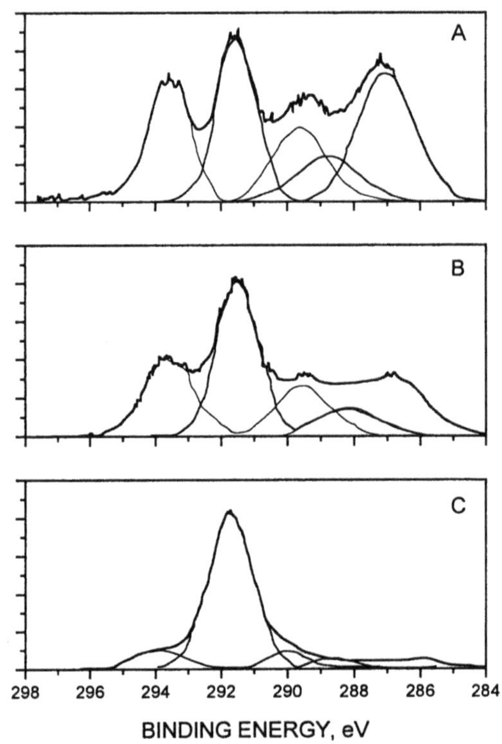

Figure 6.1. C(1s) X-ray photoelectron spectra obtained from plasma depositions of C_3F_6O on silicon substrates as a function of RF duty cycle. Duty cycles expressed as ratio of on:off times (in ms): (A) 10/60, (B) 10/100, and (C) 10/200 [9].

Similar results to those obtained with C_3F_6O were observed for the pulsed plasma polymerization of the cyclic monomer perfluoro-2-butyl-tetrahydro-furan ($C_8F_{18}O$) [22]. Again, the progressive variation from cross-linked to linear perfluorocarbon poly-

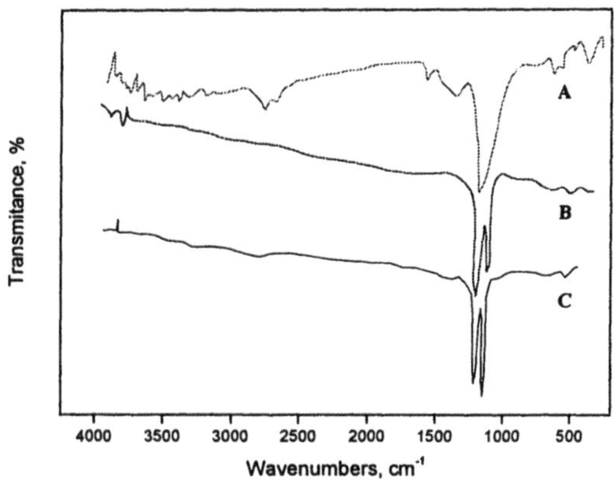

Figure 6.2. FT-IR transmission spectra of plasma deposited films of C_3F_6O on KCl disks as a function of on:off times (in ms): (A) 10/20, (B) 10/200, and (C) 10/400 [9].

mer was obtained with decreasing plasma duty cycle. Oxygen atom incorporation in the films was essentially negligible (<1%) and decreased with decreasing plasma duty cycles employed during film formation. No post plasma oxidation was observed in these films. Overall, the range of films chemistry variations obtained under pulsed plasma conditions greatly exceeds those reported by Yasuda and Wang [16] from the same monomer polymerized under CW conditions involving controlled variations of substrate temperature (from –25 °C to 75 °C) along with changes in the W/FM composite variable by a factor of 17.8.

In more recent years, the pulsed technique has been employed during plasma polymerization of a range of functionalized monomers. An illustrative

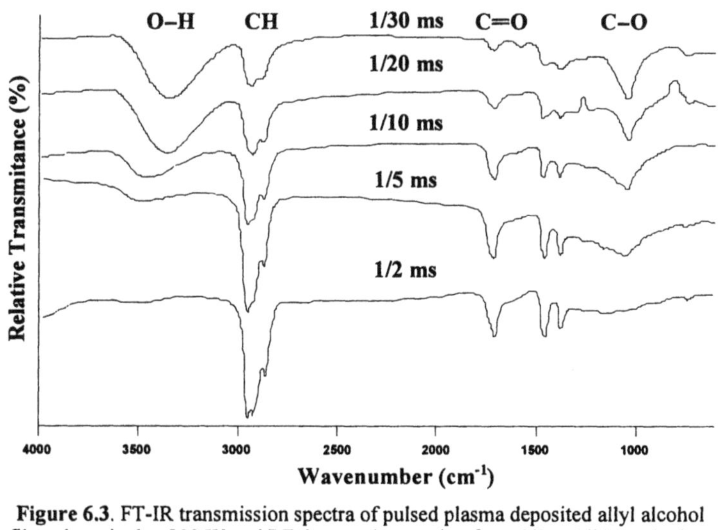

Figure 6.3. FT-IR transmission spectra of pulsed plasma deposited allyl alcohol films deposited at 300 W and RF duty cycles ranging from 1/2 to 1/30 (ms/ms) as shown [42].

example of the progressive film chemistry variations observed with changing duty cycles are results obtained with allyl alcohol [42]. Fig. 6.3 shows FT-IR transmission spectra of films obtained from allyl alcohol under different duty cycles, arranged in order of increasing duty cycles (top to bottom). The progressively controlled film variations with duty cycle changes are most easily recognized by simple comparison of the relative intensities of the O-H (~3400 cm^{-1}) and C-H (~2900 cm^{-1}) vibrational bands.

Clearly, a dramatically increased percentage of –OH groups in the starting monomer are retained in the plasma generated polymers as the duty cycle is decreased. The hydrophilic character of these films changes in an orderly fashion as shown by water contact angles which varied progressively from 87° to 50° as the ratio of plasma on/off times during films formation was changed from 1/2 (ms) to 1/30 (ms). Also notable, in terms of the spectra shown in Fig. 6.3, is

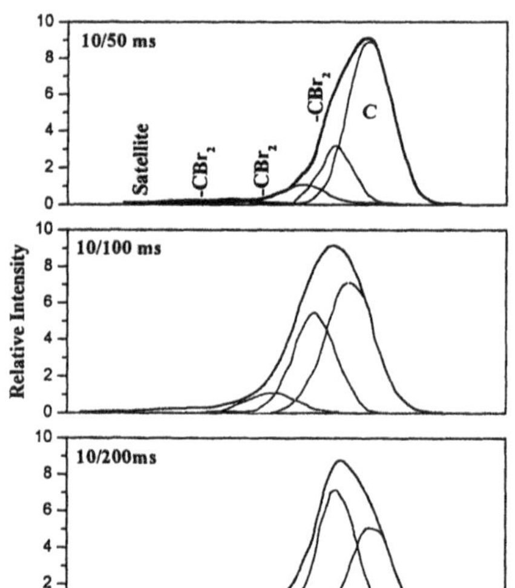

Figure 6.4. High Resolution C(1s) spectra of films obtained from the pulsed plasma polymerization of CH_2Br_2 arranged in order of decreasing duty cycle from top to bottom. All runs were carried out at a peak power of 300 W and constant 10 ms plasma on time [45].

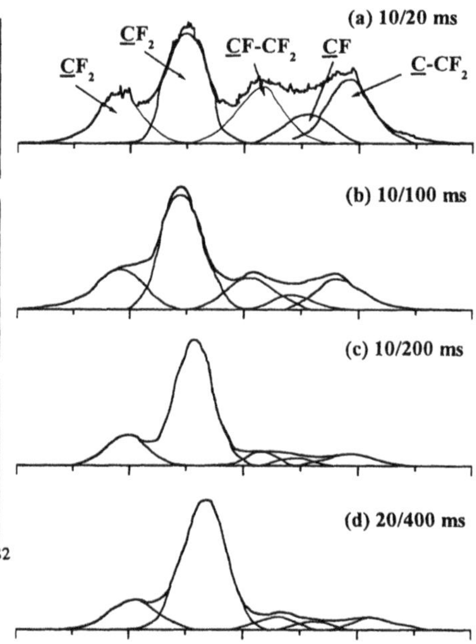

Figure 6.5. High Resolution C(1s) spectra of fluorocarbon films obtained from the pulsed plasma polymerization of perfluorohexane monomer arranged in order of decreasing plasma duty cycles during film deposition reading from top to bottom. All films were obtained at a peak power of 200 W; (a) 10/20 ms, (b) 10/200 ms, (c) 10/400 ms [29].

the sharply reduced relative contribution of the absorption band at ~1700 cm^{-1}, reading bottom to top. This band presumably reflects the presence of a carbonyl group created during the polymerization process. The reduction, in some cases elimination, of new functional groups created under plasma discharge conditions in films obtained under low duty cycle conditions is characteristic of the pulsed plasma technique.

Many other equally dramatic examples of film compositional changes with duty cycle variations to those shown above can be cited to document the generality of the pulsed plasma technique with respect to film chemistry controllability issues. These studies include retention of even extremely reactive functional groups such as anhydrides [23], acid chlorides [24] and epoxides [43] (see below). Other studies include work with the orgametallic monomers tetramethyltin [44] and ferrocene [33]. Although the majority of studies have involved unsaturated monomers, it is of interest to note that the pulsed plasma technique provides a high level of film chemistry controllability with saturated monomers. For example, XPS spectra of films obtained from the saturated monomers dibromomethane [45] and perfluorohexane [29] at various plasma duty cycles are shown in Figures 6.4 and 6.5, respectively. Clearly large scale compositional changes are occurring with duty cycle variations during film depositions from the saturated precursors.

6.3.2. Film Formation Rates

As noted above, in view of variable plasma off times, film deposition rates under pulsed conditions are better analyzed in terms of overall energy input efficiencies. One way to accomplish this goal is to compute the film thickness achieved per unit energy input (i.e. per Joule) as a function of the plasma duty cycle or average power employed. The results obtained are interesting and indicate that the films thickness/J versus duty cycle dependence is very much a function of the monomers employed. Illustrative examples of several such plots are shown in Fig. 6.6 for the monomers C_3F_6O, perfluoropropylene (C_3F_6) and vinyl ferrocene.

For the perfluoromonomers, the films were deposited at a constant plasma on time of 10 ms plasma on time and constant peak power (300 W in the case of C_3F_6O and 200 W for C_3F_6). The ferrocene films were all carried out at 10 ms on times and included runs carried out at peak power inputs ranging from 25 to 200 W, along with a range of plasma off times. These particular film formation/J curves were chosen simply to illustrate the variety of results which have been obtained in contrasting film formation efficiency curves obtained for various monomers. Although there are clearly major differences among these plots, a common feature of all curves is a region of significant increased film formation energy efficiency with plasma off times. For a number of monomers, the energy efficiency curve plateaus at longer plasma off times (e.g. C_3F_6O). Such behavior would be consistent with a polymerization chain process occurring to termination given sufficient plasma-off times. Relatively similar results were obtained from vinyl ferrocene at higher peak powers (Fig. 6.6(c)).

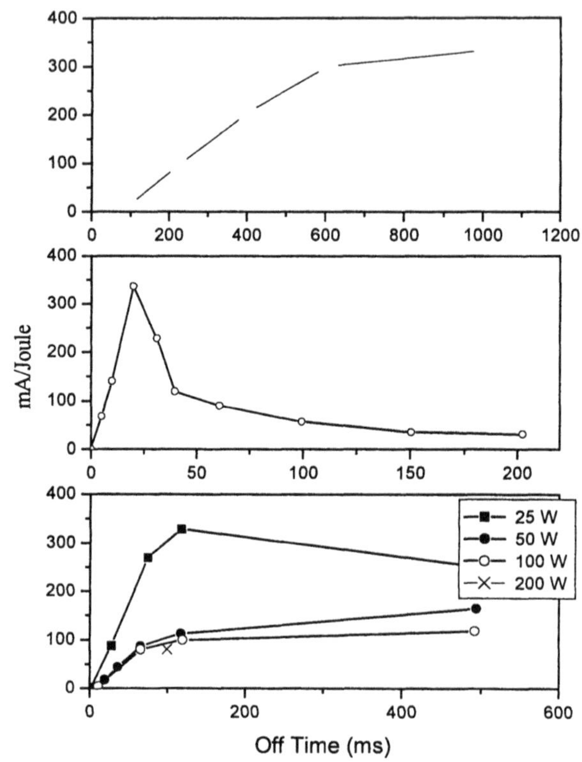

Figure 6.6. Illustrations of film formation energy efficiency curves for several monomers during pulsed plasma polymerization. The series of runs for each curve represents depositions at constant plasma on times, constant peak power and variable off times as shown. (A) C_3F_6O [22], (B) C_3F_6 [22], and (C) vinyl ferrocene [33].

However, the energy efficiency curve from vinyl ferrocene exhibited a maximum for runs carried out at lower peak power. Similarly a maximum in the energy efficiency curve was obtained in the work with C_3F_6. It is difficult to rationalize these maxima, particularly the sharpness of the extrema such as that obtained with C_3F_6, for series of experiments involving no variations in monomer flow rates and pressures. Additional studies are required on this topic since it seems reasonable to speculate that the energy efficiency curve may well provide valuable insights into plasma dynamics under pulsed conditions. In the meantime, it is significant to note that the consistent observation of at least certain regions of increasing plasma off times producing sharply increased values of film thickness/J is indicative of significant film formation during plasma off times.

6.3.3. Polymerization under Ultra Low Power Input Conditions

As noted earlier, the pulsed plasma technique permits polymerizations to be carried out under exceptionally low power input conditions. For example, runs carried out at plasma on/off times of 10/500 and 25 watt peak power correspond to average power input of only ~0.5 W. High quality films accompanied by reasonable films formation rates are readily obtained with many monomers at such low power input, even in reactors as large as 5 L in volume. In the case of less reactive monomers (e.g. saturated perfluoroalkanes), higher power inputs

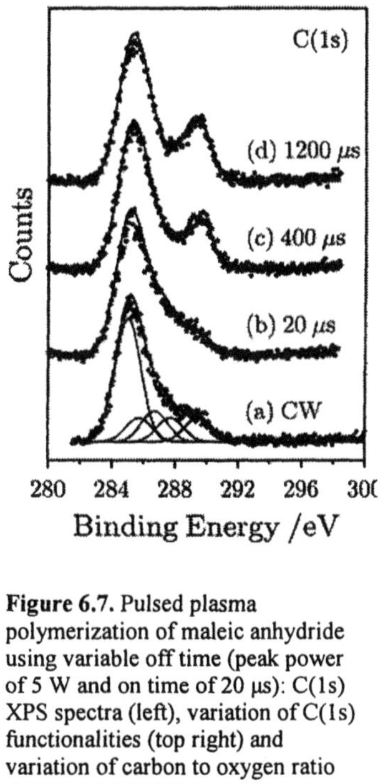

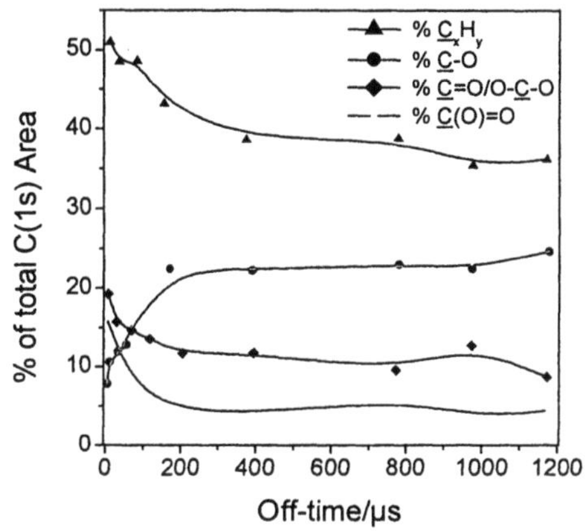

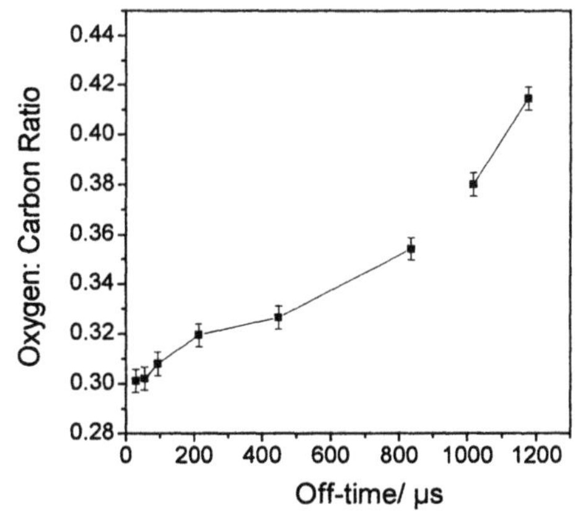

Figure 6.7. Pulsed plasma polymerization of maleic anhydride using variable off time (peak power of 5 W and on time of 20 μs): C(1s) XPS spectra (left), variation of C(1s) functionalities (top right) and variation of carbon to oxygen ratio (bottom right) [23].

are required. Nevertheless, the minimum average power inputs useable during pulsed plasma polymerization of saturated monomers such as perfluoroalkanes is significantly lower than those required under CW conditions for these same monomers [29].

As anticipated, extending the plasma polymerization process to lower power inputs provides a means of reducing monomer fragmentation during the film formation processes. However, not so expected, is the fact that highly non-linear variations in film chemistry are usually observed as one progresses through the very low power input regimes. This effect is particularly dramatic for monomers containing highly reactive groups such as anhydrides [23] or acid chlorides [24]. Results obtained from maleic anhydride are shown in Fig. 6.7. This non-linearity has been observed with a sufficient variety of monomers to suggest that it is one of general occurrence, with the magnitude of the effect being strongly monomer dependent.

The sharply increased retention of reactive functional groups achieved in plasma polymers synthesized under ultra low power input conditions provides important application opportunities, as discussed in the next section.

6.3.4 Film Composition Obtained under Equivalent Power Input Conditions during Pulsed and CW Plasma Polymerizations

An important fundamental issue is that concerning film composition from a given monomer produced using an equivalent input power in either pulsed or CW mode. Will noticeably different film compositions be obtained under equivalent power input condition? It is not possible to provide an unequivocal answer to this question at this time. This is partly because of the relative scarcity of data on this point, coupled with the fact that the answer to this question appears to be strongly highly monomer dependent.

At this time, it appears safe to generalize that if film compositional differences from pulsed plasma versus CW depositions are to be observed, it is more likely that these differences will appear under conditions of relatively low equivalent power inputs. Higher equivalent powers generally mean high plasma duty cycles (i.e. short plasma off times) thus producing a greater similarity between pulsed and CW results. However, at lower equivalent power inputs significant film compositional differences may be obtained as illustrated in the study of dichloromethane [22]. Results with CH_2Cl_2 at equivalent power inputs of 50 and 100 W under pulsed and CW conditions are shown in Fig. 6.8 in terms of C (1s) XPS spectra of the resultant films. It is clear that significant compositional differences are obtained at 50 W but not at 100 W.

There are several other examples in which no differences in film compositions are noted in contrasting data from pulsed versus CW experiments carried out at equivalent wattages spanning a fairly wide range of power inputs. For example, virtually no differences in film compositions were obtained with difluoromethane under pulsed versus CW conditions for runs carried out under equivalent powers of 100 and 15 W [46]. The large-scale differences observed in contrasting CH_2Cl_2 and CH_2F_2 results illustrate exactly how difficult it is to generalize concerning pulsed and CW films generated films at the same equivalent powers. Another example in which essentially the same film compositions are obtained under both pulsed and CW conditions is that reported by Fisher and co-workers for the monomer acetonitrile [47]. In contrast, these same workers report relatively large-scale film compositional differences in contrasting pulsed and CW studies of acrylonitrile [47]. It is interesting to note that polymerization of both the difluoromethane and acetonitrile monomers produce polymer film compositions that are essentially independent of power inputs, whether the operational mode is pulsed or CW. Thus, in view of this

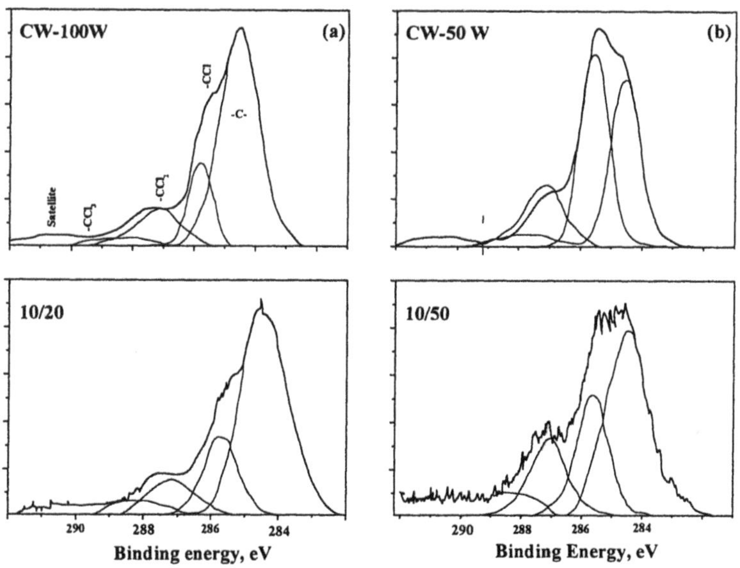

Figure 6.8. Comparison of film compositions obtained from dichloromethane monomer under equivalent power conditions during CW and pulsed plasma polymerizations. (a) 100 W equivalent power (100 W CW and 10/20 ms, 300 W); (b) 50 W equivalent peak power (50 W CW and 10/50 ms, 300 W) [22].

relative film compositional insensitivity to power input, perhaps it is not surprising that pulsed and CW runs provide essentially identical film compositions. Obviously, more studies are required on this topic.

A particularly interesting study of pulsed plasma versus CW polymerizations involves the polymerization of benzaldehyde [48]. Results obtained reveal significant differences in film compositions from pulsed compared to CW runs. Surprisingly, the higher retention of the reactive aldehyde groups were obtained under higher power inputs. This result is attributed to increased fragmentation of the aromatic ring system under higher power inputs, whereas selective plasma induced dissociation of the more reactive aldehyde group occurs at lower power inputs. It will be of interest to determine if similar results, involving selective retention of the more reactive functional groups at higher power inputs, are obtained with monomers containing more than one potentially reactive functional group.

6.4. Applications

At present, potential applications for plasma generated polymeric films span a wide range of topics. These applications involve deposition of both organic and inorganic films, as treated in various monographs and reviews. Here, we examine additional possible applications of polymeric films emphasizing those that are particularly specific for pulsed plasma generated films.

6.4.1. Biomaterials

Improved biocompatibility of many medical implants is an important goal at the present time. Surface coatings that may regulate biomolecule-surface interactions in a beneficial manner, without compromising substrate bulk properties, represent a major portion of this research. For this reason, plasma polymerization based surface modifications have received a lot of attention in biomaterials research. In particular, the conformal aspect of such coatings is an important advantage in coating unusually shaped implants vis a vis other potential coating methods.

With respect to pulsed plasma polymerization there are biomaterials application areas that may benefit from the added film chemistry controllability provided by this approach. One such application is to utilize the retention of highly reactive functional groups under pulsed polymerization for subsequent surface attachment of target molecules via simple chemical derivatization reactions. Because both the nature and surface chemistry of reactive groups are controllable under pulse plasma conditions, this approach provides an unusually simple molecular tailoring procedure resulting in three-dimensional surface structuring. The facile covalent coupling of reactant molecules to plasma-modified surfaces has been illustrated with acid chloride groups [24], epoxides [43], and brominated surfaces [31]. The latter example involved nucleophilic displacement reactions of bromide ions from surfaces by reagents such as amino acids and peptides from aqueous solutions at room temperature and neutral pH. Under such gentle reaction conditions, biomolecules presumably attach to surfaces without loss of secondary or tertiary structures The pulsed plasma technique provides an unusually wide selection of reactive functional groups, coupled with control of their surface densities, which should prove to be particularly useful in future studies, especially those involving tissue culture research.

A second important biomaterials application for plasma generated films involves synthesis of biologically non-fouling surfaces. For example, the film chemistry controllability provided by this technique can be utilized to maximize the surface densities of ethylene oxide groups during the polymerization of

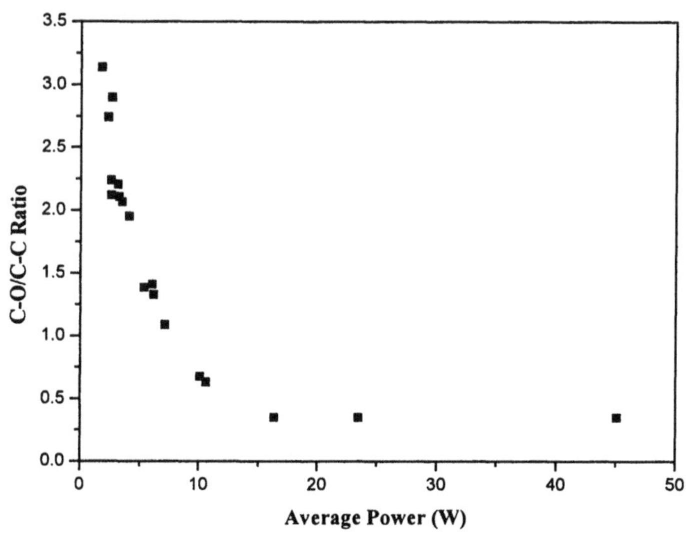

Figure 6.9. The variation of the C-O/C-C XPS peak areas of EO_2V plasma polymerized films as a function of the average power employed during deposition [30].

appropriate monomers. It is well known that polyethylene oxide (PEO) functionalized surfaces are strongly resistant to protein and other biomolecule adsorptions [49]. It has now been demonstrated that pulsed plasma polymerizations of low-molecular weight ethylene oxide monomers generates protein resistant surfaces [50]. An example of such surfaces is the polymerization of diethylene glycol vinyl ether [CH_2=$CH(OCHCH_2CH_2)_2OH$]. Sharply increased retention of monomer ether functionality is achieved as the duty cycle employed during pulsed plasma polymerization is reduced, as illustrated in Fig. 6.9 [30].

Here one notes the highly non-linear increased retention of C-O functionality as the plasma film formation is extended into the ultra-low power input regime not accessible under CW conditions. The efficacy of these coatings in reducing protein adsorption on the coated substrate surfaces has been demonstrated with decreased protein adsorption correlating with increased ethylene oxide content of these films [30].

6.4.2. Thin Layer Adhesives

As recently demonstrated by Badyal and co-workers, the pulsed plasma technique provides an effective route for applications requiring adhesive films for subsequent bonding of various substrates [43]. In this application, advantage is again taken of the ability to deposit thin films containing high surface densities of reactive functional groups, as provided under the low average power inputs available during pulsed plasma polymerizations. Surfaces containing such reactive groups can be bonded together via strong covalent bond formation between the two substrates. One such example is the pulsed plasma deposition of surface epoxide groups from glycidyl methacrylate onto two solid substrates, which were subsequently bonded together via simple application of a drop of diamine dissolved in dioxane, followed by overnight annealing at 60°C [43]. Both single lap joint and T-peel adhesion measurements showed that the adhesion bond strength was greater than the threshold failure of the host polyethylene matrix employed as the substrate. It was not possible to duplicate this strong substrate bonding behavior on films produced under CW conditions.

Clearly this particular bonding procedure should be one of great utility, which can be extended to functional groups other than epoxides. Of particular importance, in terms of potential future use, is the fact that this strong adhesion is achieved via use of extremely thin adhesive layers. There are numerous applications in which ultra thin bonding layers are of great importance in obtaining the desired substrate bonding.

6.4.3. Synthesis of Super Hydrophobic Surfaces

Recently it has been demonstrated that ultra hydrophobic surfaces can be generated using the pulsed plasma technique. This was initially demonstrated using alkene precursors, which contain relatively long perfluoroalkyl chains. With these monomers, polymer films having surface energies as low as 1.5 mN/m were obtained under pulsed plasma conditions [27,28]. To place this work in proper perspective, one notes that the surface energy of these films is about an order of magnitude lower than that of conventional Teflon. Water contact angles for these films were observed to range from 105° to 130°.

Subsequently, the pulsed plasma technique has been shown to provide superhydrophobic surfaces from polymerization of a variety of saturated perfluoroalkane monomers [29]. In this study, films having water contact angles of the order of >170° were obtained along with surface energies as low as 1.5 mN/m. In contrast with the fluoroalkenes, the superhydrophobic fluoroalkane derived polymer films exhibited essentially zero hysteresis when contrasting advancing and receding contact angles. Attempts to synthesize similar type films under CW conditions were not successful [29].

The exceptionally hydrophobic surfaces created under pulsed plasma conditions involve film formation of highly unusual morphology. These films involve highly intertwined, ribbon-like structures in which micron-sized pores are created. These rough textured materials contrast sharply with the relatively smooth films normally obtained under CW or pulsed plasma deposition of other, non-fluorocarbon monomers.

The availability of these superhydrophobic thin films has many potential applications. Already, the non-wettability [51] and non-biologically fouling aspects of these films have been utilized in practical applications [29].

6.4.4. Low Dielectric Constant Films

A major materials science challenge at present is production of lower dielectric (k) constant films for use in future smaller feature size integrated circuits (IC). These lower k films are required to reduce the RC coupling time constants between interconnects, thus improving the overall IC response times. These low k films must satisfy a host of requirements, including good thermal stability to temperature exposure as high as 400 °C, as imposed by IC fabrication technology.

It has been shown that pulsed plasma polymerization of perfluorocarbon monomers can successfully provide films having low dielectric constant as low as 1.8 [52]. These films were obtained from polymerization of fluoroaromtic monomers, utilizing the low duty cycle pulsed plasma technique to provide high retention of the fluorine and aromatic ring content of the monomer in the resultant polymeric films [52, 53]. The choice of aromatic containing precursors was done to help improve the thermal stability of the films. Although exceptionally low k films are obtainable, these materials exhibited surprisingly poor thermal stabilities, as shown by TGA studies. This low temperature weight loss has subsequently been attributed to loss of relatively volatile oligomers, trapped in the materials during film formation. Short-term temperature annealing of these films was shown to produce dramatic improvements in their temperature stabilities accompanied by only minor increases in dielectric constant values [34, 52]. The thermal annealing steps were shown to be accompanied by significant shrinkage in film thickness.

The low dielectric constants obtained under pulsed plasma conditions could not be duplicated under CW conditions, using the same precursors. Again, the low average power input capability under pulsed conditions provided higher retention of monomer structure and thus higher F/C atom ratios in the plasma deposited films. However, additional work, perhaps involving the use of a second monomer to promote increased cross-linking and thus higher film thermal stabilities, is required in order for these materials to be considered useful in smaller feature size ICs.

6.4.5 Electrically Conductive Polymers

A stringent test of film chemistry controllability made available by the pulsed plasma technique involves the synthesis of electrically conductive organic polymeric films. It is now well established using conventional polymerization techniques that films of relatively high electrical conductivity (i.e. ~10 S/m) can be achieved using various heteroaromatic monomers (e.g. pyrrole, thiophene, aniline and their substituted analogs). Electrical conductivities of these magnitudes require highly ordered polyconjugated structures, with appropriate doping to provide the electrons and holes required for conduction.

Although there have been many attempts to provide electrically conductive films via CW plasma depositions, the film conductivities achieved are many orders of magnitude lower than those obtained by conventional polymerization routes [54, 55]. Additionally, these CW generated films exhibited relatively poor stabilities in terms of atmospheric ageing measurements. However it has been recently demonstrated that the pulsed plasma technique can provide films having electrical conductivities that are orders of magnitude higher than those obtained under CW conditions [56]. In fact, the pulsed plasma films obtained from pyrrole under pulsed conditions approximate the conductivities of conventionally produced films.

Furthermore, the electrical conductivities of these films were found to increase in a sharp, non-linear manner as the duty cycle employed during film formation is reduced. The electrical conductivity increases with decreasing duty cycle during film formation are not unlike the functional groups compositional dependence on power input as shown previously in Figures 6.7 or 6.9. For example, it has recently been possible to demonstrate a long-term stable electrically conductivity of ~1.0 S/m for films produced from pyrrole at very low power inputs [57]. The conductivity of this film is at least 3 orders of magnitude higher than conductive films produced to date under CW conditions.

The higher conductivity of the pulsed plasma generated films correlate well with an increased film aromatic content [56]. Presumably this higher ring retention in the film provides a more linear, polyconjugated structure of the type required for good electrical conductivity. However under CW conditions the ring fragmentation apparently results in higher film cross-linking and insufficient double bond conjugation along the polymer chains and thus relatively low film conductivities.

6.5. Discussion

As outlined above, the pulsed operational mode introduces a number of interesting new aspects in terms of plasma polymerization science. These added considerations include film formation rates, morphological and compositional issues. In each case distinctive differences have been noted between CW and pulsed plasmas with respect to each of these factors.

Of particular importance in contrasting CW and pulsed plasma polymerization is recognition of the fact that the pulsed plasma approach provides an opportunity to generate high quality films under significantly lower power input conditions than attainable under CW conditions. This low power capability provides a convenient route to synthesis of non-highly cross-linked films that retain an exceptionally high level of monomer identity in the plasma-generated polymers. Of course, this capability may or may not be significantly dependent upon the particular application involved. However a number of applications have been identified in this chapter in which this low power input capability is of pivotal importance in producing the desired result.

The successful pulsed plasma synthesis of ultrahydrophobic films, high electrical conductivity organic films, and low k dielectrics represent specific examples in which film compositions obtained under pulsed plasma conditions could not be duplicated by CW experiments. Additionally, this low power input capability has been shown to be important in providing surfaces containing highly reactive functional groups. Utilizing these reactive surface groups for subsequent reactions with various compounds provides a promising approach to convenient surface molecular tailoring. These highly reactive groups expand considerably the range of target molecules which can subsequently be attached to the pulsed plasma modified substrate surfaces. Specific examples involving biomolecules surface attachments and synthesis of improved surfaces for adhesive bonding applications were noted in this chapter. It is anticipated that many more applications for this surface tailoring technology will be identified in future years.

In addition to its obvious utility in providing improved film chemistry compositional control, the pulsed technique may well provide valuable new insights concerning dynamic aspects of film growth in plasma discharge systems. For example, the strong monomer and duty cycle dependence of the energy efficiency in film formation under pulsed conditions, described above, should ultimately prove useful in elucidating key plasma parameters and reaction steps involved in polymer formation. We look forward to continued studies in this area in the years ahead.

Acknowledgements

The authors are pleased to acknowledge financial support for this project from the State of Texas Advanced Technology Program under Grant #00365-0052.

References

1. Lieberman M. A. and Ashida S., *Plasma Sources Sci. Tech.* **5** (1996), 145.

2. Zabeida O., Hallili A., Wertheimer M. R., Martinu L., *J. Appl. Phys.* **88** (2000), 635-642.
3. Scarsbrook G. A., Hewellyn I. P., Ojlia S. M. and Heinicke R. A., *Vacuum* **38** (1998), 627.
4. Hewellyn I. P., Rimmer N., Scarsbrook G. A., and Heinicke R. A., *Thin Solid Films* **191** (1990), 135.
5. Samukawa, S., *Appl. Phys. Lett.* **64** (1994), 3398.
6. Samukawa, S. and Mieno, T., *Plasma Sources Sci. Tech.* **5** (1996), 132.
7. Watanabe Y., Shiratani M., Kubo Y., Ogawa I., and Ogi S., *Appl. Phys. Lett.* **53** (1998), 1263.
8. Bouchoule A., Plain A., Boufendi L., Bloundeau J. and Laure C., *J. Appl. Phys.* **70** (1991), 575.
9. Savage C. R., Timmons R. B. and Lin J. W., *Chem. Mater.* **3** (1991), 575.
10. Yasuda H. and Hirotsu T., *J. Polym. Sci. Polym. Chem.* **16** (1978), 743.
11. Yasuda H., *Plasma Polymerization*, (Academic Press, New York, 1986).
12. Clark D. T., *Pure and Appl. Chem.* **54** (1982), 415.
13. Clark D. T and Abraham M. Z., *J. Polym. Sci. Polym. Chem Ed.* **22** (1994), 2083.
14. Inagaki N., Doyama M., and Igaki, H., *J. Polym. Sci. Polym. Chem Ed.* **22** (1984), 2083.
15. O'Kane D. F., and Rice, D. W., *J. Macromol. Sci. Chem.* **A10** (1976), 567.
16. Yasuda H., and Wang C. R., *J. Polym. Sci. Polym. Chem Ed.* **23** (1985), 87.
17. Lopez G. P., and Ratner B. D., *J. Appl. Polym. Sci.: Appl. Poly.* Sym. **46** (1990), 493.
18. Ying T. H., Saramadi A. M., Hop C. E., and Denes F., J. Polym. Sci. **55** (1995), 1537.
19. Clark D. T. and Dilks A., *J. Polym. Sci. Polym. Chem Ed.* **15** (1977), 2321.
20. Liston E. M., Martinu L., and Wertheimer M. R., in *Plasma Surface Modification of Polymers: Relevance in Adhesion*, Strobel M., Lyons C. S. and Mittal K. L. (eds.) (VSP: Utrecht, The Neatherlands, 1994).
21. Savage C. R., Timmons R. B. and Lin J. W., *In Structure-Properties Relations in Polymers*, Urban M. W. and Craver C. D. (eds.), *ACS Advances in Chemistry Series* **236** (1993), 745.
22. Panchalingam V., Chen X., Savage C. R., Timmons R. B. and Eberhart R., *J. Appl. Polym. Sci.: Appl. Poly. Sym.* **54** (1994), 123-141.
23. Ryan M. E., Hynes A. M. and Badyal J. P. S., *Chem. Maters.* **8** (1996), 37-42.
24. Calderon J. G. and Timmons R. B., *Macromolecules* **31** (1998), 3216.

25. Han L. M. and Timmons R. B., *J. Polym. Sci: Part A Polym. Chem.* **36** (1998), 3121.
26. Zabeida O. and Martinu L., *J. Appl. Phys.* **85** (1998), 6366.
27. Coulson S. R., Woodward I., Badyal J. P. S., Brown S. A. and Willis C., *Langmuir* **16** (2000), 6287.
28. Coulson S. R., Woodward I., Badyal J. P. S., Brown S. A. and Willis C., *J. Phys. Chem. B.* **104** (2000), 8836.
29. Qiu H., *Controlled Chemical and Morphological Surface Modifications via Pulsed Plasma Polymerizations: Synthesis of Ultrahydrophobic Surfaces*, Ph.D. Thesis, University of Texas at Arlington (2001).
30. Wu Y. J., Timmons R. B., Jen S. J. and Molock F. E., *Colloids and Surfaces B. Biointerfaces* **18** (2000), 235.
31. Timmons R. B. and Wang J. H., U.S. Patent 5, 875, 753.
32. Van Os M. T., Menges B., Foerch R., Knoll W., Timmons R. B. and Vansco G. J., *Mater. Res. Soc. Symp. Procs.* **544** (1996) 45.
33. Han L., Rajeshwar K. and Timmons R. B., *Langmuir* **13** (1997) 5941.
34. Han L. M., Timmons R. B., Lee W. W., Chen Y. and Hu Z., *J. Appl. Phys.* **84** (1998) 439.
35. Yasuda H. and Hsu T., *J. Polym. Sci. Polym. Chem Ed.* **15** (1977), 81.
36. Nakajima K., Bell A. T. and Shen M., *J. Appl. Polym. Sci.* **23** (1979), 2627.
37. Limb S. J., Edell D. J., Gleason E. F. and Gleason K. K., *J. Appl. Polym. Soc.* **67** (1998), 1489.
38. Lau K. S. K. and Gleason K. K., *J. Phys. Chem. B.* **102** (1998), 5977.
39. Butoi C. I., Mackie N. M., Gamble J., Castner D. G., Miller A. M., Barnd J. E and Fisher E. R., *Chem. Mater.* **12** (2000), 2014.
40. Clark D. T. and Abraham, M. Z., *J. Polym. Sci. Polym. Chem Ed.* **19** (1981), 2129.
41. Labelle C. B., Links S. J. and Gleason K. K., *J. Appl. Phys.* **82** (1997), 1784.
42. Rinsch C. L., Chen X., Wang J. H., Eberhart R. C. and Timmons R. B., *Langmuir* **12** (1996), 2995.
43. Tarducci C., Kinmond E. J., Badyal J. P. S., Brewer S. A. and Willis C., *Chem. Mater.* **12** (2000), 1884.
44. Chen X., Rajeshwar K., Chen J. J., Chyan O. M. R. and Timmons R. B., *Chem. Mater.* **8** (1996), 1067.
45. Wang J. H., Chen X., Calderon J. and Timmons R. B., *Plasma and Polymers* **2** (1997), 245.

46. Chen X., *Molecular Tailoring of Surfaces via Pulsed Plasma Depositions*, Ph.D. Thesis, University of Texas at Arlington (1995).
47. Lefohn A. E., Mackie N. M. and Fisher E. R., *Plasma and Polymers* **3** (1998), 197.
48. Leich M. A., Mackie N. M., Williams K. L. and Fisher E. R., *Macromolecules* **31** (1998), 7618.
49. Harris J. M. and Zalipsky S. (eds), *Poly(ethylene glycol) Chemistry and Biological Applications*, *ACS Symp. Ser.* **680** (American Chemical Society, Washington D.C., 1997).
50. Beyer D., Knoll W., Ringsdorf H., Wang J. H., Timmons R. B. and Sluka P., *J. Biomed. Mater. Res.* **36** (1997), 181.
51. Coulson S. R., Woodward I., Badyal J. P. S., Brewer S. A. and Willis C., *Chem. Mater.* **12** (2000), 2031.
52. Han L. Timmons R. B. and Lee W. W., *J. Vac. Sci. and Tech. B.* **18** (2000), 799.
53. Hynes A. and Badyal J. P. S., *Chem. Mater.* **10** (1998), 2177.
54. Gong X., Dai L., Mau A., Griessser H. J., *J. Polym. Sci A: Polym. Chem.* **36** (1998), 633.
55. Cruz G. J., Morales J., Olayo R., *Thin Solid Films* **342** (1999), 119.
56. Qui H., Sanchez-Estrada S. F., Timmons R. B., *J. Photopolym. Sci. and Tech.* **12** (2000), 29.
57. Sanchez-Estrada S. F. and Timmons R. B., (unpublished results).

CHAPTER 7

DEGRADATION AND STABILITY OF PLASMA POLYMERS

ANDREAS HOLLÄNDER AND JÖRG THOME
Fraunhofer-Institut für Angewandte Polymerforschung, Geiselbergstr. 69, 14476 Golm/ Potsdam, Germany

7.1. Introduction

Polymers as organic materials are known to degrade during long term use or due to the action of a reactive agent. In many cases it is not the regular structure where the alterations start. All polymers contain irregular structures which originate from side reactions during polymerization, from radical termination reactions, from chain transfer reactions, but also, for example, from impurities in the monomers. The weak bonds in these irregular structures are very often the starting point of reactions resulting in a degradation of the regular polymer structure and finally in a decay of the useful properties.

The scission of a chemical bond requires an energy input which is at least as high as the bond energy. The dissociation energy does not only depend on the atoms which form the bond. The electron density at these atoms plays an important role too. It is influenced by the chemical environment of this particular atom. Some values for typical organic bonds are listed in the table 7.1. For example, the dissociation energy of a C-H bond can vary by more than 30 % depending on the substituents on the carbon atom. The neighboring π electron system of a double bond or an aromatic ring decreases the dissociation energy of a C-C bond by 67 kJ/mol [1]. Consequently, bonds can connect the same atoms but they break upon energy input with a very different probability.

The structure of plasma polymers is much more heterogeneous compared with traditional polymers. As a result of the high energies involved in the formation of these materials, many crosslinks and some rather "strange" structural features can be found. Their analysis can be very difficult. A number of research groups have shown that plasma polymers undergo structural changes over different time scales. Due to the complex structure it is often hard to investigate mechanisms and to generalize the findings.

In this chapter we will give a summary of what is known about degradation mechanisms of well investigated traditional polymers. We will emphasize critical structures which can also be expected to appear in plasma polymers. Then we review the literature on plasma polymer degradation. Finally, we will present some general conclusions on how to make more stable plasma polymers.

Table 7.1. Bond (E_{bond}) and dissociation (E_{dis}) energies of bonds in organic molecules [1,2].

Bond	E_{bond} kJ/mol	E_{dis} kJ/mol	eV
$H–CH_3$		436	4.52
$H–CH_2CH_3$		411	4.26
$H–CH-(CH_3)_2$		394	4.09
$H–C-(CH_3)_3$		381	3.95
$H–CH_2-C_6H_5$		322	3.34
C–C	348		3.61
C=C	594	246[a]	2.55
C≡C	778	184[b]	1.91
C–C–O		302	3.13
C–C=O – aldehyde		314	3.26
C–C=O – ketone		344	3.57
CH_3-OH		381	3.91
C–O–C		335	3.48
C=O – aldehyde	741	364[a]	3.77
C=O – ketone	749	372[a]	3.86
$CH_3–NH_2$		335	3.47
$CH_3–F$		453	4.70

[a] calculated as difference between the bond energy of the double bond and the single bond.
[b] calculated as difference between the bond energy of the triple bond and the double bond.

7.2. Mechanisms of Polymer Degradation

Since the large scale application of synthetic polymers started some 70 years ago, there have been steady efforts to understand why and how these materials degrade. Most of them are rather stable and can be used for many years. However, over an extended period of time or under extreme environmental conditions they are destroyed more rapidly. An overview on the degradation of polymers can be found for example in [3,4].

The degradation mechanisms of a polymer are determined by its formal chemical structure but also, and this is sometimes even more important than the regular structure, by irregular structures. In this chapter we will summarize what is known from the polymer degradation research about the stability of certain structural features and point out what are the weak structures. We will, of course, emphasize what is important from the viewpoint of plasma polymers.

7.2.1. General Considerations

Depending on the way of the energy input we have to distinguish thermal degradation, oxidation, photodegradation, and hydrolysis. The first three are mostly based on radical reactions and show some common features. Radical reactions in general comprise of three types of reactions:

1) initiation: the radicals are formed.
2) propagation: the radicals react with other structures and promote the degradation.
3) termination: the radicals are destroyed.

The initiation reactions are specific for thermal, oxidative, and photodegradation. Besides the actual propagation there are also radical transfer reactions on one and the same molecule (intramolecular) or between two molecules (intermolecular). The termination can be a recombination of two radicals or a disproportionation which follows this general scheme 7.1.

$$R\bullet + \bullet R \longrightarrow R{-}CH_3 + H_2C{=}\overset{H}{C}{-}R \qquad (7.1)$$

The recombination increases the molecular mass while the disproportionation decreases it. The ratio between the two reaction rates determines whether the material crosslinks or whether it loses mass. In general, disproportionation is favored for structures with linkages or if one radical is stabilized e.g. by a delocalization at an adjacent double bond.

7.2.2. Thermal Degradation

Purely due to the influence of elevated temperature organic materials decompose. Depending on the chemical structure, this degradation starts at different temperatures and proceeds with different rates. In the next paragraphs we will give an overview about the sensitivity of various structures to thermal stress.

7.2.2.1. Hydrocarbon Polymers

Linear long chain polyethylene PE $(CH_2)_n$ undergoes a major weight loss above 420 °C [5] which is governed by a random scission mechanism [6]. The temperature sensitivity rises if the material is branched. Low density LDPE which contains a higher concentration of branches starts to lose its mass at a temperature which is 20 K lower than for the linear high density HDPE.

The lowered stability can be explained by two reasons: (1) The hydrogen atom at a tertiary carbon atom (branch) is bonded weaker than the one in a CH_2 group (see Table 7.1) so that an initiation proceeds more readily at this site. (2) If a main chain bond is broken, the disproportionation/ recombination ratio is higher for the more branched structures. Consequently, the branched structures

Table 7.2. Dependence of the thermal stability of polyolefins on the chain branching [5].

	Polymer	Temperature of rapid weight loss[°C]
Polyethylene HDPE	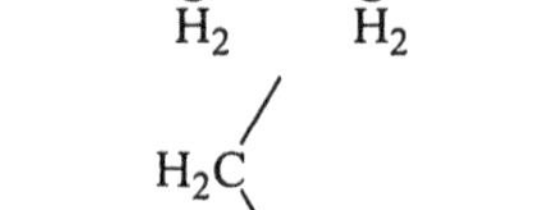	420
Polypropylene PP	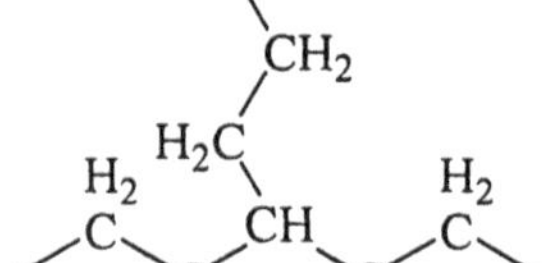	390
Branched LDPE	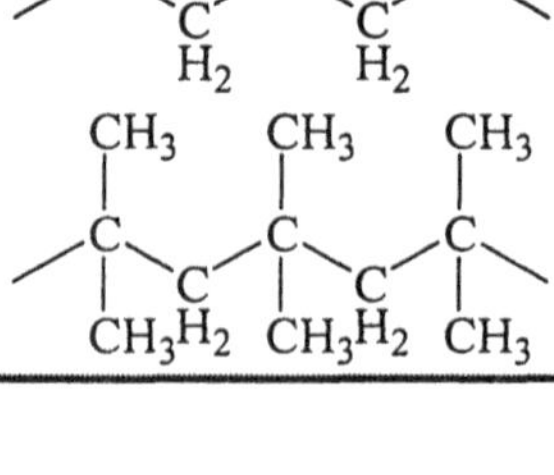	380
Polyisobuthylene		350

undergo main chain scission more readily.

While a rapid weight loss is observed at the quite high temperatures listed in table 7.2, the degradation of hydrocarbon polymers starts at much lower temperatures. PE and PP start to reduce their molecular weight at a temperature of 290 °C [7] and 250 °C [8], respectively, as a result of the scission of weak bonds (scheme 7.2).

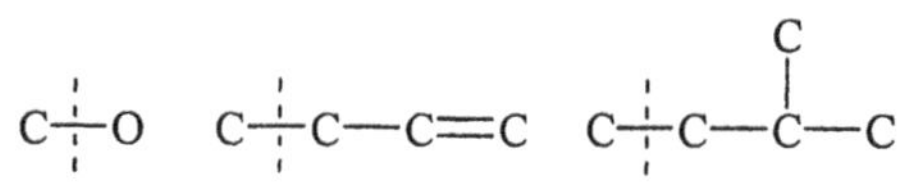

(7.2)

The role of tertiary carbon atoms and branches as in the latter structure was investigated by the analysis of the thermal degradation products of polymethylpentene (scheme 7.3) [9].

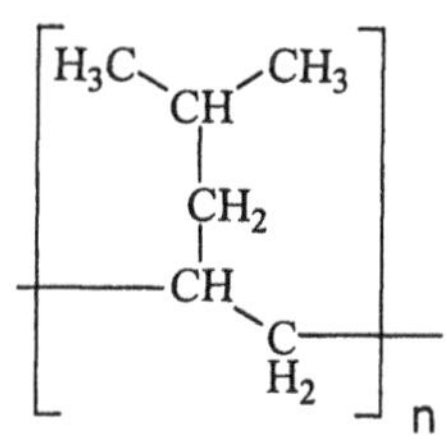

(7.3)

The dominating thermal degradation mechanism of this polymer is a radical transfer to the tertiary carbon atom followed by the scission of a C-C bond in the β-position. This mechanism is important for all branched hydrocarbon polymers and for crosslinked polymers [10]. It follows the general reaction route (scheme 7.4).

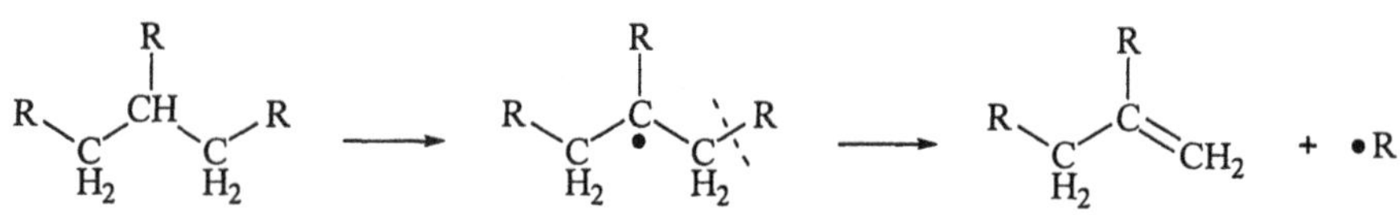

(7.4)

Double bond containing polymers like natural rubber or polybutadiene (7.5) start to crosslink above 200 °C. They also form polycyclic aromatic structures which are quite stable. The degradation mechanisms change dramatically at higher temperature. Above 400 °C a rapid volatilization is observed [11,12].

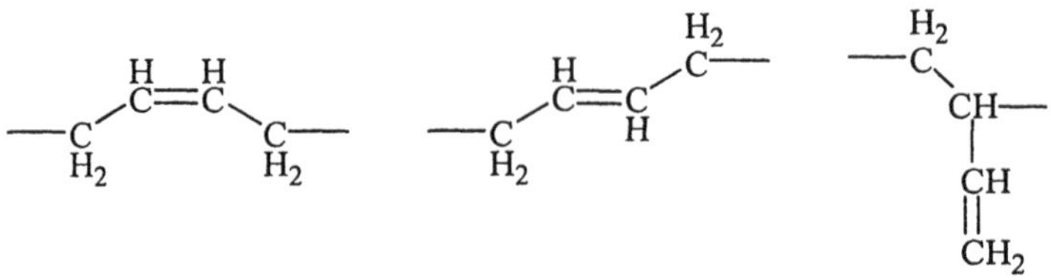

(7.5)

1,4 cis 1,4 trans 1,2

Aromatic structures are much more stable than aliphatic structures. The heating of polystyrene (PS) above 300 °C results in a volatilization [13]. Monomeric styrene accounts for about one half of the products. The rest are oligomers. Poly-α–methylstyrene decomposes above 240 °C forming almost pure monomer [14,15]. In both polymers the degradation is based purely on the scission of aliphatic main chain bonds. Polymers with aromatic rings in the main chain like poly(p-phenylene) or poly (p-xylylene) show only a small amount of volatilization below 400 °C [16,17].

7.2.2.2. Oxygen Containing Polymers

Simple oxygen containing polymers are usually less stable than hydrocarbon polymers. The dissociation energy of a C-O bond is lower than the value for most C-H bonds. C-O bonds are more polar and the lone electron pairs can undergo acid/ base interactions which open new reaction paths.

Polyoxymethylene $(-CH_2-O-)_n$ decomposes mainly by a non-radical depolymerization process which starts at the $-CH_2-OH$ chain ends and forms formaldehyde [18]. At a temperature as low as 170 °C the degradation is observed. Other end groups such as $-(C{=}O)-CH_3$ or $-C-O-CH_3$ render the material more stable and force a random initiation [19]. For higher polyethers with more carbon atoms in the chain the degradation is initiated by a C-H scission at the carbon atom in β position to an oxygen atom followed by the scission of the backbone C-O bond [20,21] (scheme 7.6).

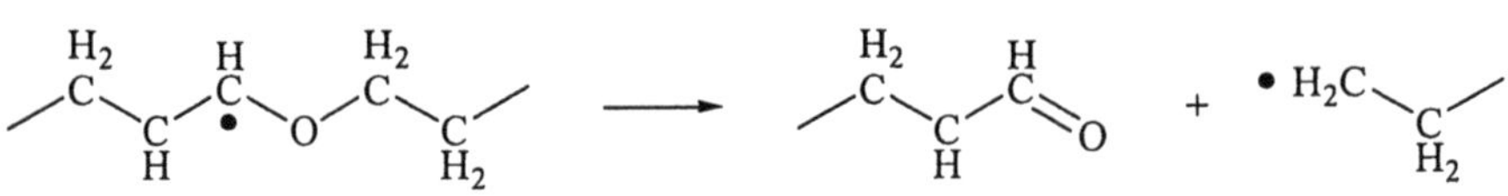

(7.6)

Polyvinylalcohol (PVA) starts to eliminate water above 220 °C in a chain reaction forming conjugated polyenes [22]. Polyvinylacetate (PVAc) degrades in a similar way forming acetic acid and double bonds at slightly higher temperatures [22].

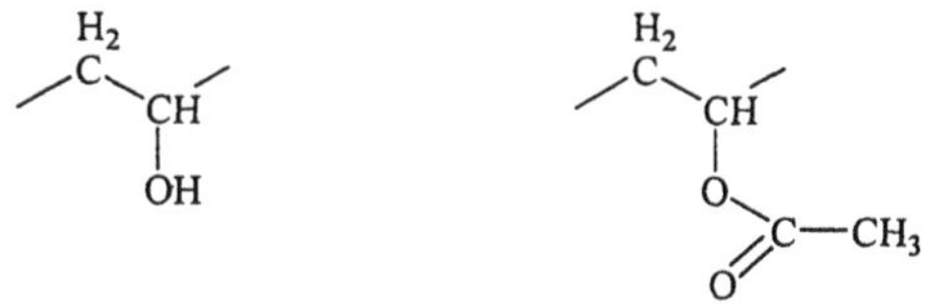

(7.7)

polyvinylalcohol (PVA) polyvinylacetate (PVAc)

It must be considered as a general rule that side groups which are bonded weakly to the backbone and which form a very stable small molecule split off easily (7.8).

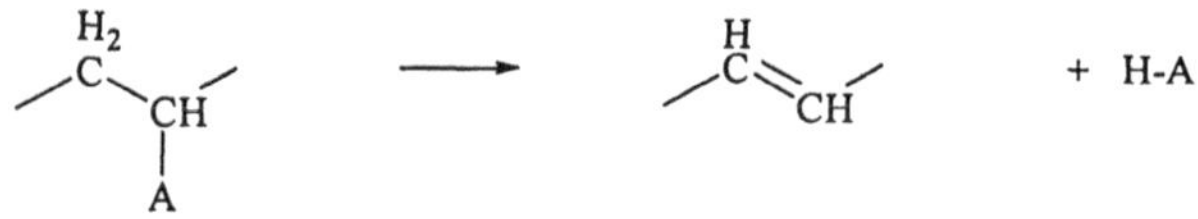

(7.8)

Polymethylmethacrylate (PMMA) is a typical example for a polymer that decomposes by a depolymerization mechanism. If it was prepared by a radical reaction many molecules have double bonds at the chain end (radical termination by disproportionation) where above about 220 °C the depolymerization starts [23]. At higher temperatures (> 270 °C), chain end initiation and random initiation occur simultaneously. If the temperature is increased above 300 °C, the molecules with double bond chain end degrade very rapidly [24].

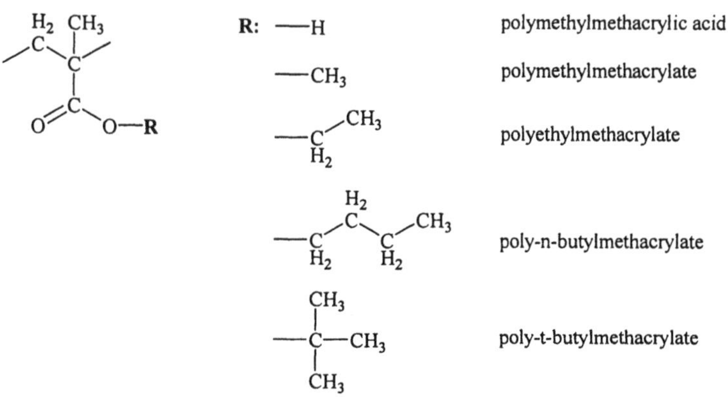

(7.9)

Polymers from other methacrylic acid esters (7.9) often degrade in a more complex scheme. The ester decomposition can take place if there is a hydrogen atom at the β carbon atom of the alcoholic residue [25] (7.10). For polyethylmethacrylate and poly-n-butylmethacrylate the ester decomposition competes with the depolymerization while the ester scission is the main degradation route for poly-t-butylmethacrylate [26]:

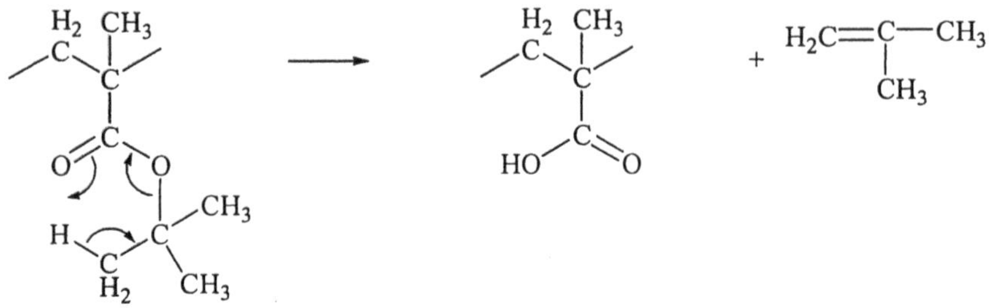

(7.10)

At a high conversion carboxylic acid groups can react in the same way as polymethacrylic acid (PMAA) [27]. Two neighboring groups split off water and form an anhydride. In the case of PMAA water formation starts below 200 °C.

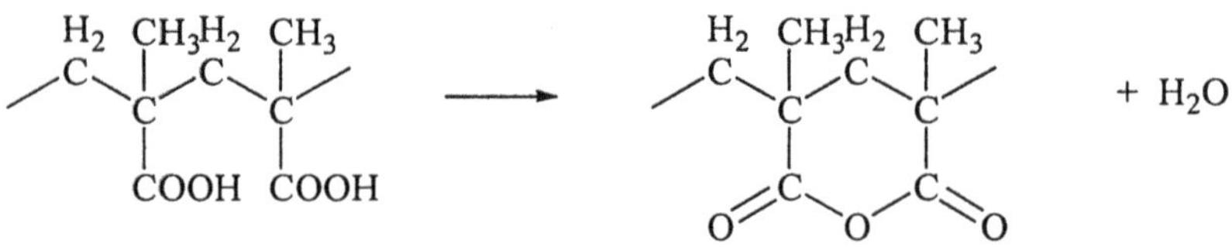

(7.11)

Polyacrylates lack the methyl side groups. Depolymerization does not take place and the reaction paths are more diverse than in the case of polymethacrylates. The tertiary carbon atoms tend to split off the hydrogen atom randomly giving rise to β scission reactions as discussed for the hydrocarbon polymers [28]. The scission of the ester group results in the formation an C=C end group and an acid group as shown in scheme 7.10. Carbon dioxide can be evolved from the acid group which forms a main chain radical [29-31]. In the case of poly-t-butylacrylate the ester scission starts well below 200 °C [32]. The volatilization of other polyacrylates typically starts at about 250 °C.

Polyesters and polyamides are produced by a polycondensation which is an equilibrium reaction. The presence of even small amounts of water results in a lowering of the molecular weight upon heating due to a hydrolysis reaction of the ester group and the amide group, respectively. Usually these reactions start to become important around 300 °C.

Beside the hydrolysis reactions there are numerous thermally initiated reactions resulting in the degradation of the material. Most of the reactions involve the ester and the amide group, respectively [33,34]. The amide is thermally more stable than the ester group. Poly (ethylene terephthalate) (PET, scheme 7.12), for example, degrades during processing (280 °C – 300 °C). The most labile bonds are marked in scheme 7.12. Acetaldehyde, water, carbon dioxide, and carboxyl end groups [35,36] are among the most abundant reaction products.

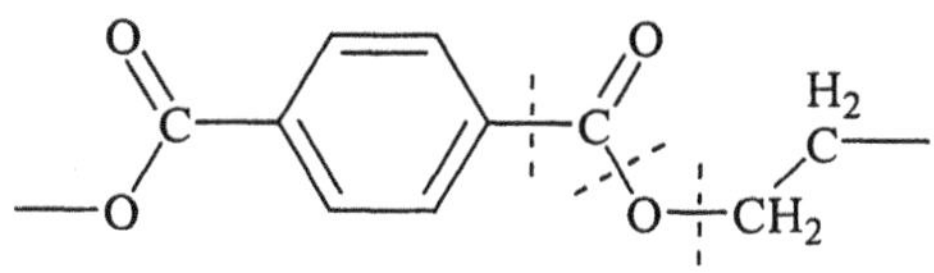

(7.12)

7.2.2.3. Aromatic Structures

Aromatic rings in the polymer back bone improve the thermal stability. Polycarbonate, for example, starts to decompose extensively between 375 °C and 450 °C depending on the structure [17]. The $-C(CH_3)_2-$ group is considered as the weakest link. The carbonate group degrades forming carbon dioxide [37].

Commercial polyimide can be used for more than 200 h at 360 °C. At 600 °C carbon monoxide, carbon dioxide, and water are formed after the breaking of the imide group [38]. The degradation of such high temperature polymers was reviewed by Arnold [39].

7.2.2.4. Fluoropolymers

The exceptional thermal stability of polytetrafluoroethylene (PTFE, Teflon) inspired an extensive research on fluoropolymers and their degradation behavior [40,41]. For a review see reference [42].

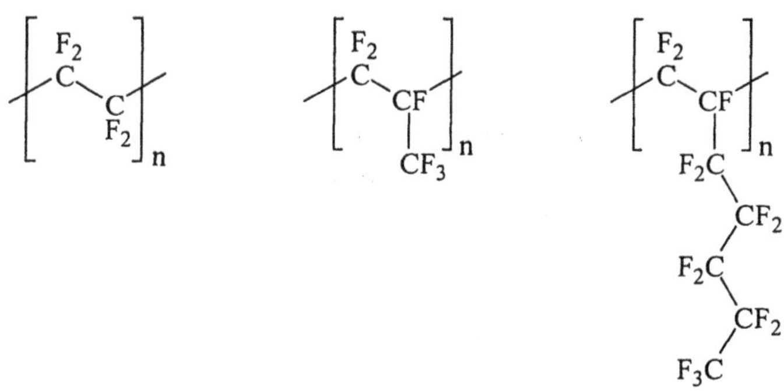

(7.13)

PTFE PHFP PPFH

The mechanism proposed in reference [43] for the degradation of PTFE is based on a random chain scission followed by depolymerization. The weight loss increases to 1 %/ min at 510 °C [40]. Polyhexafluoropropene (PHFP) and polyperfluorohept-1-ene (PPFH) comprise tertiary carbon atoms in their structure (scheme 7.13) which are responsible for the much lower thermal stability of these polymers. Weight loss is observed well below 300 °C [44]. The major reactions products are low molecular weight fluorocarbons [45]. For polymers which also contain hydrogen (polyvinylfluoride -$(CH_2\text{-}CHF)_n$-, polyvinylidenefluoride -$(CH_2\text{-}CF_2)_n$-), hydrogen fluoride is among the most abundant reaction products [46,47]. The dehydrofluorination starts at 240 °C and at 320 °C it proceeds rapidly [48]. Also an ether group like in the perfluoromethyl vinyl ether polymers reduces the degradation temperature [49].

7.2.2.5. Silicone Polymers

The thermal degradation of polydimethysiloxane starts at about 200 °C. Si-OH groups at the chain end first undergo a condensation increasing the molecular weight [50]. The weight loss at higher temperature results from the depolymerization to a cyclic trisiloxane [51]. The reduction of the Si-OH concentration increases the stability of the polymer [50].

7.2.3. Oxidation

The oxidation of organic substances follows a general scheme which will be described in the following paragraphs. The fundamental mechanisms were found by Bolland and Gee in the late 1940s [52]. A more detailed discussion can be found in [4, 53, 54].

As a radical reaction it comprises initiation, propagation, and termination reactions. The initiation reaction produces radicals either by physical factors (absorption of radiation, heating) or by chemical means like for example by oxygen atoms (scheme 7.14).

$$PH \longrightarrow P^{\bullet} + H^{\bullet} \qquad (7.14)$$

These processes require an energy input. Most of the following radical reactions, however, proceed without additional activation or energy input very quickly.

For the understanding of the reactivity of oxygen atoms and molecules their electronic structure is very important which has been fully discussed in reference [55]. The oxygen molecule in its ground state is paramagnetic and, therefore, it has to be considered as a biradical. This is extremely important for oxidation reactions because oxygen reacts rapidly with radicals forming peroxy radicals (scheme 7.15).

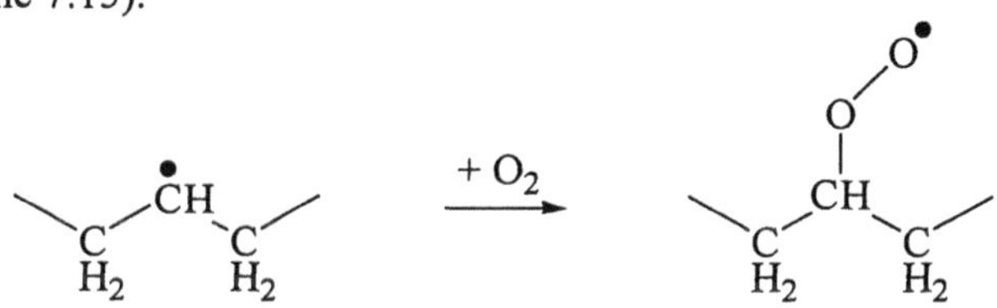

(7.15)

Because molecular oxygen is usually the most abundant oxidizing species, this reaction is the first propagation step in polymer oxidation. The peroxy radical abstracts a hydrogen atom from a neighboring carbon atom of the same chain (intramolecular) or from another chain (intermolecular) forming a hydroperoxide group. This hydrogen abstraction occurs preferably from a tertiary carbon atom (scheme 7.16).

(7.16)

For polymers with a high number of tertiary carbon atoms in their structure the formation of peroxy radicals and hydroperoxides can occur several times. In the thermal oxidation of polypropylene only a small fraction of the hydroperoxides was found to be isolated; the majority occurs in dimers, trimers, and even in longer sequences [56].

Hydroperoxides decompose thermally, by the influence of UV radiation, or by the action of chemicals like for example radicals or traces of metal catalysts (schemes 7.17 –7.19).

$$\mathrm{POOH} \longrightarrow \mathrm{PO}^{\bullet} + \mathrm{HO}^{\bullet} \qquad (7.17)$$

$$\mathrm{POOH} + \mathrm{RO}^{\bullet} \longrightarrow \mathrm{POO}^{\bullet} + \mathrm{ROH} \tag{7.18}$$

$$\mathrm{POOH} + \mathrm{HO}^{\bullet} \longrightarrow \mathrm{POO}^{\bullet} + \mathrm{H_2O} \tag{7.19}$$

Hydroperoxides and alkoxy radicals are the most important intermediates in the oxidation process. They are the starting point for the formation of a rather large number of different functional groups, of main chain scissions, and the formation of low molecular weight products. The β scission of the polymer's backbone beside an alkoxy radicals lowers the molecular weight and results in aldehydes (R = H) and carbonyl groups (ketones) depending on the polymer structure (scheme 7.20).

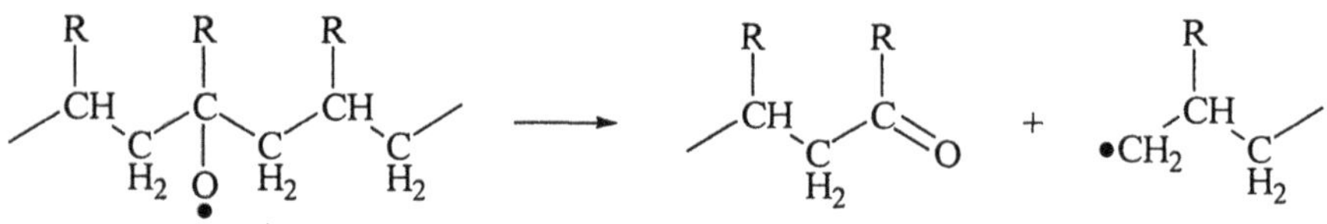

(7.20)

Carbonyl groups can also be formed without breaking the polymer backbone by the scission of hydroperoxides adjacent to a radical (scheme 7.21).

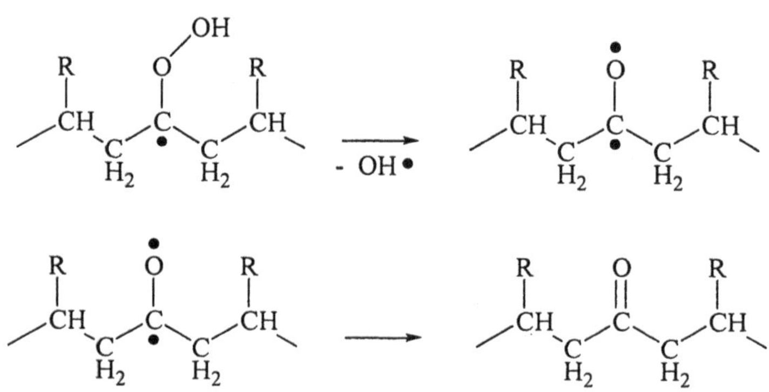

(7.21)

Hydroxyl groups are formed from alkoxy radicals by hydrogen abstraction as shown in scheme 7.18 or in scheme 7.22.

$$\mathrm{PH} + \mathrm{PO}^{\bullet} \longrightarrow \mathrm{P}^{\bullet} + \mathrm{POH} \tag{7.22}$$

Besides these decay reactions the radicals also undergo bimolecular reaction. The recombination of two radicals results in a deactivation and forms linkages. The crosslinking is mainly a result of radical recombination. But also structures like ethers, ester, and lactones are formed this way.

Oxidation starts at temperatures well below the temperature at which thermal degradation is observed in vacuum or in an oxygen free atmosphere. For example, PP and PE start to get oxidized in air at a temperature slightly above 100 °C with a measurable rate [57,58]. The two polymers show an induction period in the oxygen uptake which is a measure for the oxidation yield. At 110 °C it takes 4.5 h until atactic PP takes up oxygen. For LDPE this period is 70 h and for PS it is longer than 10000 h [59]. The preferred location for the first attack is a tertiary carbon atom. Consequently, branching makes the oxidation proceed more readily.

7.2.4. Photooxidation

Photochemical processes start with the excitation of a molecule by the absorption of light. The excited state results directly (primary process) or indirectly (secondary process after transition in a triplet state) in the decomposition of the molecule. Pure saturated hydrocarbons [60] and fluorocarbons [61] do not absorb ultraviolet radiation with wavelengths longer than 160 nm. Molecules with isolated double bonds and with oxygen or nitrogen single bonds absorb mainly below 200 nm [62]. For the photochemistry in the near UV and in the visible region these structures are not relevant. Most of the photochemistry involved in the degradation of polymers is based on carbonyl (C=O) structures.

In 1956 Guillet and Norrish [63] found the two basic carbonyl scission reactions when they investigated the photodegradation of polyvinylmethylketone (PVMK) with 313 nm radiation. It is the Norrish type I reaction (scheme 7.23) which is a primary photoreaction with a quantum yield of 0.04 at 25 °C and 0.07 at 50 °C [63].

(7.23)

The Norrish type II reaction is more complicated and follows the general scheme 7.24.

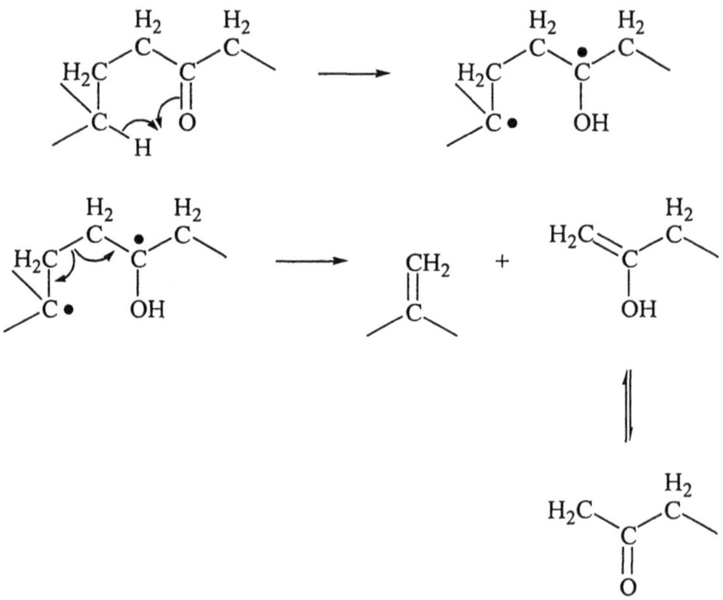

(7.24)

A hydrogen atom at the γ carbon atom is necessary for this reaction. In the case of PVMK, a quantum yield of 0,04 was determined which is independent of temperature [64]. The Norrish scission reactions take place with other carbonyl containing functional groups like esters and amides too. For example, the UV induced main chain scission of PMMA is largely based on Norrish type II reactions (quantum yield at 253.7 nm 0.037 [65]).

7.3. The Degradation of Plasma Polymers

7.3.1. Structural Characteristics of Plasma Polymers

7.3.1.1. General Considerations

The structure of a plasma polymer depends on the structure of the monomer, the conditions of the plasma polymerization, including the design of the plasma reactor, and on the properties of the substrate on which the plasma polymer is deposited. The dependence on the process parameters is much stronger for plasma polymers than for traditional polymers. There are, however, general characteristics which distinguish plasma polymers from conventional polymers: [66-70]

- Due to the high energy input plasma polymerization often involves monomer fragmentation i.e. the loss of functional groups (e.g. methyl, ester or carboxylic groups), hydrogen abstraction from alkyl groups or ring opening of cyclic and aromatic structures.
- Plasma polymerization proceeds by rather complex mechanisms which are different from the chain addition (step growth) mechanism of many conventional polymers.
- Plasma polymers are not composed of regular repeat units. Their chains are in general highly branched and crosslinked.
- Carbon based plasma polymers often contain numerous double bonds.
- As a result of their structural inhomogeneity plasma polymers are in general amorphous.
- Considerable amounts of free radicals are trapped in the deposited plasma polymers. These radicals are still reactive after plasma polymerization.

The high degree of crosslinking in the plasma polymers reduces the mobility of the chains and makes a recombination of the trapped radicals rather difficult. The quantity of free radicals trapped in a plasma polymer is largely determined by the chemical structure of the monomer. The amount of monomer fragmentation and of irregular structures in the plasma polymer can be controlled by the deposition parameters. As a general trend the "disorder" in the deposited polymer increases with the power input in the plasma.

As a promising alternative to the continuous plasma polymerization Yasuda et al. developed the pulsed plasma polymerization [68,69]. Free radicals are only created during the "on period" of the pulsed discharge. During the "off period" unactivated monomer molecules can react with previously formed radicals and radicals can recombine. Studies by Yasuda and Hsu [68,69] showed that the use of a pulsed discharge reduces the fragmentation of monomers such as carboxylic acids. For most monomers, with the exception of acetylene and ethylene, the pulsed discharge also reduces the concentration of trapped free radicals.

7.3.1.2. Plasma Polymer Coatings

In the last section we discussed the intrinsic properties of plasma polymers and we considered them as structurally homogeneous. However, a real plasma polymer coating usually does not grow constantly. The deposition conditions change during the process solely by the presence of the substrate and its changing properties.

A metal surface is very often covered by a layer of oxides, hydroxides, carbonates and other compounds formed during the storage of the material. A pure metal has a very high surface free energy and, therefore, it is covered with a layer of adsorbed gases and vapors (water, carbon dioxide, hydrocarbons etc.) which adheres strongly. The surface free energy of oxides like, for example, glass and the compounds mentioned above is usually not as high as for metals. But it is still sufficiently high to form an adsorbed layer which can not be removed completely simply by evacuation. The energy input from a plasma, however, gives rise to at least a partial desorption. Consequently, in the initial stage of a plasma deposition the composition of the process gas is altered by the substances evolved from the surface.

For most polymers the situation is different. Their surface free energy is rather low and, therefore, only a weak adsorbed layer is found. Polymers, however, readily react under the action of a plasma. They often form low molecular weight decomposition products which volatilize and mix with the process gas.

In both cases, the highly energetic surfaces and the polymers, the influence of the substances contaminating the process gas gradually declines as the layer grows*. The chemical composition of the plasma polymers continuously changes during that stage of the deposition process and we observe something which is called an interphase [71]. The properties of the interphase can differ substantially from the properties of the actual plasma polymer.

The action of the plasma vacuum ultraviolet (VUV) radiation (wavelength < 200 nm) also results in a heterogeneous stacking of the plasma polymer layer. While the particle interactions at the plasma-solid interface are confined to a few

* This is true, actually, only in batch processes. In continuous deposition processes these contaminations are present more or less during all the layer formation.

nanometers, the radiation penetrates deeper. 90 % of the VUV photons are absorbed within a layer which is some ten nanometers thick. These photons initiate photolysis reactions in the layer much more efficiently than near UV radiation [72].

Considering these phenomena, we can generalize the structural stacking of a plasma polymer coating: Directly on the surface of the substrate there is an interphase layer. It follows a plasma polymer layer which was photolyzed by the plasma VUV radiation. And on the top of all there is a fresh plasma polymer. The thickness and the properties of the layers and of the transitions between them largely depends on the deposition conditions.

7.3.2. Aging of Plasma Polymers in Air

When a plasma polymer is exposed to air, the free radicals trapped in the polymer can react with oxygen and water vapor and to some extent they also recombine. The oxidation reactions follow the same general scheme as it was discussed for traditional polymers in section 7.2.3. These processes lead to a gradual rearrangement of the polymer network and are often referred to as "aging" of the plasma polymer [66-70]. The oxygen uptake results in the formation of oxygen-containing functional groups which make the surface of the plasma polymer more hydrophilic.

7.3.2.1. Hydrocarbon Plasma Polymers

For all hydrocarbon plasma polymers aging in air is caused by the attack of oxygen or water vapor at carbon-centered radicals. Yasuda investigated the concentration of trapped radicals in hydrocarbon polymers with respect to the monomer structure [68]. Monomers with a triple bond such as acetylene yield plasma polymers with a very high radical concentration. For monomers with a double bond, e.g. ethylene the concentration of trapped radicals is moderate and it is the lowest for saturated monomers (scheme 7.25)

$$HC{\equiv}CH \gg H_2C{=}CH_2 > RCH_2\text{-}CH_2R' \qquad (7.25)$$

Acetylene easily forms diradicals by opening the triple bond and complete reaction of the diradicals is very unlikely once the molecule is built in the layer.

Saturated monomers, however, can only form radicals by hydrogen abstraction, which requires a higher energy input (see Table 7.1). As a result of the high radical concentration, acetylene plasma polymer films have random structures [70] and are gradually oxidized in air [68].

Gengenbach et al. investigated the aging of hexane plasma polymers in air over a period of more than one year [73-75]. In contrast to the oxidation of conventional polyolefins in air there is no induction period in the oxygen uptake. The oxygen incorporation is even very rapid within the first few hours of exposure to air, then it gradually slows down over a period of several months until the oxygen to carbon ratio almost levels out approximately 6-7 months after the plasma deposition. In comparison with the oxidation of conventional polyolefins at elevated temperature, highly oxidized carbon such as in carbonyl, carboxylic and ester groups is formed to a much lower extent. Gengenbach et al. identified C-O structures as the main oxidation product of n-hexane plasma polymers. The plasma polymer coatings do not show any mass loss (reduction in film thickness) or cracking of the films. They retain their physical integrity. It is also reported that the surface is restructured in a similar way like plasma oxidized polymers surface [76].

Based on their experimental data Gengenbach et al. proposed a model for the oxidation process of hexane plasma polymers [73]. They assign the rapid oxygen uptake within the first few hours to the fast formation of peroxy radicals (C-O-O·) by oxygen addition to carbon-centered radicals (scheme 7.15). In the second oxidation stage the much less reactive peroxy radicals form hydroperoxides by hydrogen abstraction (scheme 7.16), which creates new carbon-centered radicals and eventually depletes the reservoir of available C-H groups. The metastable hydroperoxides slowly decay and start reactions that form stable products (scheme 7.17 – 7.22).

7.3.2.2. Oxygen Containing Plasma Polymers

Similar to hydrocarbon plasma polymers the aging of methyl methacrylate (MMA) plasma polymers in air is due to oxidative processes [77]. The oxygen uptake of the plasma coatings is very rapid during the first week of exposure to air, then slows down significantly and stops within one year of exposure to air, leaving the coatings physically intact. Oxidation results mainly in the formation of hydroxyl groups and to a smaller extent of carbonyl and ester groups.

Gengenbach and Griesser studied the influence of the plasma deposition parameters on the initial structure of the PMMA coatings and on their aging behavior [77]. For a low power plasma input (5 W) they observed a high degree of retention of the original monomer structure in the freshly deposited PMMA coatings. A higher plasma power input (40 W), however, results in substantial monomer fragmentation (loss of ester groups) and in the formation of double bonds by hydrogen abstraction. Aging of the coatings even increases the diversity of their structure. Whereas oxidative aging is very moderate for the PMMA coatings deposited at low power input, the PMMA coatings prepared at high power input oxidize very rapidly especially during the first few days of exposure to air. A wide variety of carbon-oxygen functionalities was found after aging.

Plasma polymers from various aldehydes can uptake oxygen upon storage in air and decrease the water contact angle with time or right the opposite can happen depending on the monomer structure and the actual deposition conditions [78].

7.3.2.3. Nitrogen Containing Plasma Polymers

Gengenbach and Griesser extensively studied the aging behavior of plasma polymers which contain oxygen and nitrogen beside the carbon. In reference [79] they discussed some general features of the aging of this type of plasma polymer, namely the finding that the post plasma oxidation occurs preferably at carbon atoms bonded to nitrogen atoms and that the nitrogen ends up in amide and imide structures after a long aging.

The same authors investigated earlier the plasma polymers from heptylamine [80] and from alkyldiamines (1,3-diaminopropane [81] and 1,2-diaminoethane [82]) and a number of other amines [83]. Analogous to hydrocarbon plasma polymers, alkylamine plasma polymers show a rapid initial oxygen uptake during the first hours of exposure to air. The oxygen incorporation then slows down within some months, but it does not stop within one year. No oxidized N groups, such as nitroso, nitro or nitrate groups, were detected in the films. Like in hydrocarbon plasma polymer films, oxidation of carbon atoms not bonded to nitrogen atoms results in the formation of hydroxyl (OH) groups, carbonyl (C=O) groups and to a smaller extent carboxyl (COOH) groups are formed too. The nitrogen content of freshly deposited films is already lower than in the monomers, which can be explained by a scission of C-N bonds during the plasma polymerization. During aging in air the N content, as determined by surface analysis, further decreases slightly, in the case of heptylamine plasma polymers e.g. by about 25 % within one year [80]. According to Gengenbach and Griesser [82] several processes can account for this loss of nitrogen at the film surface: an oxidative scission of C-N bonds, a hydrolysis of imide groups formed during plasma polymerization and a surface rearrangement during aging, which results in the migration of polar amine groups from the surface to the less polar bulk of the plasma polymer.

7.3.2.4. Fluoroaromatic Plasma Polymers

Clark and Abu-Shbak studied the aging behavior of pentafluoropyridine and pentafluorobenzene plasma polymers in air [84]. Whereas plasma polymers produced from per- or pentafluorobenzene keep their hydrophobic nature upon exposure to air, pentafluoropyridine plasma polymer films show a rapid change from nonwetting to wetting. This effect is explained with a hydrolysis reaction. In comparison with fluorobenzenes, fluoropyridines exhibit a higher reactivity for a nucleophilic attack of water molecules due to the presence of the N atom in the aromatic ring. This accounts for the increasing hydrolytic instability of fluorinated heteroaromatic polymers with increasing number of nitrogen atoms. Munro and Allaker stored plasma polymers from 2,4,6-trifluoro-1,3,5-triazine (scheme 7.26)

(7.26)

for 10 days in air in glass bottles and observed significant etching of the storage bottles due to hydrogen fluoride formed during the aging [85].

7.3.2.5. Siloxane and Silazane Plasma Polymers

One very early paper on the aging of plasma polymers was published by Wrobel [86] and dealt with organosilicons prepared from hexamethyldisilane, hexamethylcyclotrisilazane. These plasma polymers react with oxygen and water in a rather complex mechanism. The reaction is controlled by the diffusion of oxygen and water into the film structure. Organosilicon plasma polymers prepared from hexamethyldisiloxane (HMDSO) show a very good long-term stability in air. Gengenbach and Griesser studied the structure of freshly deposited HMDSO plasma polymers and their structural changes during storage in air over a period of several years [87]. Whereas conventionally polymerized polydimethylsiloxane (PMDS) consists of linear chains, the plasma polymerization process leads to the formation of crosslinks by a partial cleavage of Si-CH_3 bonds in the HMDSO monomers. During the first days of exposure to air the organosilicone plasma polymers show no measurable oxygen uptake which is in contrast to the rapid oxidation of most other plasma polymers. Long-term storage of the HMDO plasma polymers in air results only in minor structural changes such as a small extent of methyl group abstraction and the formation of additional siloxane crosslinks.

As opposed to the high stability of the siloxane (Si-O-Si) group, the structure of disilazane (Si-N-Si) based plasma polymers is extremely unstable in air [73]. Already during plasma polymerization of hexamethyldisilazane (HMDS) there is extensive monomer fragmentation which includes not only methyl group abstraction but also the scission of Si-N bonds. As a result freshly deposited HMDS plasma polymers show a much higher structural diversity than HMDSO plasma coatings. During storage of HMDS plasma polymers in air about 50 % of the silazane groups are lost within 6 months and 90 % within one year. The silazane groups are highly susceptible to hydrolysis which leads to their conversion into crosslinked siloxane structures. The structure of the aged HMDS plasma polymers appears to stabilize after 1-2 years and their final structure is rather similar to the HMDSO based plasma polymers.

7.3.2.6. Thermal Degradation of Plasma Polymers

There have been very few studies in literature addressing the thermal and thermo-oxidative stability of plasma polymers. Most of these studies concern fluorine containing plasma polymers [88-91] and almost all were performed as thermogravimetric analysis (TGA) experiments under a nitrogen atmosphere [88,90-92].

Behnisch at al., however, studied the thermo-oxidative degradation of fluorohydrocarbon plasma polymers in air [89]. Three plasma polymers with different fluorine to carbon ratios from 0 to 2 were subjected to a thermooxidative cycling, i.e. they were repeatedly heated to 200 °C. For the "teflon-like" plasma polymer with a F/C ratio of 1.93 the oxygen incorporation was minimal, while a higher hydrogen content (or lower F/C ratio) reduces the stability. However, all three fluorohydrocarbon plasma polymers showed a low thermal long-term stability. After 10-20 h of thermooxidative cycling the mass loss was 50 %.

The thermal stability of perfluoroallylbenzene [90,91] and pentafluorostyrene [91] polymers deposited under pulsed plasma conditions was investigated by Han, Timmons and Lee. In contrast to the corresponding plasma polymers synthesized under continuous discharge conditions the pulsed plasma polymerization results in mostly linear polymers in which the monomer structure is retained to high degree. However, the plasma polymer films as deposited showed only a poor thermal stability in TGA experiments. Both the perfluoroallylbenzene and the pentafluorostyrene polymers films suffered considerable weight loss and thickness reduction above 300 °C.

For perfluorobenzene plasma polymer coatings and for powdered pentafluoropyridine plasma polymers Clark and Abu-Shbak reported a fast thermal decomposition under a nitrogen atmosphere at about 600 K [88]. Below 600 K the rate of weight loss was low.

Chou and Chen analyzed the thermal stability of plasma polymers from four different organosilicone monomers: hexamethylcyclotrichlorotrisilazane (HMCTSN), hexamethyldisilazane (HMDSN), hexamethyldisiloxane (HMDSO) and tetramethylsilane (TMS) [92]. Under a nitrogen atmosphere all these plasma polymers had good thermal stability. Up to 350 °C there were only minor differences between the four plasma polymers. At higher temperature the thermal stability decreased in the order HMCTSN > HMDSO > HMDSN > TMS. The thermal stability of hexamethyldisiloxane (HMDSO) plasma polymers is higher than that of hexamethyldisilazane (HMDSN) plasma polymers due to the higher binding energy of the Si-O bonds.

7.3.2.7. Photooxidation of Plasma Polymers

In contrast to the extensive research on the photostability of traditional polymers, very little has been reported concerning the photooxidation of plasma polymers. Potter, Ward and Short studied the photostability of styrene and methyl methacrylate (MMA) plasma polymers [93]. Both plasma polymers were deposited during a continuous discharge and irradiated in air for up to 8 weeks with a UV lamp in the range λ = 300 – 410 nm which was similar to the UV radiation of natural sunlight.

Unlike conventional polystyrene that undergoes photooxidation after an induction period of about 8 weeks the styrene plasma polymer immediately incorporates oxygen without an induction period. The rate of oxygen uptake in the styrene plasma polymer decreases only after about 10 days of irradiation. During the oxidation process C-O, C=O and O-C=O groups are formed in the plasma polymer. The MMA plasma polymer was prepared under conditions that ensured considerable fragmentation of the monomer, leaving only about 25 % of the ester groups intact. Under these preparation conditions the MMA plasma polymer also suffered rapid photooxidation without any induction time.

This behavior is in sharp contrast to conventional PMMA. However, the authors did not report result of control samples which were not irradiated. Gengenbach et al. compared their results of the atmospheric aging of methyl methacrylate plasma polymers [77] with the results of Potter at al. and state that the radiation has only a minor effect.

7.3.3. General Features of Plasma Polymer Degradation

The investigations which have been summarized in the preceding sections reveal two types of processes that result in plasma polymer aging and degradation. The first type of process is caused by free radicals which are buried in the layer after the deposition often in substantial concentrations. These radicals react readily with oxygen when the layers are exposed to air. This reaction results in the formation of transient peroxy radicals and hydroperoxides which then decompose slowly as described in Section 7.2.3. Because one single radical can result in the incorporation of many oxygen atoms, the alterations in the plasma polymer layer can be tremendous.

The second type of processes comprises reactions of weak structures. Depending on the nature of these structures and on the storage conditions, the reaction can be very different and the reaction rates can cover a wide range. One can find similarities with traditional polymers.

7.4. Preparation of Stable Plasma Polymers

7.4.1. Reduction of the Radical Concentration

The first and most important step towards more stable plasma polymers is the reduction of the radical concentration in the layer. There were reported different strategies in literature which will be discussed in the following sections.

7.4.1.1. Copolymerization

It was found that the concentration of free radicals trapped in the layer during plasma polymerization can be significantly reduced by a copolymerization with a suitable comonomer. As mentioned in section 7.3.2.1. acetylene plasma polymers contain a large amount of free radicals and gradually age in air. Adding nitrogen or carbon monoxid to the acetylene plasma results in layers with a slightly lower radical concentration. The addition of water to an acetylene/ nitrogen plasma or an acetylene/ carbon dioxide plasma reduces the concentration of free radicals drastically and the chemical structure of the layers remains more or less unchanged during a 15-month period in air [94].

7.4.1.2. Thermal Annealing

Plasma polymers can be stabilized by thermal annealing. The thermal treatment increases the mobility of the crosslinked polymer chains and therefore favors the recombination of trapped free radicals [66,95]. Morita, Sawa and Ieda determined the concentration of trapped radicals in freshly deposited styrene plasma polymers as a function of temperature [95]. They observed a significant decrease of the radical concentration up to 200 °C. Excessive heat treatment, however, leads to the formation of new free radicals due to the breaking of chemical bonds. In the case of the styrene plasma polymer, there was an abrupt increase in the concentration of free radicals above 200 °C.

Han, Timmons and Lee reported a dramatically increased thermal stability of pulsed plasma polymers from perfluoroallylbenzene [90,91] and pentafluorostyrene [91] by thermal annealing under nitrogen at 400 °C and 280 °C, respectively. After thermal annealing both plasma polymers showed excellent thermal stability up to a temperature of more than 400 °C. However, the thermal treatment resulted in a considerable weight loss.

7.4.1.3. Quenching of Trapped Radicals

A third strategy of reducing the radical concentration in plasma polymer films can be a controlled reaction. Friedrich et al. reported the stabilization of an acetylene pulsed plasma film by introducing nitric oxide (NO) into the plasma reactor at the end of the deposition [70]. NO is a relatively stable radical. It reacts readily with carbon alkyl and alkoxy radicals forming nitroso groups, oximes, nitro groups or nitrated depending on the constitution of the original radical [96]. As a small molecule NO can saturate all trapped radicals also in the bulk of a thin plasma polymer film. After 5 weeks of storage in air the NO quenched plasma polymer showed no sign of oxidation.

7.4.2. Design Rules for Stable Plasma Polymers

An essentially radical free plasma polymer reacts very much like a traditional polymer with the same chemical structure. However, as discussed in section 7.3.1. , most plasma polymers exhibit a number of structural features which are rarely found in traditional polymers and which comprise weak bonds.

One of the most pronounced characteristics of a plasma polymer is its high degree of crosslinking. A linkage in a plasma polymer very often consists of a tertiary carbon atom. A hydrogen atom at this kind of carbon is bonded weakly and it is susceptible to the abstraction by a thermal energy input or by the action of a radical. The β-C-C bond is weak too in such a structure. Many carbon based plasma polymers comprise a substantial amount of multiple bonds which result in a weakening of the C-C bond in the β position (see Section 7.1. and scheme 7.2). Consequently, a lower concentration of these structures will increase the stability of plasma polymers. Reducing the concentration of linkages and multiple bonds actually means to prepare a plasma polymer with a more regular structure. However, in many cases the crosslinking is an integral part of the structural characteristics of a plasma polymer coating which the particular application needs.

Beside these general features, there can be many more weak structures in plasma polymers. Depending on the monomer and on the intended application, the critical structures can vary and with them the limiting application conditions. For example: A good hydrophilic film can be prepared if it contains ether groups, hydroxyl groups, or carboxyl groups. However, all these groups have necessarily weak bonds in their neighborhood which limit the maximum application temperature. Consequently, these films are hydrophilic but they can not be used at high temperatures. In high temperature applications the films have to follow other rules. The concentrations of oxygen, of tertiary carbon atoms, and of double bonds (isolated or in small sequence) must be as low as possible. Aromatic rings and/ or perfluorinated structures have to be preferred.

The stability of a plasma polymer coating does not only depend on the plasma polymer itself but also on the structural properties of the interphase (see section 7.3.1.2.). A coating on metals and oxides comprises metal organic groups which can undergo hydrolysis reactions much more readily than the overlying plasma polymer. The surface of polymers can be weakened by inappropriate conditions during the plasma deposition process. Also in such a case the coating is less stable than the plasma polymer structure would suggest.

The design of a stable plasma polymer coating asks also for the consideration of the substrate in order to obtain a good and durable adhesion. In a number of cases it could be necessary to adjust the deposition conditions in the first stage of the coating formation to the needs of the substrate. Because the combinations of substrates and coatings are numerous we will only give some examples. The coating of polymers which tend to decompose quickly due to the action of a high power plasma like e.g. PP or PMMA results in more stable products if the energy input is reduced in the beginning of the deposition process. Because the nitrogen containing metal compounds are often less stable than the corresponding oxygen structures it could be advantageous to deposit a nitrogen containing layer not directly on a metal substrate but on a underlayer which promises to be less susceptible to the attack of agents from the application environment.

7.5. Conclusions

The preparation of stable plasma polymer coatings asks for more control over the chemical structure of the films. It is not primarily a question of retaining the monomer structure. The minimal concentration of free radicals in the layer is the most crucial point. Weak bonds can be the starting point of the coatings degradation by decomposing the plasma polymer itself or by resulting in an adhesive failure. Therefore, the desired long term stability can only be obtained if the formation of certain groups is avoided.

References

1. Van Krevelen D. W., *Properties of Polymers* (Elsevier Science, Amsterdam, 1997) 635.
2. Christen H. R. and Vögtle F., *Organische Chemie*, Band 1, (Otto Salle Verlag, Frankfurt, 1988) 411.
3. Bamford C. H. and Tipper C. F. H. (eds.), *Degradation of Polymers* in *Comprehensive chemical kinetics* Vol. 14 (Elsevier Scientific, Amsterdam, 1975).
4. Grassie N. and Scott G., *Polymer Degradation and Stabilization* (Cambridge University Press, Cambridge, 1985).
5. Winslow F. H. and Hawkins W. L. in *Crystalline Olefin Polymers Part 1*, ed. Raff R. A. V., Doak K. W. (Interscience, New York, 1965).
6. Bockhorn H., Hornung A., Hornung U. and Schawaller D., *J. Anal. Appl. Pyrolysis* **48** (1999) 93.
7. Wall L. A., Brown D. W., Straus S. and Simha R., *J. Amer. Chem. Soc.* **76** (1954) 3430.
8. Davis T. E., Tobias R. L. and Peterli E. B., *J. Polym. Sci.* **56** (1962) 1487.
9. Reginato L., *Makromol. Chem.* **132** (1970) 113.
10. Yao Q. and Wilkie C. A., *Polym. Degr. Stab.* **69** (2000) 287.
11. Madorsky F. H., *Thermal Degradation of Organic Polymers*, (Interscience, New York, 1964) 100.
12. Belavicqua E. M. in *Thermal Stability of Polymers,* ed. Conley R. T. (Marcel Dekker, New York, 1970).
13. Madorsky S. L., *Thermal Degradation of Organic Polymers,* (Interscience, New York, 1964) 26.
14. Brown D. W. and Wall L. A., *J. Phys. Chem.* **62** (1958) 848.
15. Jellinek H. H. G. and Kachi H., *J. Polym. Sci., Part C* **23** (1968) 97.
16. Wright W. in *Progress in high polymers, Vol. 2*, ed. Robb J. C. and Peaker F. W. (Heywood Books, London, 1968) 189.
17. Korshak V. V., Manucharova I. F., Vinogradova S. V. and Salazkin S. N., *Vysokomol. Soedin.*, 7 (1965) 1614.
18. Grassie N. and Roche R.S., *Makromol. Chem.* **112** (1968) 16.
19. Dudina L. A., Karmilova L. V., Tryapitsyna E. N. and Enikolopyan N. S., *J. Polym. Sci., Part C* **16** (1967) 2277.
20. Madorsky S. L. and Straus S., *J. Polym. Sci.*, **36** (1959) 183.

21. Blyumenfeld A. B. and Kovatskaya B. M., *Vysokomol. Soedin., Ser. A* **12** (1970) 633.
22. Gardner D. L. and McNeill I. C., *J. Therm. Anal.* **1** (1969) 389.
23. McNeill I. C., *Eur. Polym. J.* **4** (1968) 21.
24. Cameron G. G. and Kerr G. P., *Makromol. Chem.* **115** (1968) 268.
25. Grassie N. and McCallum J. R., *J. Polym. Sci. A* **2** (1964) 983.
26. Grant D. M. and Grassie N., *Polymer* **1** (1960) 125.
27. Geuskens G., Hellinckx E. and David C., *Eur. Polym. J.* **7** (1971) 561.
28. Cameron G. G. and Kane D. R., *Makromol. Chem.* **109** (1967) 194.
29. Grassie N. and Speaksman J. G., *J. Polym. Sci. A 1* **9** (1971) 919.
30. Grassie N. and Speaksman J. G., *J. Polym. Sci. A 1* **9** (1971) 931.
31. Grassie N. and Speaksman J. G., *J. Polym. Sci. A 1* **9** (1971) 949.
32. Schaefgen J. R. and Sarosohn I. M., *J. Polym. Sci.* **58** (1962) 1049.
33. Wiloth F., *Makromol. Chem.* **144** (1971) 283.
34. Peebles L. H. and Huffman M. W., *J. Polym. Sci., Part A-1* **9** (1971) 1807.
35. Pohl H. A., *J. Am Chem. Soc.*, **73** (1951) 5660.
36. Birladeanu C., Vasile C. and Schneider I. A., *Mokromol. Chem.* **177** (1976) 121.
37. Davis A., *Makromol. Chem.* **132** (1970) 23.
38. Bruck S. D., *Polymer* **6** (1965) 49.
39. Arnold C. J., J. *Polym Sci: Macromolecular Reviews* **14** (1979) 265.
40. Wright W. in *Progress in High Polymers, Vol. 2*, ed. Robb J. C. and Peaker F. W. (Heywood Books, London, 1968) 193.
41. Wall L. A. (ed.), *Fluoropolymers* (Interscience, New York, 1972).
42. Scheirs J. (ed.), *Modern Fluoropolymers* (Wiley, New York, 1997).
43. Jellinek H. H. G. and Kachi H., *Polym. Eng. Sci* **5** (1965) 200.
44. Wright W.W. in *Thermal Stability of High Polymers, Vol. 1*, ed. Conley R. T., (Marcel Dekker, New York, 1970) 287.
45. Madorsky F. H., *Thermal Degradation of Organic Polymers*, (Interscience, New York, 1964) 138.
46. Usmanov Kh. U., Yulchibaev A. A. and Khalikov K. R., *Vysokomol. Soedin., Ser. B* **12** (1970) 327.
47. Oksentevich L. A and Pravednikov, *Vysokomol. Soedin., Ser. B* **10** (1968) 49.
48. Knight C. J. and Wright W. W., *J. Appl. Polym. Sci.* **16** (1972) 683.
49. Ameduri B., Boutevin B. and Kostov G., *Prog. Polym Sci,* **26** (2001) 105.

50. Rodé V. V., Verkhotin M. A. and Rafikov S. R., *Eur. Polym. J.* (Supplement) (1969) 401.
51. Thomas T. H. and Kendrick T. C., *J. Polym. Sci., Part A-2* 7 (1969) 537.
52. Bolland J. L., *Quart. Rev. Chem. Soc.*, **3** (1949) 1.
53. Conley R. T. (ed.) *Thermal Stability of Polymers*, (Marcel Dekker, New York, 1970).
54. Reich L. Stivala S. S., *Auto-Oxidation of Hydrocarbons and Polyolefins*, (Marcel Dekker, New York, 1969).
55. Herzberg G., *Molecular Spectra and Molecular Structure*, ((Van Nostrad, New York, 1950) 560.
56. Chien J. C. W., Vandenberg E. J. and Jabloner H., *J. Polym. Sci., Part A-1* **6** (1968) 381.
57. Adams J. H., *J. Polym. Sci., Part A-1* **8** (1970) 1077.
58. Adams J. H. and Goodrich J. E., *J. Polym. Sci., Part A-1* **8** (1970) 1269.
59. Hansen R. H. in *Thermal Stability of Polymers,* ed. Conley R. T., (Marcel Dekker, New York, 1970) 153.
60. Painter R., Arakawa E. T., Williams M. W. and Ashley J. C., *Radiation Research* **83** (1980) 1.
61. Seki K., Tanaka H., Ohta T., Aoki Y., Imamura A. and Fujimoto H., *Phys. Scr.* **41** (1990) 167.
62. Onari S., *J. Phys. Soc. Japan* **26** (1969) 500.
63. Guillet J. E., Norrish R. G. W., *Proc. Roy. Soc., Ser. A* **233** (1956) 153.
64. Amerik Y., Guillet J. E., *Macromolecules* **4** (1971) 375.
65. Fox R. B., Isaacs L. G., Stokes S., *J. Polym. Sci., Part A* **1** (1963) 1079.
66. Yasuda H., *Plasma Polymerization* (Academic Press, Orlando, 1985) 334.
67. Biederman H. and Osada Y., *Plasma Technology 3: Plasma Polymerization Processes* (Elsevier, Amsterdam, 1992) 3-7, 73, 151.
68. Yasuda H., *J. Macromol. Sci. -Chem.* **A 10**, (1976) 383.
69. Yasuda H. and Hsu T., *J. Polym. Sci., Part A: Polym. Chem.* **15**, (1977) 81.
70. Friedrich J. et al., *Proc. 7th Conference on Plasma Surface Engineering* (2000) 230.
71. Martinu L., Wertheimer M. R. and Klemberg-Sapieha J. E., *Mat. Res. Soc. Symp. Proc.* Vol. **544** (1999) 251.
72. Wertheimer M. R., Fozza A. C. and Holländer A., *Nucl. Instr. and Meth. in Phys. Res.* **B 151** (1999) 65.

73. Gengenbach T. R. et al., *J. Polym. Sci., Part A: Polym. Chem.* **32**, (1994) 1399.
74. Gengenbach T. R., Chatelier, R. C. and Griesser H. J., *Polymer Preprints* **38** (1997) 1045.
75. Gengenbach T. R., Chatelier R. C. and Griesser, H. J., *Polymer Preprints* **38** (1997) 1004.
76. Gengenbach T. R., Zoran R. V., Chatelier R. C. and Griesser, H. J., *Plasmas and Polymers*, **1** (1996) 207.
77. Gengenbach T. R. and Griesser H. J., *J. Polym. Sci., Part A: Polym. Chem.* **36** (1998) 985.
78. Gong X. and Griesser H. J., *Plasmas and Polymers* **2** (1997) 261.
79. Gengenbach T. R., Chatelier R. C. and Griesser H. J., *Plasmas and Polymers* **4** (1999) 283.
80. Gengenbach T. R., Chatelier R. C. and Griesser, H. J., *Surf. Interf. Anal.* **24** (1996) 271.
81. Gengenbach T. R. and Griesser H. J., *J. Polym. Sci., Part A: Polym. Chem.* **37**, (1999) 2191.
82. Gengenbach T. R., Chatelier R. C. and Griesser, H. J., *Surf. Interf. Anal.* **24** (1996) 611.
83. Gengenbach T. R., Zoran R. V., Li S., Chatelier R. C. and Griesser, H. J., *Plasmas and Polymers* **2** (1997) 91.
84. Clark D. T. and Abu-Shbak M. M., *J. Polym. Sci., Part A: Polym. Chem.* **21**, (1983) 2907.
85. Munro H. S. and Allaker R. S., *J. Polym. Sci., Part A: Polym. Chem.* **22**, (1984) 2661.
86. Wrobel A. M., *J. Macromol. Sci. Chem.*, **A22** (1985) 1089.
87. Gengenbach T. R. and Griesser H. *J., Polymer* **40** (1999) 5079.
88. Clark D. T. and Abu-Shbak M. M., *J. Polym. Sci., Part A: Polym. Chem.* **21**, (1983) 2907.
89. Behnisch J. and Holländer A., *Proc. International Symposium on Plasma Chemistry ISPC13* (1999) 1287.
90. Han L. M., Timmons R. B. and Lee W. W., *J. Vac. Sci. Technol. B* **18** (2000) 799.
91. Han L. M., Timmons R. B. and Lee W. W., *Mat. Res. Soc. Symp. Proc.* **511** (1998) 93.
92. Chou S. and Chen S.-H., *Polym. Polym. Compos.* **8** (2000) 333.

93. Potter W., Ward A. J. and Short R. D., *Polym. Degrad. Stab.* **43** (1994) 385.
94. Yasuda H., Marsh H. C., Bumgarner M. O. and Morosoff N., *J. Appl. Polym. Sci.* **19** (1975) 2845.
95. Morita S., Sawa G. and Ieda M., *J. Macromol. Sci.-Chem.* **A 10** (1976) 501.
96. Rees Y. and Williams G. H. in *Advances in Free Radical Chemistry, Vol. 3*, ed. Williams G. H. (Logon Press New York, 1969) 199.

CHAPTER 8

APPLICATION OF ATMOSPHERIC PRESSURE DISCHARGE FOR PLASMA POLYMER PROCESSES

MASUHIRO KOGOMA
Faculty of Science and Technology
Sophia University
Tokyo, Japan

8.1. Introduction

Because a high quality process is possible, the surface modification and polymer deposition in low temperature plasma are very useful for the high technology industry. Most of such processes have been done at a low–pressure of a few torrs, and therefore it is difficult to apply plasma processing for large-scale objects, especially by continuous treatment. We previously reported that surface treatment and thin film deposition could be carried out with the atmospheric pressure glow plasma (APG) process. The APG process has been developed by M. Kogoma and coworkers since 1987 [1,2]. This approach can reduce apparatus costs and can also be applied to high vapor pressure substances such as gum, powders and bio-materials [3,4]. A stable glow discharge at atmospheric pressure requires: (i) The use of helium or argon for dilution of carrier gas; (ii) The insertion of a dielectric insulator between the electrodes in the discharge gap; (iii) The use of a high frequency, at least higher than 1kHz. Normally the processing rate is dependent on the applied frequency because the discharge current should have one or two pulses per half cycle in low frequency APG discharge [2]. However, the temperature rise inside the discharge zone due to the increase of the discharge frequency restricts the maximum applied frequency.

In this article, we report the application of APG discharge to deposit mainly such organic polymers as C_2H_4 (Et), C_2F_4 (TFE), and $Si(OC_2H_5)_4$ (TEOS). The films are investigated by SEM, ESCA, and FTIR. The study of the deposition process using emission spectrometry, IR absorption spectra and high-pressure mass spectrometry will be described.

8.2. Experimental

Figure 8-1a shows the schematics of our APG discharge chamber with parallel plate-type electrodes covered by glass insulator. The system was used for the measurement of deposition rate, for the emission spectrometry, and for the surface analysis by ESCA (SSI, SSX-100).

Figure 8-1b shows the schematic representation of the system that we used for the direct mass spectrometry in the APG discharge. In the latter system we used a Pyrex glass discharge tube of 6 mm inner diameter and 8 mm outer diameter. Double ring electrodes, each 5mm wide and with 1 cm separation between the two electrodes, cover the outer surface of the discharge tube. The

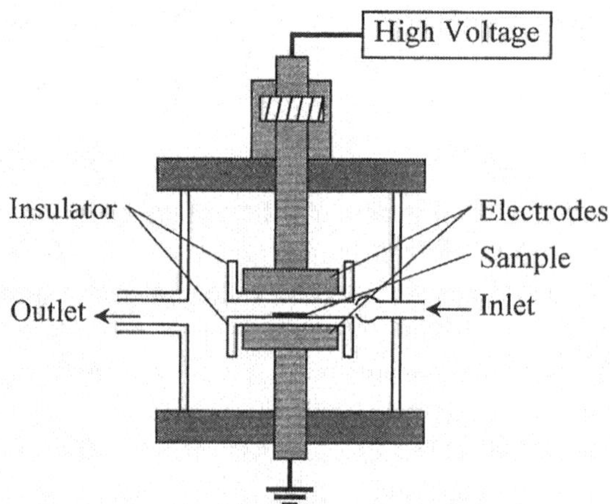

Figure 8.1a. Schematic representation of parallel type discharge apparatus.

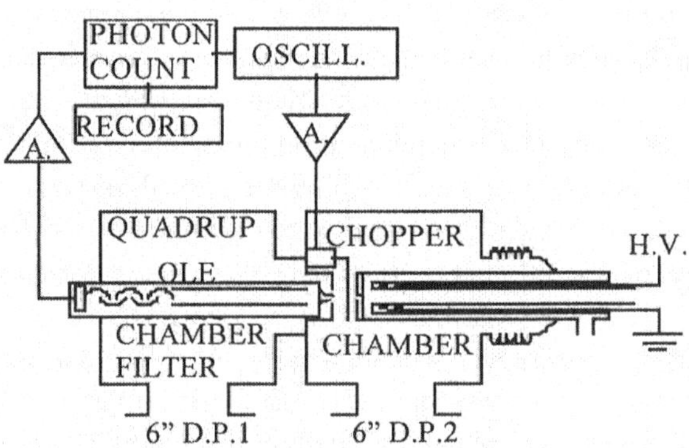

Figure 8.1b. Schematic representation of AP mass spectrometer with APG discharge tube.

grounded electrode is very closely faced with a Teflon sampling cone which has a 50 μm sampling hole. A molecular beam extracted from the hole passes through a differential pumping system; it is then introduced to the main chamber. A quadrupole mass spectrometer is set in the main chamber where the pressure is kept lower than 5×10^{-6} torr with dual 1300 l min^{-1} diffusion pumps. The carrier gas mixture of He and the monomers such as TFE, Et or CF_4 passes through the flow meters and is introduced into the discharge tube. The total pressure of the discharge gas is 1atm. The flow rates of the mixed gas are 2 l/min to 4 l/min; each includes 1 – 2 % of the monomer. We used the discharge sources of 3 kHz, 15 kHz and 120 kHz in frequency.

For the analysis, we used FTIR (FT7000, Japan Optics Co. LTD), ESCA (SSX-100 Surface Science Inst.), thickness meter (DEKTAK IIA, Slorn Co. LTD) and quadrupole mass spectrometer (MSQ190, ULVAC Co. LTD).

8.3. Results and Discussion

8.3.1 FTIR and Emission Spectra

Fig. 8.2 shows the FTIR absorption spectrum of the plasma-polymerized TFE (PPTFE) film deposited on the silicon wafer. In the figure, we found the typical CF stretching absorption peak that is very similar to that for the chemically

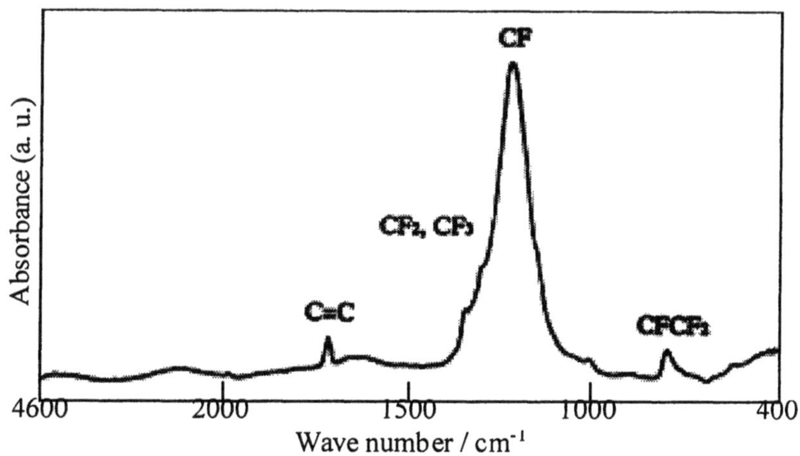

Figure 8.2. FTIR absorption spectrum of APG plasma - polymerized PPTFE deposited on the silicon wafer.

synthesized PTFE (PTFE) film. The deposited film also has similar physical properties to those of PTFE film, for examples, soft-white film and low friction surface material.

In the case of Et plasma-polymerized film (PPEt), we got a similar FTIR absorption spectrum to that of chemically synthesized Et (PEt) film. The physical properties of plasma-synthesized Et film are also similar to those of PEt film. Using SEM photography, we obtained very smooth surfaces for PPTFE and PPEt films on the substrates.

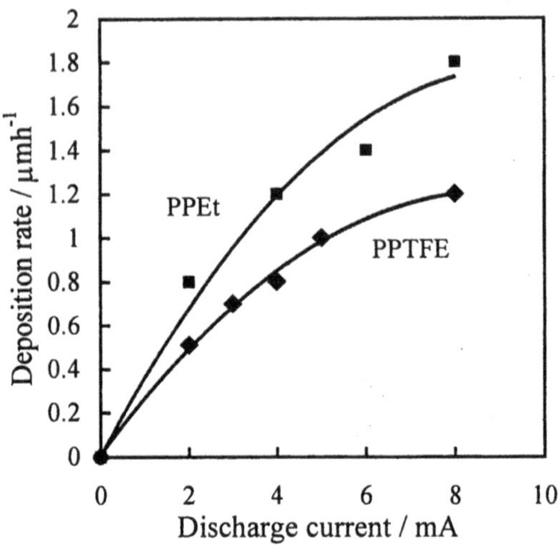

Figure 8.3. Deposition rates of PPEt and PPTFE as a function of discharge current.

Fig. 8.3 shows the deposition rates of PPTFE and PPEt as a function of the discharge current. Both deposition rates increase with increasing of the discharge current. The deposition rates are about the same order as that for the low-pressure glow plasma deposition. In the PPTFE film, the ratios of atomic composition of F/C measured by ESCA are between 1.4 – 1.7, depending on the discharge current and the monomer flow rate.

In Fig. 8.4, the ESCA C1s spectra of PPTFE contains many peaks, for example, CH, CF, CF_2 and CF_3. It seems that the film contains a 3-D network (cross-linking chain) with their fragments.

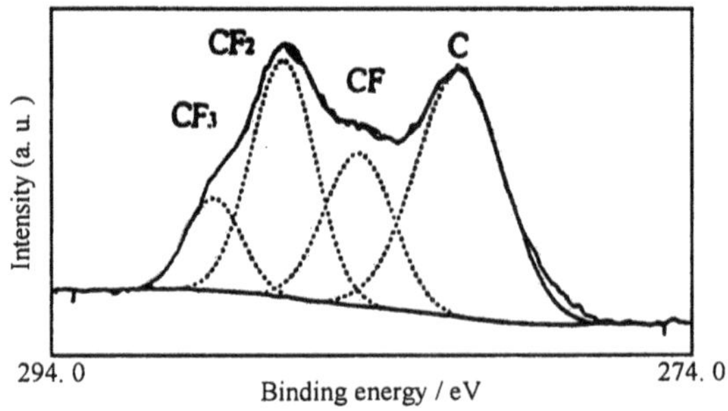

Figure 8.4. ESCA spectrum of PPTFE. He, 5 l min^{-1}; C_2F_4, 8.4 cm^{3}min^{-1}; 3 kHz; 3 mA.

Fig. 8.5 shows UV photo emission spectra of TFE/He plasma in 120 kHz APG discharge. In the spectra, we have strong emissions assigned for the CF_2 molecule. The same emission spectra were found in the C_2F_4 low-pressure glow plasma. But we found nothing in the emission

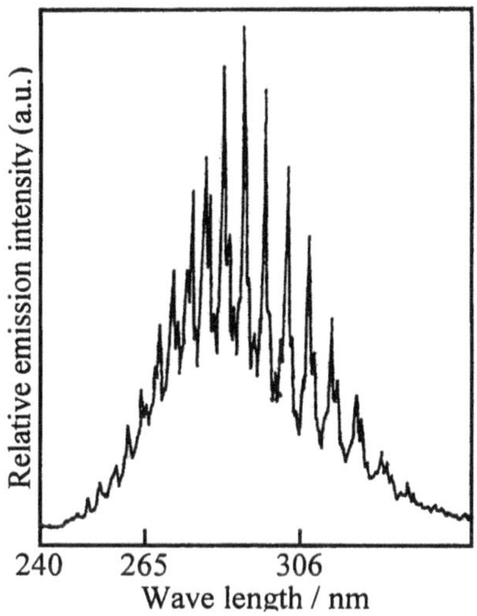

Figure 8.5. UV photo emission spectra of C_2F_4/He plasma. He, 4 l min^{-1}; C_2F_4, 4 cm^3min^{-1}; 120 kHz.

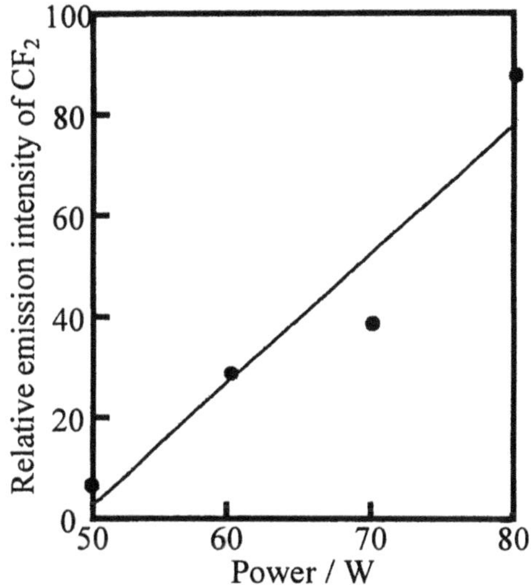

Figure 8.6. Relative emission intensity of CF_2 (265 nm) as a function of discharge power He, 2 l min^{-1}; C_2F_4, 1 cm^3min^{-1}; 120 kHz.

spectra related with CF_n in the visible area.

The relative emission intensity of CF_2 band spectra (~265 nm) increases with increasing of the discharge power, as shown in Fig. 8.6. We suppose that the CF_2 radical is one of the important precursors of the deposited polymer in C_2F_4/He APG discharge.

8.3.2 Mass Spectrometry of C_2H_4 and C_2F_4 in He APG Discharge

Because some kinds of radicals and fragments do not radiate the photo-emissions, we tried to use the direct mass spectrometry of the plasma zone. In the C_2F_4/He APG plasma, we found C_2F_4, HF and the radicals such as CF_2 and C_2F_3 in the mass spectrum. When we cut the discharge, we could not detect any radicals generated in the ionization chamber in vacuum. We used an ionization voltage lower than 18 V to restrain the fragmentation of monomer molecules.

Fig. 8.7 shows the relative intensities of $C_2F_4^+$, CF_2^+ and $C_2F_3^+$ as a function of the discharge current. In Fig. 8.7, CF_2^+ and $C_2F_3^+$ increase and C_2F_4 decrease with increasing of the current. On the other hand, in Fig. 8.8, the intensity of CF_2^+ decreases and that of $C_2F_3^+$increases with increasing of the monomer (C_2F_4) flow rate. Many previous reports suggest that the important

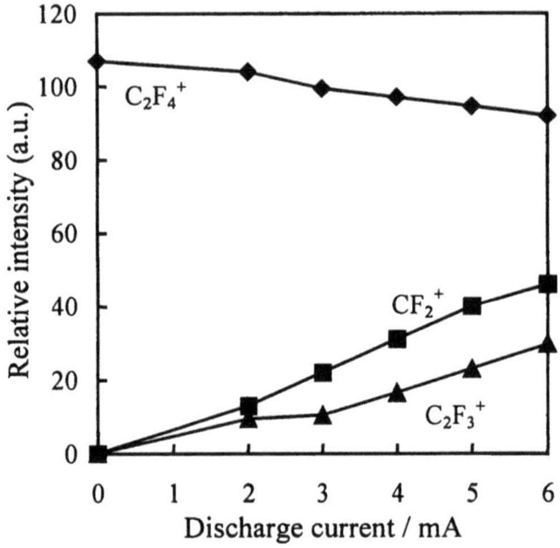

Figure 8.7. Relative intensity of C_2F_4, CF_2 and C_2F_3 ions as a function of discharge current. He, 2 l min^{-1}; C_2F_4, 1 cm^3min^{-1}; 120 kHz; 70 W.

precursors of PPTFE are CF_2 and C_2F_3 radicals in TFE discharge. The mechanism is as follows:

$$C_2F_4 + e \rightarrow 2\,CF_2 + e \tag{8.1}$$

$$C_2F_4 + CF_2 \rightarrow C_3F_6 \tag{8.2}$$

$$C_3F_6 \rightarrow CF_3 + CF_2{=}CF \rightarrow \text{polymerization} \tag{8.3}$$

So, the deposition mechanism of PPTFE in APG plasma has almost the same scheme except for the electron exciting mechanism. In He APG discharge, high energy and long life meta-stable He atoms can dissociate the monomer, so then the later chemical reaction will be started [5]. A C_2H_4 molecule has the same molecular structure as C2F4 does, so we tried to compare the mass spectra of C2F4/He and C_2H_4/He extracted from APG plasma. In the C_2F_4/He plasma, we found only $C_2F_4^+$, H_2^+, $C_2H_2^+$, and He^+ in the mass spectrum. Figure 8-9 shows the relative intensity of $C_2H_2^+$ vs. discharge current. The C_2H_2 concentration is almost linearly dependent on the discharge current. A previous report suggests that the acetylene molecule is one of the important precursors in the ethylene plasma polymerization scheme. So, Et-APG plasma will have an analogous polymerization mechanism with that in the low pressure Et plasma, but not one like that in the TFE deposition mechanism. Table 8.1. shows the dissociation energy of C_2F_4 and C_2H_4.

The dissociation energy of C_2H_4 to produce CH_2 is much higher than that of C_2H_2 production reaction. This is the reason why C_2H_4 does not produce CH_2 radical. So, TFE and Et do not have analogous schemes to produce polymers in APG plasma. As a consequence, the chemical processes of the polymerization of the C_2F_4 and of the C_2H_4 in the APG plasma are exactly the same as those in the low-pressure plasma.

Table 8.1. Dissociation energy of monomers.

Reactions	ΔH / kcalmol^{-1}	Reactions	ΔH / kcalmol^{-1}
$C_2F_4 \rightarrow 2CF_2$	70.4	$C_2H_4 \rightarrow 2CH_2$	171.7
$C_2F_4 \rightarrow C_2F_2 + F_2$	162.4	$C_2H_4 \rightarrow C_2H_2 + H_2$	41.7
$C_2F_4 \rightarrow C_2F_2 + 2F$	200.2	$C_2H_4 \rightarrow C_2H_2 + 2H$	145.0

8.3.3 TEOS Deposition by APG plasma

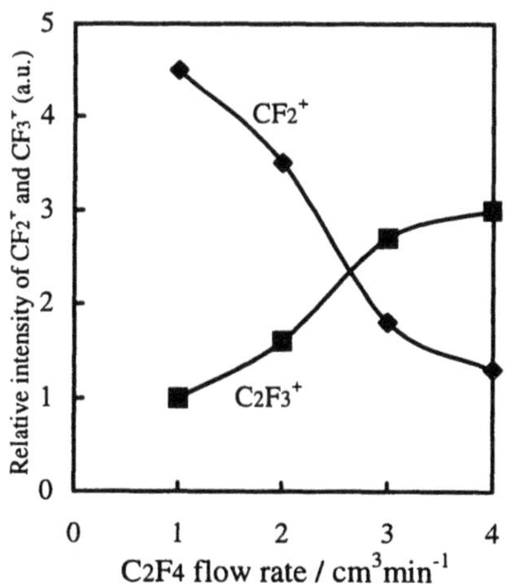

Figure 8.8. Relative intensity of CF_2^+ and $C_2F_3^+$ as a function of C_2F_4 flow rate. He, 2 l min^{-1}; C_2F_4, 1 cm^{3}min^{-1}; 3kHz.

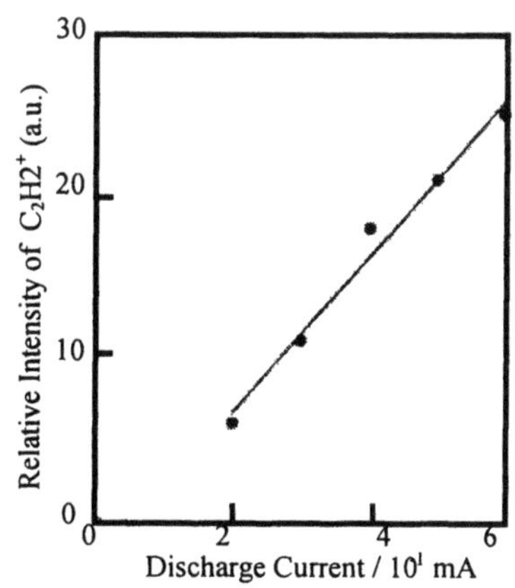

Figure 8.9. Relative intensity of $C_2H_2^+$as a function of discharge current. He, 2 l min^{-1}; C_2H_4, 2 cm^{3}min^{-1}; 3 kHz.

A silica-coated film has many important applications, for example, insulation layer on the silicon wafer, gas barrier coating, water protection and hard coating.

Fig. 8.10 shows FTIR spectra of TEOS monomer and TEOS deposition film on the silicon wafers prepared in TEOS/He or TEOS-O_2/He. It is clear that TEOS films deposited without oxygen retained the structure of the starting material. But the TEOS polymer obtained by adding oxygen has inorganic characteristics even at 50 °C of the substrate temperature. Due to the OH groups, the new broadband absorption peaks at 3650, 3400 and 930 cm^{-1} are created.

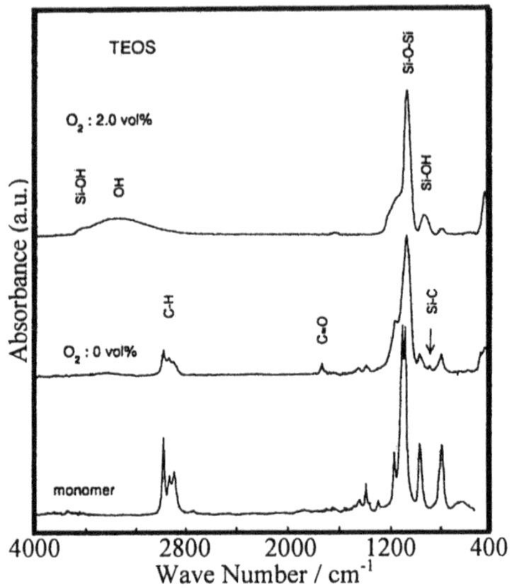

Figure 8.10. FTIR spectra of TEOS monomer and TEOS deposition film on the silicon wafer prepared in TEOS/He and TEOS-O_2/He.

The SEM observation of the deposited film shows that the surface is smooth and uniform [6].

Table 8.2 shows the atomic compositions of C, Si and O in the films analyzed by ESCA. The film prepared in TEOS/He contains a large amount of carbon component. However, we do not have any carbon component in the deposited material obtained in TEOS-O_2/He system. The ratio of O/Si in the latter films is about 1.8 - 1.9. The film density in the oxygen-rich condition is very near to that of the pure silica. It seems that the former one is a polymer-like film but the latter one is a silica film.

Table 8.2. The results of ESCA; Power, 50 W; TEOS flow rate, 10.3 mg/min; Substrate temperature, 50 °C.

Oxygen	Film composition/%		
concentration / %	C	Si	O
0	33	29	38
0.3	0	36	64

Fig. 8.11 shows the deposition rate of the film and the film density as a function of oxygen % in the mixture. The initial deposition rate first increases

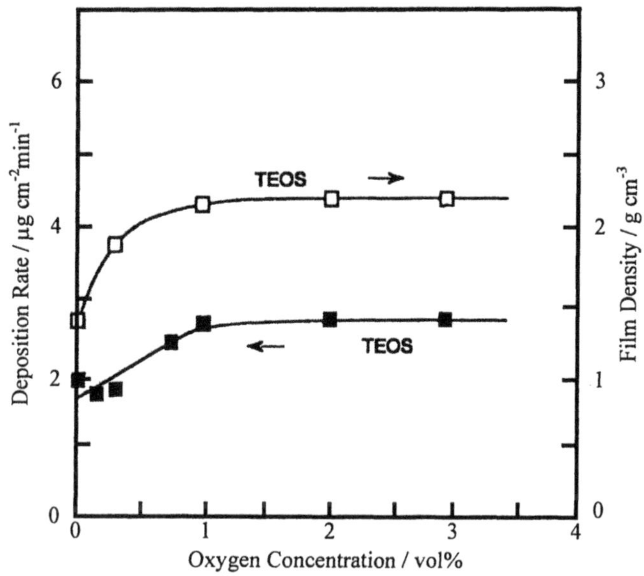

Figure 8.11. Deposition rate and film density of deposited polymers as a function of oxygen concentration.

with increase of oxygen %, then reaches a plateau. The film density also increased and approached a value of 2.2 gcm^{-1}, the density of amorphous silica. It is known that, in the siloxane-based polymer deposition in low-pressure plasma, the excited oxygen promotes gas-phase decomposition of monomer molecules. In the APG plasma, oxygen is assumed to play the same role. Therefore, adding of oxygen increases the deposition rate. However an excessive amount of oxygen inhibits the polymerization reaction and restrains further increase of the rate.

From the point of view of the cost and simplicity of the plasma systems, the APG method has many advantages over the low-pressure method.

References

1. Kanazawa S., Kogoma M., Moriwaki T. and Okazaki S., *Proc. Japan Symp. Plasma Chemistry*, Tokyo, **3** (1987) 1839.
2. Yokoyama T., Kogoma M., Kanazawa S., Moriwaki T. and Okazaki S., *J. Phys. D: Appl. Phys.*, **23** (1990) 374.
3. Sawada Y. and Kogoma M., *Powder Technology*, **90** (1997) 245.

4. Mori T., Tanaka K., Inomata T., Takeda A. and Kogoma M., *Thin Solid Films*, **316** (1998) 89.
5. Kobayashi H., Bell A. T. and Shen M., *Macromolecules*, **7** (1974) 277.
6. Sawada Y., Ogawa S. and Kogoma M., *J. Appl. Phys.*, **28** (1995) 1661.

CHAPTER 9

HARD PLASMA POLYMERS, COMPOSITES AND PLASMA POLYMER FILMS PREPARED BY RF SPUTTERING OF CONVENTIONAL POLYMERS

HYNEK BIEDERMAN, PAVEL KUDRNA AND DANKA SLAVÍNSKÁ
Department of Macromolecular Physics
Faculty of Mathematics and Physics, Charles University,
Prague, Czech Republic

9.1. Introduction

The energy flux into the growing plasma polymer film is decisive for the film structure and final properties. An increase of this energy results in films that are usually harder becomes more disordered and crosslinked with an excess of carbon. If the energy flux is very low, a plasma polymer is created that retains more of the molecular structure of the monomer and resembles more of a conventional polymer.

Within about the last two decades, the three directions of the plasma polymer field appeared:

1) new materials including hard coatings and composite films

In this case the energy flux is often realised by some kind of irradiation, usually by means of energetic positive ions or neutrals with the mean energy more than 10 eV up to 100 eV, while the substrate temperature is kept at moderate values (< 300 °C)

2) plasma polymers resembling conventional polymers

These are usually prepared by substantially reducing the energy flux into the growing film e.g. by preventing the growing film from energetic particle (corpuscules and photons) bombardment and keeping the substrate at room and below room temperatures.

3) plasma polymers prepared under moderate energetic conditions

In this case no special arrangements are done in the plasma polymerization process as regarding the energy of particles bombarding the substrate and its temperature. The substrate is at a floating potential or it is earthed and because the plasma potential, in usual plasma reactors, is not extremely high, the mean energy of e.g. positive ions remains below 10 eV and the substrate temperature below 300 °C. This third direction of plasma polymerization field is likely its mainstream.

Plasma polymers discussed in this chapter will fall in the first and in the third directions mentioned above. Several reviews have appeared in the first direction e.g. [1-3]. In the third direction, where we focus on sputtering of polymer targets, reviews [3,4] may be also helpful to the reader.

9.2. Hard Plasma Polymer and Hard Carbon Films

If a growing hydrocarbon plasma polymer is subjected to energetic positive ion bombardment, then its structure, composition and properties are changed. It becomes more disordered, the ratio C/H grows and the hardness increases.

Hard plasma polymer films were first prepared by König and Helwig [5] who used a DC glow discharge operated in benzene and grew the films on the anode. They characterized the films by IR spectroscopy [6]. The first a-C:H films were found by Schmellenmeier [7]. Review on these early stages of carbonaceous films was published by Pagnia [8].

Low temperature plasma deposition using a DC glow discharge in a hydrocarbon gas was suggested in order to grow the carbonaceous film on the ion bombarded cathode [9]. The most simple way was proposed by Holland who suggested using an RF glow discharge in a hydrocarbon gas [10]. He deposited a carbonaceous films on the negatively biased probe capacitively coupled to a RF power supply. In later publication Nyaiesh and Holland [11] related the growth of the film to the RF power input per unit area of the receiver (RF excitation electrode) and unit gas pressure (for methane) and claimed that the soft polymer was deposited from zero up to 5 $Wcm^{-2}Torr^{-1}$. From this value up to 9 $Wcm^{-2}Torr^{-1}$, hard plasma polymers were obtained while the temperature of the growing film did not exceed 175 °C. In the range from 10 to 23 $Wcm^{-2}Torr^{-1}$, a-C:H film was grown. Above 23 $Wcm^{-2}Torr^{-1}$, a graphite like film was created.

The example of hard carbonaceous film preparation using the simplest way – i.e. according to Holland [10] may be seen in Fig. 9.1. Any kind of substrate resting on the RF excitation electrode that is capacitively coupled to an RF power supply develops a DC negative self bias. Then positive ions with the energy that corresponds to the value of the bias are bombarding the growing film. It was shown by Biederman et al. [12] that in the case of an RF (20 MHz) discharge operated in benzene at 2 Pa, soft plasma polymer films were deposited for bias values within 0 and -100 V. For bias values ranging from -100 V to -180 V (-200 V), hard plasma polymer films were obtained and for even more increased negative bias values, a-C:H films were grown. Amorphous hydrogenated carbon films (a-C:H) have been considered as a special class of films and subjected to vast investigations because of their special properties: high electrical resistivity, hardness higher than sapphire, optical transparency in the infrared, and chemical inertness. For more information, the reader should look for e.g. a special journal (Diamond and Related Materials) published by Elsevier.

Further studies by Wagner et al. [13] investigated in detail the deposition process using benzene at a pressure of 3 Pa and a DC negative bias $V_B = -1000$ V. Mass spectroscopy has shown that the dominant specie among the positive ions is $C_6H_6^+$. Spatially resolved optical emission spectroscopy has shown that $C_6H_6^+$ ions are in part extracted from the plasma glow and accelerated by a sheath potential (close to V_B); at impact on the substrate they are broken into fragments. The majority of them serve for a-C:H film growth, but a part of them are back-scattered. These fragments, e.g. C_2 and especially CH, do not depend on the kind of hydrocarbon employed as the working gas.

Process parameters such as bias voltage, V_B, and pressure, p, determine the a-C:H films structure and their physical properties [14]. Bubenzer et al. [15], in the earlier work where they used 2.3 MHz power, stressed the importance of the energy of positive ions bombarding the growing deposit. The energy of positive ions is given by the sheath potential difference, V_s, between the plasma and powered electrode i.e.: $V_s = V_p + V_B$ where V_p is plasma potential and V_B is self negative bias voltage of the powered electrode. Usually V_p is small and as a processing parameter self negative bias, V_B, is taken. However, the energy of the ions depend also on their mean free path i.e. on the working gas pressure, p. They deduced an empirical relation for the average energy of positive ions, $\bar{E} \sim V_B / p^{0.5}$. For an RF discharge operated in benzene at 3 Pa, hydrogen contents (determined by NRA), refractive index and optical (Tauc) gap (determined from absorption measurements), mass density and microhardness were found to change with bias ranging from -100 V to -1000 V as follows:

[H]	from 42 at% to 20 at%
opt. Gap	from 2.2 V to 1.0 eV

refr. Index	from 1.7 to 2.2
density	from 1.45 g/cm^3 to 2 g/cm^3
microhardness	from 4 GPa to 10 Gpa

These authors therefore conclude that below $V_B = 100$ V, the films are "polymerlike", transparent (optical gap > 2 eV) and soft. They contain more than 40 at%H. At biases above 300 V, the films are dense, hard and strongly crosslinked. They noted a trasition region between the two types of carbonaceous films (soft plasma polymer and a-C:H) for V_B of 100 – 300 V that correspond to the average ion impact energies between 40 and 120 eV. The average ion energy was found to be 0.4·($eV_{B)}$ eV [14]. In this region (more likely between 100 and 200 eV), so called hard plasma polymers are deposited that is denoted C:H in this chapter. It was found by IR spectroscopy that at $V_B = 100$ V, prepared films depend strongly on the hydrocarbon used [16]. At higher bias ($V_B = 400$ V) there is no dependence. Actually, refractive index is about 2, opt. gap between 1.44 and 1.36, hydrogen contents about 31 at% and sp3:sp2 ratio is about 72:28. However, deposition rate ($V_B = 400$ V, pressure $p = 3$ Pa) decreases from 85 nm/min for benzene to 25 nm/min for n-hexane, and to 5 nm/min for methane [14].

The concept of Koidl et al. [14] of the deposition of hard a-C:H films because of the impact induced fragmentation of the energetic hydrocarbon ions at the growing surface is oversimplified. Möller et al. [17] summarized previous models regarding a methane plasma (mostly ECR) and concluded that the most abundant reactive species delivered by these plasmas are hydrocarbon radicals (mainly CH_3 below 1 Pa), atomic hydrogen as neutrals and CH_3^+ and CH_4^+ as ions. During the development of their model they moved from the adsorbed layer model, that assumes a saturable adsorption of radicals at the surface [18-19], to the surface activation model (radicals may be adsorbed only on surface sites activated by ion damage [17]) and also incorporated hydrogen re-etching. Finally, von Keudell and W. Möller proposed a combined plasma-surface model [20].

Several other deposition systems were used for the preparation of carbon films by plasma polymerization processes (PECVD) e.g. dual frequency (MW/RF) apparatus [21] and a lower frequency (50 kHz) plan parallel plate system [22]. The MW/RF system, which became very popular, consists of vacuum chamber equipped with a fused silica window, through which the MW power (2.45 MHz) is applied using a linear slow wave applicator. An insulated substrate holder is capacitively coupled to the RF (13.56 MHz) for development of a negative self bias. The MW/RF deposition enables independent control of the ion mean energy, E_i, and the ion-to-condensing particle flux ratio.

Intensive ion bombardment causes a number of growth-related effects, such as interfacial atom mixing, higher surface mobility of arriving species, knock-in effects (forward sputtering) and resputtering of loosely bound species. It is very well known that energetic ion bombardment causes that a-C:H films develop, during the growth, compressive stress. If the film is grown to a higher thickness and has a low adhesion to the substrate, then this film peals off. The example is an a-C:H film of thickness more than 200 nm on glass. However, the adhesion on Si and Ge is very good because of the carbide interface layer that is created. Therefore, a-C:H films were successfully used as optical coatings for the IR region of light. They were also used as hard, non-abrasive (low friction) coatings on various materials, including polymers (for review please see Ref. 23).

A DC glow discharge in a hydrocarbon gas has also been employed in a diode arrangement to deposit a-C:H films on the conducting substrates that are placed on the cathode [9]. DC sputtering became more popular for the production of this type of coating, which was first reported using post-cathode cylindrical and planar magnetrons [24]. The cathode is covered by a-C:H that is sputtered on an electrically floating substrate, where a film is also growing due to the plasma polymerization process. Simultaneous energetic bombardment is provided by neutrals reflected from the cathode when low working gas pressure is used. For the production of hard carbonaceous coatings also, so called, an unbalanced planar magnetron is used [25-28]. In this case, an amorphous hard carbon (a-C) containing no hydrogen was deposited by sputtering of the graphite in argon at a pressure of 1 Pa.

Researchers attempted materials such as a-C:F:Cl [12] and a-C:F [29] using r.f. diode arrangements. Films of a-C:F:H type received more attention [30]. Leidlich et al. [31] reported DC magnetron sputtering from a glassy carbon target in an argon-hydrogen-hexafluoroethane atmosphere. The films possessed low conductivity, good transparency, great hardness and low wettability. They exhibited higher density and refractive index than the films deposited in a glow discharge [30,32]. In [33], a-C:H:F films were prepared in an RF discharge operated in H_2/C_2F_6 mixture using the reactor with a triode electrode arrangement. These films were obtained when they were deposited on the substrate resting on the third electrode possessing a temperature below 200 °C and a DC negative bias from 100 to 150 V. The feed gas mixture contained 5%-12% C_2F_6.

Since the first report on nitrogen doped a-C:H films by Jones and Stewart [34], the hydrogenated carbon nitride films (a-C:H:N) were prepared by plasma deposition from CH_4 and N_2 gases and also using an cyklohexan-N_2-Ar mixture.

Recently, a-C:H:N films were fabricated in an RF diode system in a methane/nitrogen mixture with nitrogen up to 25 % at a total pressure 8 Pa and negative self bias of 370 V. Films were deposited on Si substrates with the thickness up to 350 nm and also comparative a-C:H films were grown at the same conditions. In these, the composition was 86.1 at% C and 13.9 at% H while for a-C:H:N, the concentrations of C, H, and N were 78.1,14.1 and 7.8 at%, respectively. The incorporation of 7.8 at% of nitrogen in a-C:H:N films did not change the hardness (21 GPa), however, it considerably lowered the internal stress from 2.5 GPa (a-C:H) to 1.6 GPa (a-C:H:N) [35,36].

Concluding: amorphous hydrogenated carbon films (a-C:H) contain hydrogen (20-30 at%). Hydrogen may be bounded or unbounded (up to 50 % of total H content) [37,38]. A microstructural model was developed describing a-C:H as a network of sp^3 carbon bonds with embedded sp^2 bonded clusters [39]. Fluorine, nitrogen and recently oxygen with silicon [40] have been successfully incorporated.

Hard plasma polymers that contain around 30 % and more of hydrogen will be now at the center of attention. Hard plasma polymers (C:H) can be prepared in an unbalanced planar magnetron using a mixture of a hydrocarbon gas and argon. The energetic, positive argon ions and the hydrocarbon fragment ion bombardment is caused by the accelerating electric field. This field originates from the floating potential appearing on the substrate surface when it is immersed in a plasma beam directed at it by the unbalanced magnetic field of the magnetron [25-27]. Using the described type of magnetron, graphite was sputtered in argon and propane [41,42]. Sputtering of the graphite resulted in a-C films that were electrically insulating and rather hard but their hardness did not reach that of a-C:H. An addition of a small amount of propane to the argon caused the obtained films to be more optically transparent - their optical gap was well above 2 eV and increased with the addition of more propane.

Using this technique, another class of films can be prepared: composite metal/plasma polymer films [1,2] that will be discussed in the next paragraph.

9.3. Composite Metal/Plasma Polymer Films

The metal/plasma polymer films appeared in the seventies as a new class of composite thin films [43-45]. In the following years, a number of studies devoted to this topic increased as can be seen from a number of reviews e.g.

[1,2,29,46]. One should note that the first composite metal/"organic films" were produced by vacuum co-evaporation [47,48].

Simultaneously metal/inorganic dielectric (cermets) prepared by vacuum co-evaporation and co-sputtering were investigated and these studies layed down some basic concepts for the film property explanations, including theoretical models [49].

9.3.1 Deposition techniques

9.3.1.1 Simultaneous plasma polymerization and sputter-etching using an RF discharge

The principal arrangement - RF diode-parallel plate electrode system is shown in Figure 9.1.

The metal target is placed on the excitation electrode while the substrate is attached to the earthed support (counterelectrode). Halocarbons as working gas were used e.g. CF_4, C_3F_8 , C_2F_3Cl (for summary se e.g. [1,29]) and, in this case, sputtering of a metal (Au) [50-52], sputter etching of metals (Mo,Cu,Al) [45] or their halide compounds from the target electrode takes place and the metal is thus incorporated into the composite film on the substrate. The polymer component is created from the plasma polymerization process taking place in the plasma volume and on the substrate. The metal incorporation can be simply controlled by DC negative bias of the target. However, when this is increased by increasing the r.f. power the substrate heating affects the film growth [51,52]. An RF planar magnetron has been applied in order to increase both, metal emission and

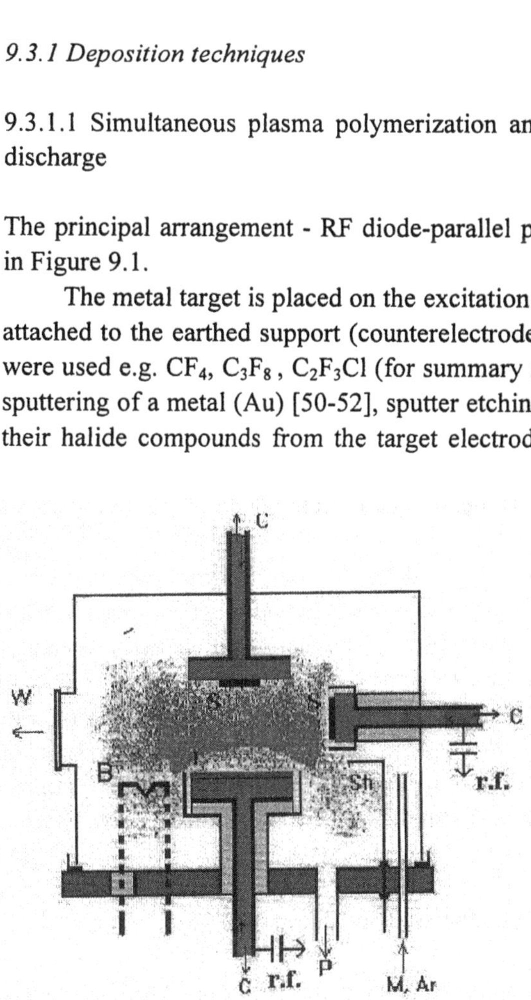

Figure 9.1. T- target, S- substrates Sh- shutter, r.f. to r.f. power supply, C-cooling, M,Ar-monomer, argon inlets, W-window, B –evaporation boat, P- to pumps.

plasma polymerization rate [50], while substrate heating is surpressed because most of the electrons are trapped at the target by the magnetic field. The structure (morphology) of this composite film is the following: metal inclusions (with some metal halide compounds on the surface of the inclusions in the case of reactive metals) embedded in the plasma polymer matrix.

9.3.1.2. Simultaneous plasma polymerization of an organic gas and evaporation of a metal

If one places a vacuum evaporation source (boat) in any plasma polymerization reactor, then this deposition system can be used for metal/plasma polymer preparation [53] (see Fig. 9.1). The deposition rate of both components - polymeric and metal can be independently varied by changing the plasma polymerization process parameters (power to the discharge, pressure or flow rate of the monomer) or heating of the evaporation boat. In the case of the internal mesh electrodes [54], the discharge can be excited by AC voltage or even DC voltage in a special case. When RF excitation is used, a coil can be placed outside the reactor, usually made of silica [55,56].

9.3.1.3. Plasma polymerization of metal organic compounds

Researchers first studied organotin compounds by plasma polymerization [43,44]. Suhr and co-workers [57,58] showed that by varying the power density, substrate temperature, and monomer flow, it is possible to change the metal contents in the composite plasma polymer films and end up with entirely metallic films. These investigators prepared composite plasma polymers with incorporated metals such as Pd, Ni, Co, Sn and Au. Plasma polymerization of ferrocene [59] and iron pentacarbonyl [60] was performed. Copper phthalocyanine [61,62] and copper acetylacetonate (and others) were plasma polymerized [63,64].

9.3.1.4. Co-sputtering from composite metal/polymer target

Biederman and Holland [50] prepared composite gold/fluorocarbon plasma polymer films by RF sputtering of PTFE/Au composite target in CF_4. Roy et al [65] sputtered, in argon, PTFE/Pt- 10% Rh, Au, Cu and Ekonol/Pt- 10% Rh, Au targets. Ekonol (trade name of Carborundum Co) is an aromatic polyester.

In this kind of composite film deposition, a gaseous monomer is not supplied to the discharge zone for the plasma polymerization process. Instead, positive ions sputter metal from metallic part of the target. Simultaneously positive ions sputter from polymeric part of the target its polymeric chain fragments that are volatile. These are further fragmented, excited and ionized in the plasma and a solid plasma polymer is deposited on the adjacent surface as in the usual polymerization process. This mechanism is dominant in the case when argon alone is used, and it is combined with the gaseous plasma polymerization mechanism when argon is mixed with a monomer or if a monomer only is applied.

9.3.1.5. Dc planar magnetron sputtering and simultaneous plasma polymerization

Both cylindrical and planar magnetrons were used by Harding and Craig [24,66] who prepared Fe/C:H composite films by "reactive sputtering" of stainless steel in an Ar/C_2H_2 plasma. Fast neutrals were considered to be responsible for deposition of C:H (hard plasma polymer instead of usual hydrocarbon (soft) plasma polymer films.

At the same time, Sikkens [67,68] reactively sputtered, in a simple DC diode sputtering arrangement, Ni in an Ar/CH_4 mixture. Weissmantel [69] reported composite films prepared by planar magnetron co-sputtering of graphite and Sn and Ti. Deposits prepared in these studies were claimed to be metal/hard carbonaceous films. Additional metals were incorporated, such as Ta and W [70-72].

The deposition process of metal/C:H composite films was improved after the unbalanced mode of planar magnetron operation was discovered [27]. An escaping beam of plasma from the near target region induces a negative bias (floating potential) from -10 to -70 V against ground of the isolated substrate.

This bias attracts mild positive ion bombardment. Biederman et al. [41,42] prepared, in this way, composite metal/C:H films from a composite metal graphite target using a mixture of Ar and propane. Gold, molybdenum and some others were attempted but mostly silver was investigated [42]. The closed field unbalanced magnetron sputtering arrangement was used for the deposition of graded and multilayered Ti:C-H/TiC thin film systems for wear-resistant application [73]. An unbalanced planar magnetron, equipped with various targets (Ag,Ni,Mo and Ge) in a DC discharge and operated in Ar/n-hexane working gas mixture was used for the deposition of Ag/C:H[74], Ni/C:H[75], Mo/C:H[76] and Ge/C:H[77]. In recent years, great attention has been paid to Ti/C:H [78–86] because of the tribological applications, optical application (selective absorber) [87] and biomedical applications [88]. Finally, composite metalCu/C:H was examined [89].

9.3.2. Structure and morphology of the composite films

In the first approach, the morphology of the metal composite films can be viewed as it is seen in Fig. 9.2. This picture corresponds very well to the TEM micrograph of Au or Ag/plasma polymer composites. Size of metallic inclusions is typically 20 nm but may vary from 1 to 100 nm.
The amount of a metal is usually described by a filling factor or volume fraction ratio defined as

$$f = \frac{V_{metal}}{V_{composite}} \tag{9.1}$$

The best determination of f is to use a quartz crystal microbalance in order to determine the mass (weight) of the composite film. By measuring the thickness using an independent method (e.g. Surfometer or Tolansky optical method) of the film deposited on the known area, one can obtain the volume of the film and therefore, its density. The filling factor is then given by:

$$f = \frac{\rho_{comp} - \rho_p}{\rho_{met} - \rho_p} \tag{9.2}$$

where ρ with the indices of *comp*, *met* and p are densities of composite, metal and plasma polymer, respectively.

The size of the metal inclusions depends on the combination of the metal - its reactivity and surface diffusion properties (usually assessed by melting point) - and the plasma polymer used. For example, a high melting point metal that is non reacting

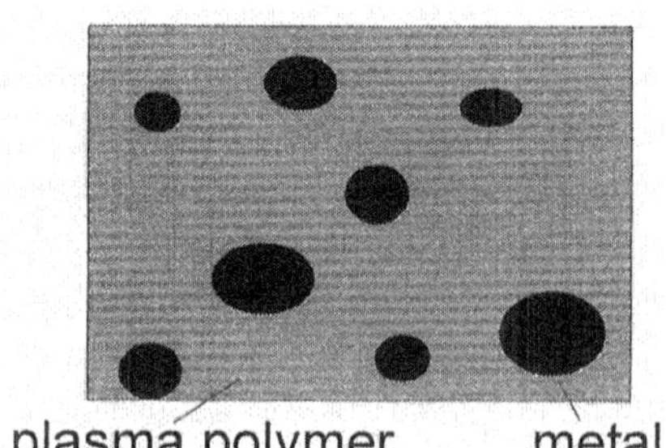

Figure 9.2. Schematic structure of metal/plasma polymer composite film.

with a plasma polymer matrix gives smaller inclusions at room temperature than a low melting point metal. Also, when the temperature of the substrate during deposition is increased, the size of the metal inclusions increases because of the increased surface mobility of the metal atoms. When we want to increase the filling factor, it is better to change the ratio of the monomer/argon working gas mixture because this does not increase the size of the inclusions by substrate heating.

Large variations of the metal inclusions are possible in the case of metalorganic compounds, as shown above. Suhr et al. reported variable size of inclusions up to atomic level using metalorganic compounds [58].

9.3.3. Deposition process and composite film growth

The deposition process can be viewed in the first approach as follows. Metal atoms are arriving to the substrate – some of them are resputtered (reemitted), some of them are adsorbed and create at first nuclei at certain substrate points usually according to the lowest energy rule. Nuclei of several atoms grow with the deposition time and are turned into islands (should be considered three dimensional) of the size from 10 to 100 nm and more. This concept is similar to the initial stages of metallic film growth [90] except the fact that, in our case, at the same time, with metal atoms, the fragments (radicals, neutrals, ions) of organic molecules of a monomer are also arriving to the substrate where they combine into a plasma polymer component of the composite. This may hamper surface diffusion of the metal atoms, slowing down the growth speed of the metallic islands. This also causes the encapsulation of the metal islands, turning them into the metallic inclusions dispersed in the three dimensional plasma polymer matrix. This holds when there is less metal than plasma polymer and no chemical reaction exists between the metal and organic (plasma polymer) phase e.g. in case of Au/C_2F_3Cl [51,52]. In the other case, e.g. Ti/C:H, the Ti inclusions (dielectric and transition regime) are in fact composed from a Ti core with the TiC/C skin embedded in C:H matrix [91].

When the substrate temperature is increased generally the size of the inclusions also increases when the other conditions remain the same. When there is energy (impact) supplied by the bombarding energetic neutrals or more often positive ions, e.g. Ar^+, the plasma polymer component changes into a hard plasma polymer (partly it may be ablated or sputtered off) and metallic inclusions become more rounded, bigger and may stick off the surface of the composite film e.g. in case of Ni/C:H [41,42,92,93]. This is realized by an unbalanced magnetron or by negatively biasing the substrate holder. In case V_b is more than -200 V, the C:H hard plasma polymer component turns into a-C:H.

The deposition process should be observed by several in situ monitoring techniques. The most important are: the quartz crystal microbalance (QCM), the optical emission spectroscopy (OES) and the quadrupole mass spectrometry (QMS). QCM enables the determination of the mass of the film sample and, when we independently determine the thickness, the density and therefore the filling factor, *f*, may be determined, as mentioned above. The other technique is optical emission spectroscopy (OES) that has been used by a number of authors e.g. in [94,95]. Martinu and Biederman [94] determined the ratio of the emission gold line at 267.6 nm to the height of the band head belonging to the CF_2 radical at 265 nm, and related this to the film filling factor (gold volume fraction).

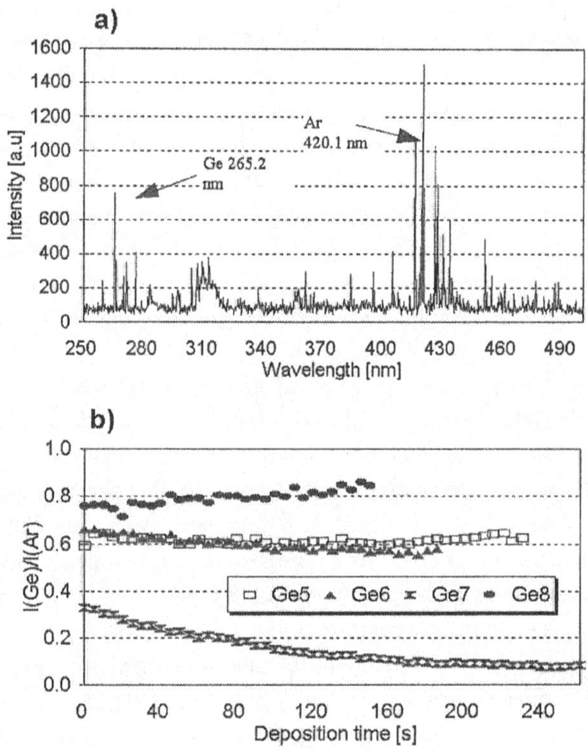

Figure 9.3. (a) Optical emission spectra during Ge/C:H composite film deposition. (b) Time dependence of the ratio of the intensities I of Ge line (265.2 nm) and Ar line (420.1 nm). After [77].

When metal/C:H films were deposited, the ratio of the metal line e.g. Ag (328 nm) to Ar line (418 nm) was employed [42]. An example for Ge/C:H is in Fig. 9.3. [77]. However, actinometric optical emission spectroscopy (AOES) must be applied for accurate determination of various species [78].

Quadrupole mass spectrometry (QMS) was used for monitoring the deposition, e.g. of composite Au/fluorocarbon plasma polymers[95]. An example for the case of Ni/C:H films is in Fig. 9.4. [93]. Langmuir probes have been used in several cases [96,97].

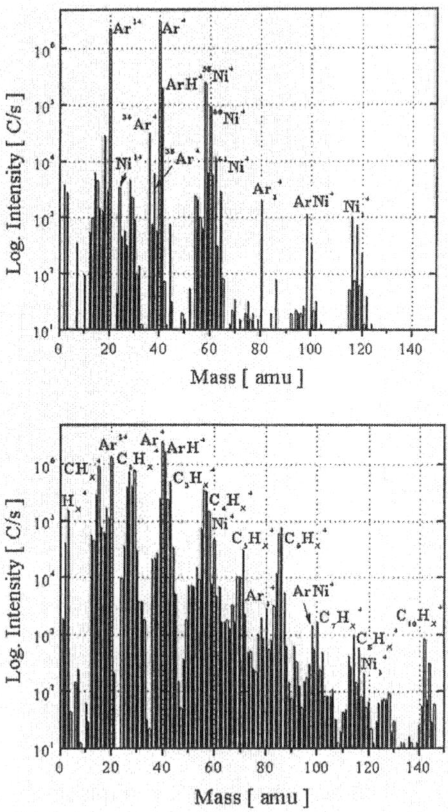

Figure 9.4. Typical ion mass spectra during Ni sputtering (After [77])
(a) Without n-hexan admixture (U_{dc} = 455 V, I_{dc} = 200 mA, p_{Ar} = 0.7 Pa, E^+ = 3 eV)
(b) With n-hexane admixture (U_{dc} = 434 V, I_{dc} = 200 mA, p_{Ar} = 0.665 Pa, $p(C_6H_{14})$ = 0.035 Pa, E^+ = 3 eV). After [93].

9.3.4. Basic physical properties – characterization

9.3.4.1. Structure, morphology and composition

A very convenient technique for determining the morphology of a composite film is transmission electron microscopy (TEM). The film is deposited on a carbon foil supported by a Cu grid (3 mm in diameter) and then the TEM observation is performed. An example of Ag/C:H composite films with filling factors ranging from 0.1 to 0.25 is shown in Fig. 9.5(a). The films were deposited by a DC unbalanced magnetron [74] with increasing power from 30 W to 100 W in order to increase the Ag contents in the film. However, the increased power delivered to the substrate caused the substrate temperature to rise and enlarge of the Ag grains (please, see further aging).

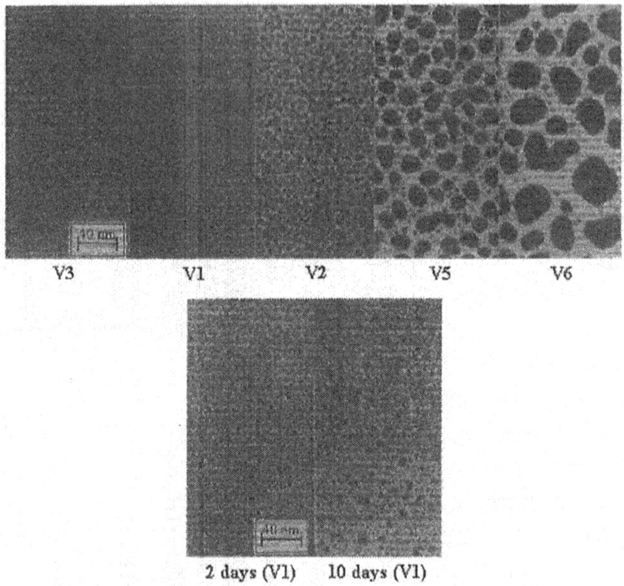

Figure 9.5. Transmission electron micrographs of samples V3,V1,V2,V5,V6 with filling factors 0.008, 0.014, 0.025, 0.08, 0.24, respectively: (a) immediately after deposition, (b) sample V_1 after 2 days and after 10 days. Thickness of V3 approx 26 nm and of the other samples 40 nm. After [74].

Another important technique is infrared absorption spectroscopy (FTIR) that enables the characterization the plasma polymer matrix. An example is in

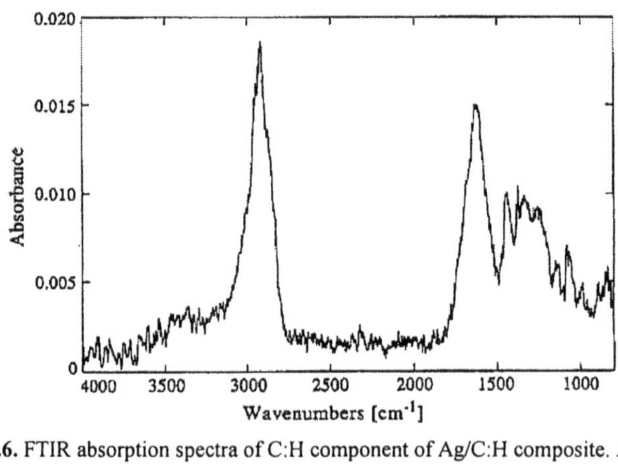

Figure 9.6. FTIR absorption spectra of C:H component of Ag/C:H composite. After [74].

Fig. 9.6. for the C:H matrix in the above described Ag/C:H composites [74]. The absorption band caused by the C-H stretching in various CH, CH_2 and CH_3 groups is peaked at 2925 cm^{-1} (near to the wavenumber related to asymetrical stretching in CH_2 (sp^3 aliphatic) group). The tail of this band near and above 3000 cm^{-1} can be ascribed to the stretching in various =C-H groups (both sp^2 and sp^3 configurations). The other absorption peaks are assigned as follows: 1460 cm^{-1} for the unresolved CH_2 scissoring and CH_3 asymetrical bending and 1375 cm^{-1} for the CH_3 symetrical bending. They witness the presence of polymeric structures (linear chains) in our films.

The incorporation of lower concentrations of silver into the C:H did not show significant changes in the FTIR absorption spectra. However, it was

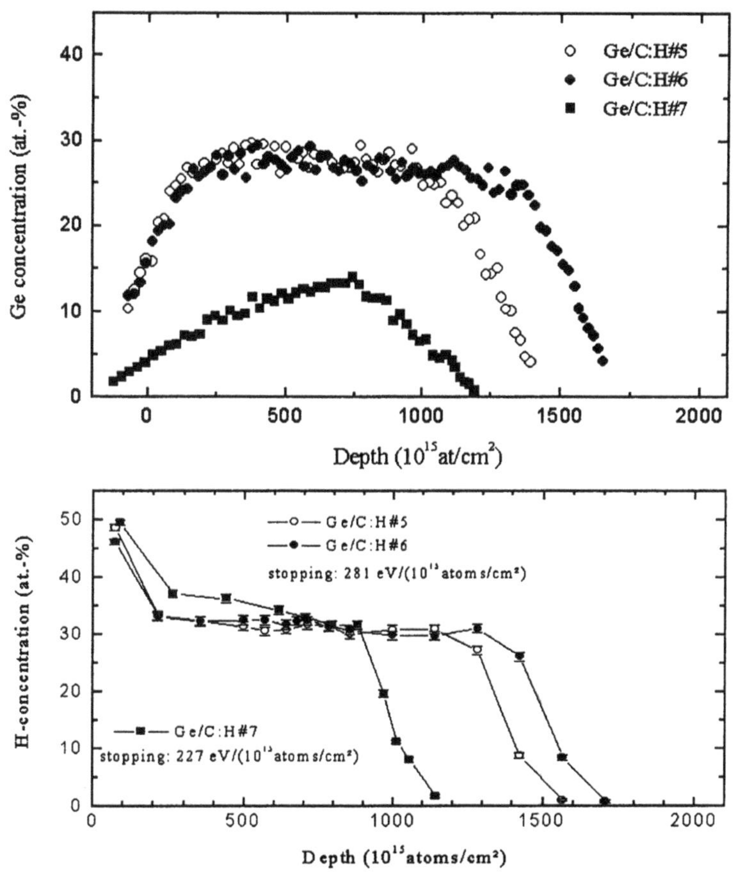

Figure 9.7. (a) Germanium concentration in the three samples measured by RBS. (b) Hydrogen concentration in the three samples measured by NRA. After [77].

shown formerly [1,52], in the case of fluoro(chloro)carbon plasma polymer matrices, that the increase of the gold contents in the film causes the decrease of the fluorine concentration in the plasma polymer matrix. The same effect was proved by XPS for gold-containing plasma polymerized C_3F_8 [98].

XPS is another technique that allows elemental analysis of the surface of the composite layer and gives information on the bonding states of the elements. For fluoro(chloro)carbon plasma polymers containing metals, profiling, with certain limitations, may be performed [99,100]. In the same manner, AES was used [99]. However, in the case of XPS and AES, hydrogen cannot be detected and therefore for metal/C:H films, profile analysis RBS (Rutherford Back Scattering) combined with ERDA (Elastic Recoil Detection) [101] or NRA (Nuclear Reaction Analysis) [77] should be used. An example is shown in Fig.

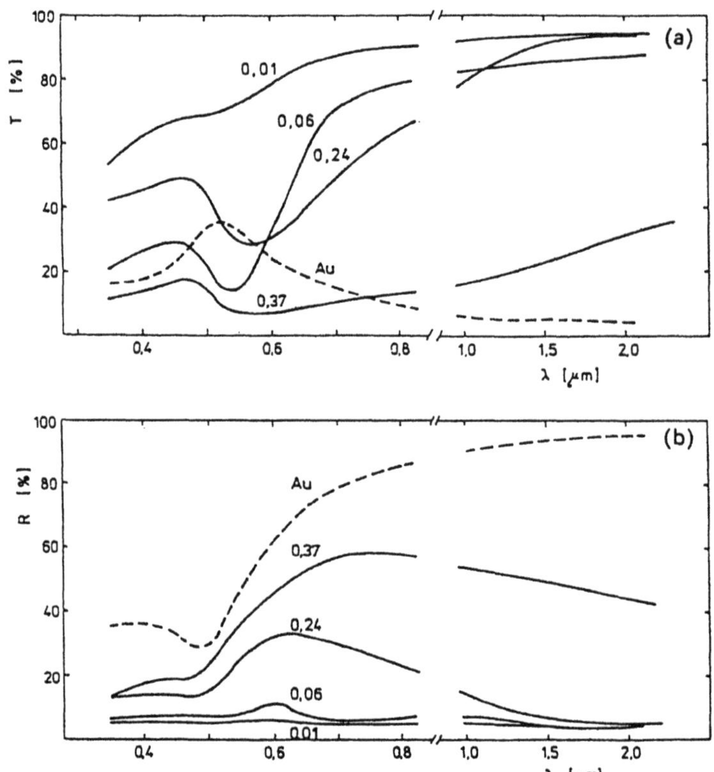

Figure 9.8. Spectral dependence of the optical transmission (a) and reflection (b) for gold/halocarbon plasma polymer films, numbers at the curves show filling factor values. After [102].

9.7. for Ge/C:H films. Please note that the information from *in situ* OES (Fig. 9.3.) is perfectly in coincidence with the elemental (especially for Ge) profiles.

9.3.4.2. Optical properties

When metal inclusions, such as gold, are dispersed in a low absorbing matrix of a plasma polymer, such as a fluorocarbon, so called anomalous optical absorption takes place because of the optical resonance. An example – transmission and reflection spectra in the visible and near infrared light region – is shown in Fig. 9.8. [102]. When the amount of gold increases over a filling factor value of 0.01, an anomalous absorption region develops at 0.55 μm. The colour of the composite film in the transmitted light changes from pink (f = 0.01) to red (f = 0.06), to violet (f = 0.24) to blue (f = 0.37). Generally, the position, size and width of the (anomalous) absorption maximum depends on the kind of metal and plasma polymer matrix, filling factor f, the size and shape of the metal inclusions and their distribution in the plasma polymer matrix. When a C:H matrix with an increased carbon contents is used, optical absorption in the visible region cannot be neglected. Colour appearence of these films in the visible light is therefore modified. Theoretical description of the optical behaviour has been done using effective medium theory. For details, please see review [1]. Recently, advanced studies of the optical properties versus microstructure were carried out by Martinu et al. [103-105].

9.3.4.3. Electrical properties

The electrical conductivity of a metal/plasma polymer composite film is generally determined by its metal contents and its arrangement on a microstructural level. Metals such as Au, Ag, Pt, Ni, Mo and etc. are dispersed in the matrix in the form of inclusions of the size on the order of 1 nm to 100 nm. The DC electrical conductivity is then strongly dependent on the filling factor, f. An example may be seen in Fig. 9.9. for Ag/C:H where this is presented as the dependence of the sheet resistance (resistance per square) $R_{\square}$ on the metal filling factor (volume fraction ratio). When this is small (less than 0.2), the resistance is very high because the metal inclusions are separated by the dielectric-plasma polymer and the dominant conductivity mechanism is then electric charge transport by the electrons and holes tunneling from one metal

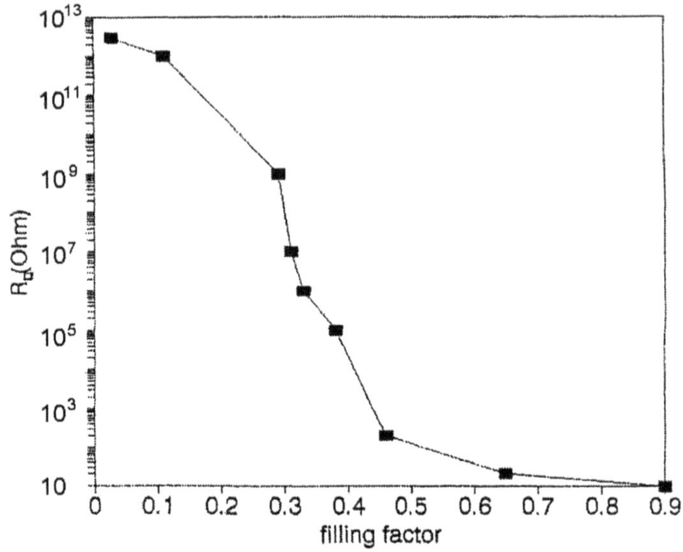

Figure 9.9. Sheet resistance as a function of filling factor for Ag/C:H. After [109].

inclusion to the other. The first part of the curve in Fig. 9.9. correspond to the so called structural "dielectric regime" as established in accord with cermet films [49] (please, see also review [1]). This is followed by the "transition regime" at an increased filling factor when metallic inclusions are approaching each other. When they start to touch, a very steep decrease of the resistance is observed – so called "percolation threshold. The third and last part of the curve in Fig. 9.9. corresponds to the structural "metallic regime" when the filling factor is so high that the plasma polymer inclusions are in a metallic continuum. For Au and Ag (relatively large inclusions-several nm in diameter when deposited at room temperature), the percolation threshold is at about $f = 0.4$ as seen from Fig.9.9. for Ag/C:H and as found for Au/plasma polymer films [106,107].

For smaller clusters, as shown for Co/plasma hydrocarbon polymer [108], the percolation threshold was found to be 0.22. It has been formerly observed [57,58] that the percolation threshold is less and less abrupt as the dimension of the inclusions decrease. For example, for Ni/C:H the percolation threshold extends between 0.2 – 0.3 [75] and for Mo/C:H between 0.3 – 0.6 [109]. Theoretical concepts follow Abeles and Ping Sheng and have been summarized in several reviews e.g. in [1,49,71].

9.3.4.4. Mechanical properties

As mentioned in the first part of this chapter, a-C:H films have a microhardness that was found to be between 20 to 50 GPa. When a metal, such as Au or Ag, is co-deposited, e.g. co-evaporated and incorporated into the a-C:H grown on the RF excitation electrode possessing DC negative bias below 200 V, then the hardness of the composite decreases [110]. The same effect was also found for some other metals and ion beam techniques [111]. Dimigen et al. [70] reported that composite films of W, Ti, Ta, and NiCr dispersed in a C:H matrix and prepared by means of a DC magnetron possess small abrasive wear and sliding friction coefficients ranging from 0.1 to 0.2. Great attention was devoted to Ti/C:H films [78–86] as mentioned in paragraph 9.3.1.5. Graded and multilayered system such as Ti/TiC:H/TiC/a-C:H [73] were deposited on steel substrates with success for tribological application and this concept was later often followed. Several authors also studied Ta/C:H (e.g. [71,112]) composite films. Structure of Ta/C:H was discussed based on gas evolution studies [112], pointing out the thermal degradation effects of the C:H component of the composite.

9.3.4.5. Aging

When composite metal/plasma polymer films are deposited at temperatures close to room temperature, they are in a metastable state as a consequence of a non-equilibrium deposition processes. Metal is incorporated in the form of inclusions of various sizes that may be composed of a few atoms or even composed of a single atom. Single atoms or even small "clusters" start to move by diffusion towards bigger inclusions, increasing the size of these inclusions that sometimes may further increase by coalescence, sintering and Ostwald ripening effects [90]. These processes were noticed as the changes of the optical transmission (colour in transmitted visible light) and the DC electrical resistance [42,113]. These effects have recently been observed in case of Ag/C:H, Ni/C:H and Mo/C:H [74–76] and an example of the sheet resistance change with the storage time in the open air is given in Fig. 9.9. [76].

In addition to the metal inclusions, a "relaxation phenomena", of the plasma polymer matrix, which contains a lot of free radicals take place in post-deposition reactions, including oxidation in the open air. These processes have been observed by means of AC electrical properties and FTIR [114]. It has been proved that it is advisable to leave the composite metal/C:H for some time in vacuum before opening the reactor (deposition chamber) to the atmosphere. The dielectric properties of the C:H pointed out that it has a very rigid network (highly crosslinked) [115].

Aging effects can be decreased by depositing composite films on substrates at elevated temperatures [2,113] or, in the case of metal/C:H, by increasing the energy of the bombarding ions [75]. However, even through the fluxes to the substrate responsible for the plasma polymer component and the metallic component of the composite remain unchanged, the change in the conditions at the substrate will change the structure, morphology and other physical properties. For example, in the case of a substrate at an elevated temperature, up to 100 °C, in the case of Au/fluorocarbon plasma polymer [113], the composite film is deposited with bigger gold inclusions and with their increased spacing. Post deposition annealing, in the case of "room temperature" substrates, produces a similar effect. In the case of ion energetic bombardment, the size of the inclusions slightly increases but the plasma polymer matrix undergoes more severe changes than in the case of a raised substrate temperature – it becomes more hard and rigid. This stops any rearrangements of the composite – i.e. aging. For example, when a Ni/C:H composite was deposited on the receiver negatively biased to -200 V, no change of the sheet resistance with time in the open air was observed [75].

9.3.5. Complex composite films

Using the plasma polymerization processes, more complex composite films were prepared such as a-SiC:H, a-GeC:H, a-Si:C:H:O [40,116] and a-Si:C:H:N [40,117] and especially their electrical properties were investigated (Please, see Chapter 3 and Chapter 5). Recently, composite SiOx/fluorocarbon plasma polymer films have been prepared and studied [118-121]. Takai and co-workers [118,119] prepared composite SiO_x/fluorocarbon plasma polymer films by a plasma polymerization process using an RF discharge operated in FAS-17 ($CF_3(CF_2)_7CH_2CH_2Si(OCH_3)_3$) vapours in a tubular reactor [118]. In the other study [119], ultra water-repellent films deposited by microwave plasma polymerization processes were reported from the mixture of the three gases: TMS, FAS-17 and Ar [119]. Composite SiO_x/fluorocarbon plasma polymer films on various substrates (microscope slides, silica, PE and PTFE) were obtained also by co-sputtering of PTFE and silica using an RF planar magnetron

[120]. Static contact angle of the water ranges from 45 to 65°, hardness from 3500 to 500 N/mm^2 when the ratio of the composite components: fluorocarbon plasma polymer (deficient in fluorine)/ SiO_x, changes from 1/1 to an almost pure fluorocarbon plasma polymer only [121]. These films were also prepared by ion beam sputtering [122]. These so-called fluoropolymer-filled SiO_2 thin coatings were deposited on polymeric substrates in order to provide for them atomic oxygen protection in the low Earth orbit in space. Some other applications were proposed, such as moisture and oxygen barrier coatings for food packaging, water repellent coatings etc. [122]. Special composite films, $Au/C_x(F,H)S_y$, were prepared by simultaneous plasma polymerization using a gas mixture of $Ar/CH_4/SF_6$ and co-sputtering gold from the target electrode for optoelectronic applications [123].

9.4. Plasma Polymer Films Prepared by RF Sputtering

If a polymeric target immersed in a plasma is subjected to an energetic positive ion bombardment, then the polymer chains get broken and volatile fragments are emitted into the plasma volume and serve as precursors of the plasma polymerization process. This is especially the case of RF sputtering of polytetrafluoroethylene (PTFE) [124-127] that has been studied since the

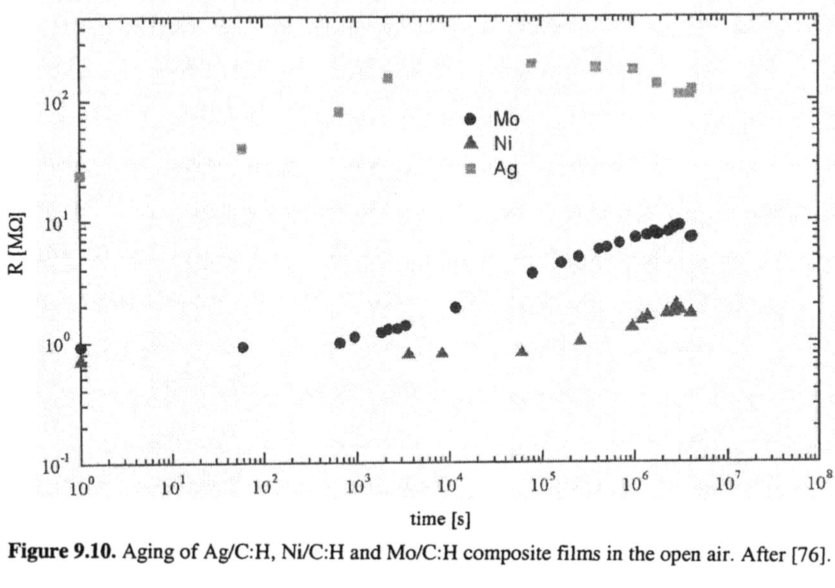

Figure 9.10. Aging of Ag/C:H, Ni/C:H and Mo/C:H composite films in the open air. After [76].

seventies of the last century with the motivation apparently to prepare good dielectric films [125], low friction coatings [124,128] and the optical applications were also considered [128].

However, other polymers were also investigated such as polyimide (PI), polyethylene (PE), polypropylene (PP) and others.

9.4.1. Fluorocarbon plasma polymers

Fluorocarbon plasma polymers were usually deposited in so called self sputtering mode [124], or they were RF sputtered in argon [125,128,129]. In the former case, argon was used to initiate the discharge that continued to be maintained by the volatile fragments when the argon supply was shut off [128]. Because, in this case, the sputtered fluorocarbon films were found deficient in fluorine, RF sputtering of PTFE in a CF_4 – Ar mixture was used and shown to give stoichiometric films [130]. As well as argon, nitrogen also was used as a working gas [131]. Marechal and Pauleau [132] compared RF PTFE sputtering in the self-sputtering mode, in Ar and in an Ar + C_3F_8 gas mixture. They worked at a fixed power of 250 W with a target 160 mm in diameter. With an Ar pressure of 1 Pa and in self sputtering mode 5 – 6 Pa, they reached high sputtering rates (0.6 nm/s). The surface of the films prepared in self sputtering mode or in Ar + C_3F_8 mixture were rather smooth and the cross section of the films were fully dense as revealed by scanning electron microscopy (SEM). The films sputtered in Ar were slightly rough and in the cross section they revealed fine grain (fiber) structures. Sugimoto and Miyake [133] used PFPE (low

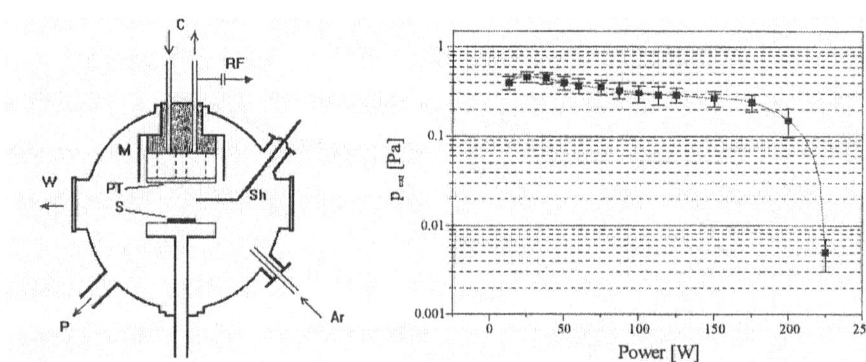

Figure 9.11. (a) The deposition arrangement (PT- polymer target, M-magnetron, Sh-shutter, W-window, Ar-gas inlet, P-to pumps), (b) self-sputtering mode development represented by p_{ext} (pressure when plasma extinguishes) versus delivered power. After [135].

vapour pressure oil) and PTFE-PFPE grease as targets in a RF sputtering for obtaining hard, lubricating and electrically conductive films. Wydeven et al. [134] compared ion beam sputtered PTFE films with plasma polymerized films in respect to $O(^3P)$ etching.

In recent years, fluorocarbon plasma polymers have been prepared in a parallel plate electrode arrangement using a balanced RF magnetron target (78 mm in diameter) electrodes (Fig. 9.11.) [135,136].

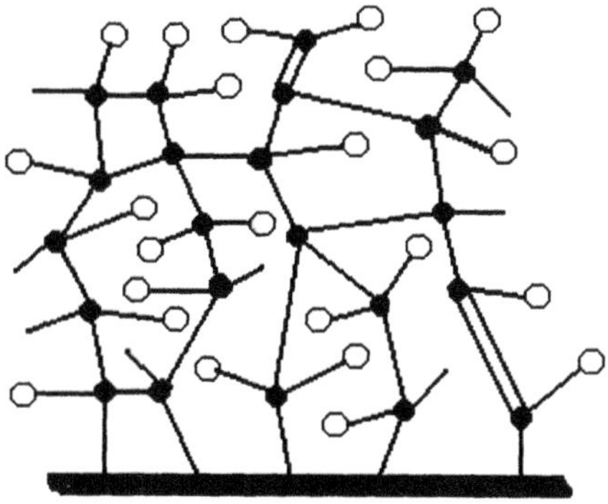

Figure 9.12. Model of a fluorocarbon plasma polymer obtained by PTFE sputtering (o- fluorine, •- carbon). After [135].

The films were deposited (sputtered) on glass substrates precoated with an Al layer. An FTIR spectrometer (Nicolet Impact 400) was used for polarized specular reflectivity spectra measurements. Shifts of the absorption band wave numbers with polarization were observed: (1250 ± 5) cm^{-1} for the radiation electric field (vector) perpendicular to the sample plane (reflectivity p-polarization) and (1215 ± 5) cm^{-1} for the electric vector in the sample plane (s-polarization in reflectivity and transmitivity measurements), respectively. This anisotropy is probably caused by an orientation of the C-C bonds predominantly perpendicular to the sample plane and C-F bonds parallel to the sample plane. Based on these findings a hypothetical model of the fluorocarbon polymer film is shown in Fig. 9.12.

This agrees well with other structural models e.g. [137]. Morphology of the films depends on the substrate temperature [138] and working gas total

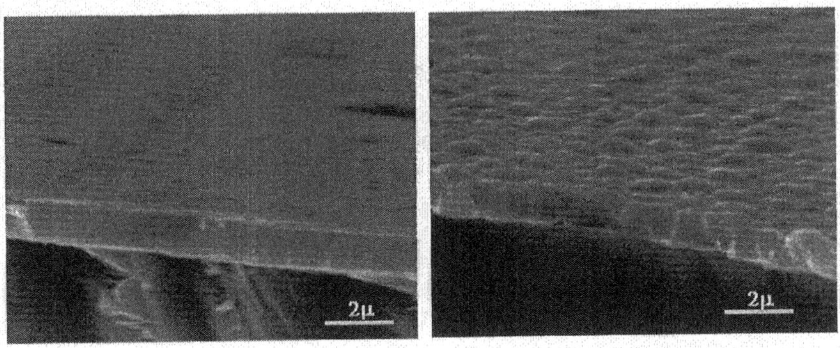

Figure 9.13. Films deposited on Si at 26 °C (left) and at –36 °C (right). After [139].

pressure [139]. Above room temperature, the film is compact (amorphous – continuous in the cross section) see Fig. 9.13. [138,139].

At a room temperature (23 °C), closely packed fibrilar structure appears. Columnar structures are clearly seen below 0 °C. In this case, the structure zone models can be applied [136]. Columnar structures mean an increased roughness, however, the static contact angle of water is increased only to 122 °, which is less than expected (static contact angle of the film at room temperature is 105 °C [140]). This is likely because of the changed chemistry of the film. However, when the roughness was obtained by increased the working gas (Ar) pressure, the contact angle increased up to 160 °C (RMS roughness about 100 nm). This way superhydrophobic films can be produced. However, more often they are prepared by pulsed plasma polymerization of C_2F_4 (See Chapter 5 and [141]).

Sputtering has been also performed in nitrogen [131], resulting in a good films. Later, a glass tubular reactor, with the inner wall lined with PTFE foil and with the inductive (by coil) RF plasma excitation was used for the sputter effect of He, Ne, Ar, N_2 and H_2 on the resulting fluorocarbon plasma polymers [142]. With nitrogen, the highest concentration of CF_2 groups was found. Fluorine content decreased with the increased atomic mass of the noble gas used. No plasma polymer deposit was obtained with hydrogen but formation of volatile byproducts (HF, CH_4 etc.) was supposed. Sputtering in nitrogen was investigated in a plan parallel plate system using an RF magnetron [140]. The

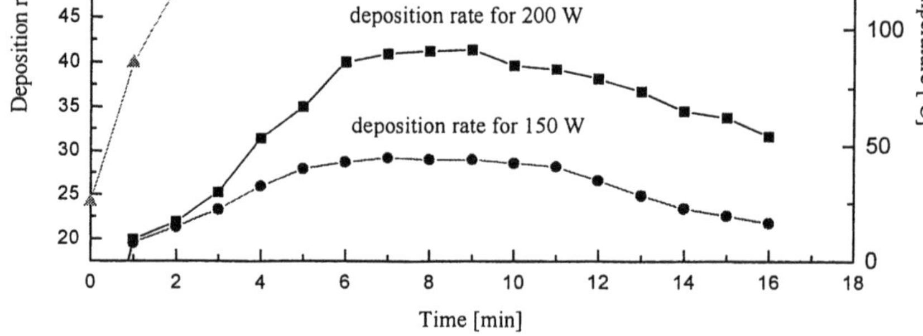

Figure 9.14. Time dependences of the deposition rate and the target temperature sputtering in N_2. Pressure 3 Pa. Target temperature at 200 W. After [140].

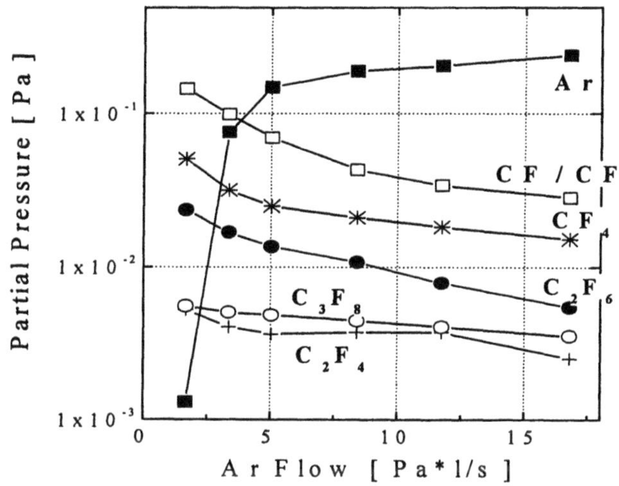

Figure 9.15. Composition of the discharge gas for different Ar flow rates ($p = 0.3$ Pa, $P_{RF} = 130$ W). After [144].

deposition rate in nitrogen and its changes during sputtering are shown in Fig. 9.14.

During the first 6 min, the target temperature (measured by inserted thermocouple) rises to 170 °C at 200 W in nitrogen and so does the deposition rate. Creation of carbon enriched (altered) layer causes a decrease of the deposition rate after the plateau. In the case of nitrogen, a higher CF_2 group presence was confirmed as compared to Ar and additionally CN and CNF groups are incorporated into the film. The static contact angle of water is 100 °C for these films.

A tubular type reactor, according to [142], was used by Wydeven et al. [143] for PTFE sputtering in Ar and He at a pressure of 13 Pa and a flow rate of 3 ccm(STP)/min at lower powers, 10 W. They found the F/C ratio in these films to be 1.51 for Ar and 1.56 for He that is better than 1.33 found for the plasma polymer of TFE. These results are in contradiction for sputtered films in a parallel plate electrode arrangement [127,128] as discussed in more detail in [4]. Using energy selective mass spectrometer Hiden Analytical EQP 300 the sputtering process was analyzed for parallel plate electrode arrangement as shown in Fig. 9.11(a). The grounded aperture of the spectrometer was 100 mm under the magnetron on its axis. In Fig. 9.15, the neutral gas composition in the discharge, calculated from the mass spectra using gas specific sensitivity factors, is shown [4,144]. A much higher contribution of CF and CF_2 fragments is observed than expected from the cracking patterns of stable C_xF_y fluorocarbons or suggested in the past from experiments with microwave discharge decomposition of PTFE in He [145]. Ar, CF_4, C_2F_4, C_2F_6 and C_3F_8 are shown as the most important stable gases. For more details, please, see Chapter 4.

9.4.2. Hydrocarbon plasma polymers

The most popular polymer after PTFE for RF sputtering seems to be PI (polyimide), also known by trade name as Kapton. Yamada et al. [146] investigated the tribological performance of sputtered PTFE in comparison to sputtered PI films in Ar at a pressure of $6.65 \cdot 10^{-1}$ Pa on glass substrates. At a constant power density of 1.27 W cm^{-2}, the deposition rate for PTFE was 70 nm/min, while for PI it was only 2 nm/min. Sputtering targets made of polymer sheets 50 μm thick were bonded to copper backing plates. The friction coefficient (sliding steel ball) was about 0.3 and 0.2 for both films. Kitoh and Honda [147] again compared fluorocarbon and polyimide film sputtering and found (at 1 Pa of Ar and at RF power 240 W) the deposition rate to be 25 nm/min and 3 to 4 nm/min, respectively. No polyimide groups at 724 cm^{-1} or 1777 cm^{-1} were observed in the FTIR spectra of sputtered PI films. Only the ether group at 1100 cm^{-1}, alkane at 1400 cm^{-1} and amino group at 1600 cm^{-1} revealed distinguishing peaks. The atomic ratio of carbon, oxygen and nitrogen for PI sputtered films (determined from XPS spectra) was 64/32/4 while for the PI parent material (target) it was 56.5/34.5/9. Lubricational properties of very thin films (20 nm) of sputtered PI. Sputtered PI films on silicon wafer were compared to MoS_2, C and BN films prepared in the same way. The PI films had the lowest friction coefficient and the highest abrasion resistance. The abrasion resistance means the number of passes of a sapphire slider on the film before its breakage. The slider reciprocates a distance of 10 mm at a velocity of 80 mm/s, is 40 mm in diameter and is loaded by a 0.49 N weight. For sputtered PI films, the abrasion resistance increases 10 times when the RF sputtering power increases from about 0.3 to 0.6 W/cm^2 at an argon pressure of 1.33 Pa. The coefficient of friction is almost constant in the range of 0.2 –0.22 while for the silicon wafer it is 0.4. The authors conclude that sputtered polyimide can serve as an effective solid lubricant.

Hishmeh et al. [148] RF sputtered in a simple diode set-up, PTFE, PI and, using co-sputtering (DC magnetron) of Al, also formed their composites with

Al. The rate of sputtering of PI and PTFE agreed with results presented above. The self-sputtering of PI was attempted, however, it required powers over 1 kW using an 8 in.-diam target. The authors concluded that the emission of PI fragments from the target was much lower than in the case of PTFE. Neutral species observed by a residual gas analyzer (quadrupole mass spectroscopy) included CO, CN, CNO, HCNO, while in the case of PTFE, more polymer building groups were found: CF, C_2F_2, CF_3, C_2F_4, CF_4. It was concluded from FTIR and XPS analyzes that PI sputtered films have no resemblance to the original parent PI (Kapton) material of the target. The imide group and the ether group of the original molecule were destroyed. Fukushima at al. [149] added during the RF sputtering process of PI, nitrogen into argon and improved stoichiometry of the obtained films. The usual blackish colour (in transmitted visible light) of PI sputtered films changed into usual yellow colour when 100% nitrogen was used as a working gas.

Rf sputtering of EKONOL, (the trade name of Carborundum Co), that is an aromatic polyester based on p-hydroxybenzoic acid [65], were performed. Rf sputtering took place in Ar at a pressure of 0.7 – 9 Pa and an RF power of 25-100 W from a target - disc 80 mm in diameter, 6 mm thick mounted on an Al backing plate. The deposition rate was 1-1.5 nm/min. Compressive stress built up when the films were deposited below 0.7 Pa and they peeled off the substrate. Obtained films were found to be amorphous and disordered hydrocarbon plasma polymer.

Polyethylene (PE) was RF sputtered by means of an unbalanced magnetron of the same dimensions as described above [135]. The deposition rate was 0.04 nm $(\text{min W})^{-1}$. FTIR absorption spectra showed that the obtained film was very different from the polyethylene target material and was revealed to be a typical hydrocarbon plasma polymer. From SEM micrographs the films deposited at room temperature were amorphous and fully dense in the cross section [135]. In recent studies [150], RF sputtering of polyethylene (HDPE Goodfellow) in Ar was studied using a balanced magnetron (80 mm in diameter). The above findings were confirmed for powers below 100 W. Above these powers, melting and evaporation of the target within the erosion zone took place, deposition rate increased and FTIR (DRIFT) spectra showed much more resemblance in shape to the spectra of the target. Similar results, as far as the deposited films are concerned, were obtained for polypropylene [151]. It was pointed out that the history of the used target is important and may be the cause of a relatively large concentration of OH groups in the deposited film. Several times of sputtering and exposure of the target to the atmosphere incorporates a

Figure 9.16. Sputtered polypropylene, Ar pressure 10 Pa, power 60 W, U_{bias} 280 V, leak rate 2.4 ccmSTP/min, time 35 min, d=120 nm. After [151].

lot of OH groups because the created free radicals react with oxygen and water vapour. The OH groups are then transferred during subsequent sputtering process into the film. The film, as observed in AFM, is in Fig. 9.16. Recently, Youngblood and McCarthy [152] studied, so called, non clasical sputtering of polymers in a tubular reactor. The target and substrate rested next to each other (5 cm apart) in the centre facing up (in the same direction) – no line of sight. A mechanism is proposed that polymer degradation to the monomer must occur at the target and then this monomer must plasma (re)polymerize on the substrate. The authors show that this does not pay only for PTFE but also for PIB(poly(isobuthylene)), PαMS (poly(α-methyl styrene)), and PCTFE (poly(chlorotrifluoroethylene)).

Composite films of metal/plasma polymers prepared by sputtering have already been discussed in Paragraph *9.3.1.4*, where composite Au /fluorocarbon plasma polymer films prepared by RF planar magnetron sputtering from a composite Au-PTFE target in CF_4 gas [50] and RF co-sputtering of Au and Pt-10 % Rh metal with EKONOL from the composite target [65] was mentioned.

Co-sputtering using an RF diode for PTFE and/or PI and a DC magnetron for Al to form the composite films were used by Hishmeh et al. [148]. It was observed that the plasma polymer component was less damaged and more like conventional polymeric materials than in the case that the polymer was sputtered without Al.

9.5. Conclusions

In this chapter we focused on plasma polymers prepared at moderate and extreme (from the point of view of energy) conditions. Hard plasma polymers extending up to DLC materials, are recognized as very important for many present and future applications. As it has been shown, not only hydrocarbon materials but also their modifications including other elements such as F, N, Si, O, Ge are investigated and applied. Incorporation of various metals led to the appearence of composite metal/plasma polymer and metal/hard carbon films.

RF sputtering of polymeric targets in a parallel plate electrode system is another area of plasma polymers – in this case prepared at moderate conditions. Instead of using gaseous feeds as it is in "conventional" plasma polymerization, the RF sputtering of a polymer target results in the emission of volatile target fragments that serve as precursors of the plasma polymerization process. Fluorocarbons were greatly studied with a number of applications proposed. Considerable attention has also been paid to polyimide in order to deposit protective coatings. Additional polymers such as polyethylene, polypropylene and others are now under consideration.

Acknowledgements

The authors are grateful to many young colleagues for their co-operation and help throughout the research reported in this chapter. The financial support of the Research Program BM MSM 113200002 and the grant OC 527.10 (COST 527), both from the Ministry of Education, Youth and Sports of the Czech Republic, is greatly acknowledged.

References

1. Biederman H. and Martinů L., in *Plasma deposition, treatment and etching of polymers*, ed. d'Agostino R. (Academic Press, Boston, 1990), 269.
2. Biederman H. and Osada Y., *Plasma Polymerisation Processes* (ELSEVIER, Amsterdam, 1992), 210.
3. Biederman H., Slavinska D., *Surf. and Coat. Technol.* **125** (2000) 371.
4. Biederman H., *J. Vac. Sci. Technol.*, **18**(4) (2000) 1641.
5. König H.and Helwig G., *Z Phys.* **129** (1951) 491.

6. Brockes A. and König H., *Z Phys.* **152** (1958) 75.
7. Schmellenmeier H., Exp. Techn. Phys. **1** (1953) 49.
8. Pagnia H., *Progress in Colloid & Polym. Sci.* **78** (1988) 143.
9. Whittmell D.S. and Williamson R., *Thin Solid Films* **48** (1976) 255.
10. Holland L., *UK Pat 1582231* (appl. Aug. 1976).
11. Nyaiesh A.R. and Holland L., *Vacuum* **34** (1984) 519.
12. Biederman H., Martinů L. and Zemek J., *Vacuum*, **35** (1986) 447.
13. Wagner J., Wild C.H., Pohl F. and Koidl P., *Appl. Phys. Lett.* **48** (1986) 106.
14. Koidl P., Wild C.H., Locher R. and Sah R.E., in *Amorphous Hydrogenated Carbon Films and related Materials: Plasma Deposition and Film properties, Diamond and Diamond-Like Films and Coatings*, eds Clausing R.E. et.al. (Plenum Press, New York, 1991), 243.
15. Bubenzer A., Dischler B., Brandt G., and Koidl P., *J. Appl. Phys.* **54** (1983) 4590.
16. Wild C., Koidl P. and Wagner J. in *Amorphous Hydrogenated Carbon films,* eds. Koidl P. and Oelhafen P,.*Proc. European Mat. Res. Soc.*, Vol. **17** (Les Editions de Physique, Paris, 1987).
17. Möller W., Fukarek W., Lange K., von Keuedell A. and Jacob W., *Jpn. J. Appl. Phys.* **34** (1995) 2163.
18. Möller W., *Appl. Phys.* **A56** (1993) 527.
19. Kersten H. and Kroesen G.M.W., *J. Vac. Sci. Technol.* **A 8** (1990) 38.
20. von Keudell A. and Möller W., *J. Appl. Phys.* **75** (1994) 7718.
21. Wertheimer M., Martinů L., and Moisan M., in *Plasma processing of Polymers,* eds. d'Agostino R. *et al* NATO ASI ASI Series E: Applied Sciences. **346**, (Kluver Academic Publishers, Dordrecht/Boston/London, 1996), 101.
22. Catherine Y. and Couderc P., *Thin Sol. Films* **144** (1986) 265.
23. Martinů L., in *Plasma processing of Polymers,* eds. d'Agostino R. *et al* NATO ASI ASI Series E: Applied Sciences. **346**, (Kluver Academic Publishers, Dordrecht/Boston/London, 1996) 247.
24. Craig S.and Harding G.L., *Thin Sol. Films* **97** (1982) 345.
25. Savvides N.and Window B., *J. Vac. Sci. Technol*, **A3** (1985) 2386.
26. Savvides N. and Window B., *J. Vac. Sci. Technol.*, **A4** (1986) 504.
27. Window B. and Savvides N., *J. Vac. Sci.Technol.* **A4** (1986) 453.
28. Window B. and Savvides N., *J. Vac. Sci.Technol.* **A4** (1986) 196.
29. Kay E., *Z. Phys. D.- Atoms, Molecules and Clusters*, **3** (1986) 251.
30. Sah R.E., Dischler B., Bubenzer A.and Koidl P., *Appl. Phys. Lett.* **46** (1985) 739.

31. Leidlich D., Schuhenn A., Umbach H.G., Niemann E. and Beyer W., *Diamond and Related Mater.* **1** (1992) 1169.
32. Ishikawa F., K. Tamahashi, M. Chigasaki, S. Onuma, M. Wakagi, T. Ohno, Y. Shimamura and C. Yamagashi, *Mat. Res. Soc. Symp. Proc.* **118** (1988) 429.
33. d' Agostino R., Lamendola R, Favia P. and Giquel A., *J. Vac Sci. Technol,* **A 12** (1994) 308.
34. Jones D.I., Stewart A.D., *Phil. Mag.* **B46** (1982) 423.
35. Franceschini D.F., Achete C.A. and Freire F.L. Jr., *Appl. Phys. Lett.* **60** (1992) 3229.
36. Freire F.L.Jr., Achete C.A., Mariotto G., Canteri R., *J. Vac. Sci. Technol.* **A** (1994) 3048.
37. Raveh A., Martinu L., Hawthorne H.M. and Wertheimer M., *Surf. and Coat. Technol.* **58** (1993) 45.
38. Grill A. and Patel V., *Appl. Phys. Lett.* **60** (1992) 2089.
39. Robertson J., *Diamond and Related Materials* **2** (1993) 984.
40. Grischke M., Hiecke A., Morgenweck F. and Dimigen H., *Diamond and Related Materials* **7** (1998) 454.
41. Biederman H., Howson R.P. and McCabe I., in *Proc. of IPAT 87,* (Brighton, 1987) 152.
42. Biederman H., Kohoutek K., Chmel Z., Stary V. and Howson R.P., *Vacuum* **40** (1990) 251.
43. Tkachuk, B.V., Marusii, N.Y. and Laurs, E.P., *Vysokomol. Soedin.* **A15** (1973) 2046.
44. Kny E., Levenson L.L. James W.J., Auerbach R.A., *J. Vac. Sci. Technol.* **16** (1979) 359.
45. Kay E., Dilks A. and Hetzler U., *J. Macromol. Sci.-Chem.* **A12** (1978) 1393.
46. Biederman H., Slavínská D., Bílková P.and Stundžia V, in *Plasma Processing of Polymers*, eds d` Agostino R., Favia P., Fracassi F. (NATO ASI Series, Ser. E- Vol. **346** ,Kluwer Academic. Pub., Dordrech, Boston, London, 1997) 365.
47. Luff P.P. and White M., *Vacuum* **18** (1968) 437.
48. Boonthanom N. and White, M., *Thin Sol. Films* **24** (1974) 295.
49. Abeles B., *Appl. Solid State Sci.* **6,** (1976) 1.
50. Biederman H. and Holland L., *Nucl. Instr. and Methods* **212** (1983) 497.
51. Biederman H., *Vacuum* **34** (1984) 405.
52. Martinů, L., *Thin Solid Films* **140** (1986) 307.

53. Beale, H.A., *Ind. Res. Develop.* **23** (1981) 135.
54. Heilmann A., Hamann C., Kamfrath G. and Hopfe V., *Vacuum* **41** (1990) 1472.
55. Morita S. and Hattori S., in *Plasma deposition, Treatment, and Etching of Polymers,* ed. d'Agostino R. (Academic Press, New York, 1990), 423.
56. Hori M., Yoneda T., Yamada H., Morita S. and Hottori S., *Plasma Chem. and Plasma Process.* **7** (1987) 155.
57. Oehr C. and Suhr H, *Thin Solid Films* **155** (1987) 65.
58. Suhr H., Bald J., Deutschmann L., Etspüler A., Feuer E., Grunwald H., Haag C., Holzschuh H., Oehr C., Reich S., Schmid R., Traus I., Waimer B., Weber A. and Wendel H., *Journal de physique* **50** (1988) C5-739.
59. Phadke S.D., *Thin Solid. Films* **48** (1978) 319.
60. Morosoff N., Patel D.L., White A.R., Umaña M., Crumbliss A.L., Lugg P.S., Brown D.B., *Thin Solid Films* **117** (1984) 33.
61. Osada Y. and Mizumoto A., *J. Appl. Phys.* **59** (1986) 1776.
62. Osada Y., Mizumoto A. and Tsuruta H., *J. Macromol. Sci.- Chem.* **A24** (1987) 403.
63. Osada Y. and Yamada K., Yoshizawa I., *Thin Solid Films* **151** (1987) 71.
64. Yamada K. and Osada Y., in *Proc. IUPAC Int. Symp. on Plasma Chemistry, ISPC-8*, (Tokyo, 1987) 1305.
65. Roy R.A., Messier R. and Krishnaswami S.V., *Thin Solid Films* **109** (1983) 27.
66. Harding G.L. and Craig S., *J. Vac. Sci. Technol.* **16** (1979) 857.
67. Sikkens M., *Solar Energy Materials* **6** (1982) 403.
68. Sikkens M., *Solar Energy Materials* **6** (1982) 415.
69. Weissmantel C., Breuer K. and Winde B., *Thin Solid Films* **100** (1983) 383.
70. Dimigen H., Hubsch H. and Memming R., *Appl. Phys. Lett.* **50** (1987) 1056.
71. Klages C.P. and Memming R., *Mater. Sci. Forum* **52/53** (1989) 609.
72. Bewilogua K. and Dimmigen H., Surf. Coat. Technol. 61(1993) 144.
73. Monaghan D., Teer D.G., Logan P.A., Efeoglu I. and Arnell R.D., *Surface and Coatings Technology* **60** (1993) 525.
74. Biederman H., Hlídek P., Pešička J., Slavínská D. and Stundžia V., *Vacuum* **47** (1996) 1385.
75. Biederman H., Hlídek P., Pešička J., Slavínská D., Stundžia V., Zemek J., Kingdon R.J. and Howson R.P., *Vacuum* **47** (1996) 1453.
76. Biederman H., Howson R.P., Slavínská D., Stundžia V., *Vacuum* **48** (1997) 883.

77. Biederman H., Stundžia V., Slavínská D., Žalman J., Pešička J., Vaněček M., Zemek J. and Fukarek W., *Thin Solid Films* **351** (1999) 151.
78. Kvasnica S., Fallmann W., Biederman H., Boldyreva H., Slavínská D., in *15th Int. Symp. On Plasma Chemistry, Symp Proceedings Vol. VI*, Eds. Bouchoule A., Pouvesle J. M., Thomann A.L., Bauchire J.M. and Robert E., (GREMI, CNRS/University of Orleans, Orleans, France 2001) 2403.
79. Vyskočil J., Široký P., Vorlíček V., Peřina V., *Diamond and Related Materials* **5** (1996) 466.
80. LoBindo N. E., Aharonov R. R.,. Fontana R. P, *Surface and Coating Technology* **94-95** (1997) 652.
81. Kulikovsky V. Yu., Fendrych F., Jastrabik L., Chvostová D., *Surface and Coatings Technology* **91** (1997) 122.
82. Michler T., Grischke M., Bewilogua K., Dimigen H., *Thin Solid Films* **322** (1998) 206.
83. Krumpiegl T., Meerkamm H., Fruth W., Schaufler C., Erkens G., Böhner H., *Surface and Coatings Technology* **120-121** (1999) 555.
84. Donnet C., Fontaine J., T. Le Mogne, M. Belin, C. Héau, J. P. Terrat, F. Vaux, G. Pont, *Surface and Coatings Technology* **120-121** (1999) 548.
85. Zehnder T., Patscheider J.,*Surface and Coatings Technology* **133-134** (2000) 138.
86. Meng W. J., Meletis E. I., Rehn L. E., Baldo P. M., *J. Appl. Phys.* **87**, 6 (2000) 2840.
87. Schüler A., Geng J., Oelhafen P., Brunold S., Gantenbein P., Frei U., *Solar Energy Materials & Solar Cells* **60** (2000) 295.
88. Hauert R., Knoblauch-Meyer L., Franz G., Schroeder A., Wintermantel E., *Surf.Coat.Technol.* **120-121** (1999) 291.
89. Hlídek P., Biederman H., Boldyreva H., Slavínská D., in *15th Int. Symp. On Plasma Chemistry, Symp Proceedings Vol. VI*, eds. Bouchoule A., Pouvesle J. M., Thomann A.L., Bauchire J.M. and Robert E., (GREMI, CNRS/University of Orleans, Orleans, France 2001) 2373.
90. Ohring M., *The Material Science of Thin Films* (Academic Press, Inc.,Boston, 1992) 195.
91. Grischke M., Bevilogua K. and Dimmigen H., *Materials & Manufacturing Processes* **8** (1993) 407.
92. Lehmberg H., Biederman H., Slavínská D., Stundzia V., *Vacuum* **50** (1998) 27.
93. Zeuner M., Neumann H., Žalman J., Slavínská D., Biederman H., *Vacuum* **51** (1998) 417.

94. Martinů L. and Biederman H., *Plasma Chemistry and Plasma Processing* **5** (1985) 81.
95. Kay E. and Hecq M, *J.Appl. Phys* **55** (1984) 370.
96. Martinů L., Spatenka P., Biederman H. and Sicha M, *Thin Sol. Films* **141** (1986) L83.
97. Biederman H., Stundžia V., Slavínská D., Glosík J., *Vacuum* **52** (1999) 415.
98. Perrin J., Despax B. and Kay E., *Phys. Rev.* **B32** (1985) 719.
99. Martinů L., Biederman H. and Zemek J., *Vacuum* **35** (1985) 171.
100. Martinů L. and Biederman H., *J. Vac. Sci. Technol.* **A3** (6) (1985) 2639.
101. Biederman H., Stundžia V., Slavínská D., Bílková P., Pešička J., Hlídek P., Zemek J. and Martinů L., Klemberg-Sapieha J.E. and Gujrathi S.C., in *13th International Symposium on Plasma Chemistry, Symp. Proceedings Vol. III.* (Peking University Press, Beijing, 1997), 1165.
102. Martinů L. and Biederman H., *Vacuum* **36** (1986) 477.
103. Dalacu D., Martinu L., *J. Vac. Sci. Technol.* **A17**, 3 (1999) 877.
104. Dalacu D., Klemberg-Sapieha J.E., Martinu L., *Surface Science* **472** (2001) 33.
105. Klemberg-Sapieha J.E., Dahl S., Martinu L.,43rd Annual technical Conference Proceedings – Denver, April 15-20, 2000, 234-238, 2000 Society of Vacuum Coaters 505/856-7188, ISSN 0737-5921.
106. Perrin J., Despax B., Hanchett V. and Kay E., *J. Vac. Sci. Technol.* **A4** (1986) 46.
107. Martinů L., *Solar Energy Materials* **15** (1987) 135.
108. Kay E., Laurent C., Parkin S.S.P. and Mauri D., in , *Physical phenomena in Granular Metals,* eds. Cody G.D., Gebale T.H. and Ping Sheng (*Mater Res. Soc. Symp.* **195** (1990) 423.
109. Biederman H. and Slavinska D. unpublished data.
110. Biederman H., Chjok H.J., Martinů L., David J., Kadlec S. and Lukáč P., in *Proc. E-MRS Conf.,* Strasbourg, 1987 **XVII.**) 499.
111. Weismantel C., in *Proc. ISIAT 83 and IPAT 83,*Kyoto, (1983) 1257.
112. Gerstenberg K.W. and Grischke M., J. Appl. Phys. 69(1991). 736.
113. Biederman H., Misina M., Fejfar A., Ohlidal M., Pesicka J. and Slavinska D.,in *Proc. of ISPC 10, Vol 3,* eds. Ehlemann U., Lergon H.G. and Wiesemann K. (Bochum, Germany, 1991) 2.5 – 5p.1.
114. Stundžia V., Bílková P., Kvasnica S., Biederman H., Slavínská D. and Hlídek P., *Vacuum* **50** (1998) 23.
115. Stundžia V., Biederman H., Slavínská D., Nedbal J., Mačkus P.K., Poškus A. and Howson R.P., *J. Phys. D, Applied Physics* **33** (2000) 719.

116.Zajickova L., Dvorak P., Bursikova V., Perina V., Mackova A., Navratil V., Janca J., in *15th Int. Symp. On Plasma Chemistry, Symp Proceedings Vol. VI*, eds. Bouchoule A., Pouvesle J. M., Thomann A.L., Bauchire J.M. and Robert E., (GREMI, CNRS/University of Orleans, Orleans, France 2001), 133.

117.Grill A., *Diamond and Related Mater.* **8** (1999) 428.

118.Hozumi A., Sekoguchi H., Kakinoki N., Takai O., *J. Mater. Sci.* **32** (1997) 4253.

119.Hozumi A., Takai O., *Thin Sol. Films* **303** (1997) 222.

120.Choukourov A., Pihosh Y., Stelmashuk V., Biederman H., Slavinska D., Kormunda M., Zajickova L., *Surface and Coatings Technology* **151-152** (2002) 214.

121.Pihos Y., Biederman H., Choukourov A. and Trchova M., in *Proceedings of contributed papers WDS 01, Part III*, ed. Safranková J. (Matfyzpress, Praha, 2001), 602.

122.Banks B.A., Rutledge S.K., Dever J.A., Hambourger P.D., Walters P. and Bruckner E.J., in *39th Annual Technical Conf. SVC* 1996 paper 83, 431.

123.Kashem M.A., Matsushita M., Morita S., in *15th Int. Symp. On Plasma Chemistry, Symp Proceedings Vol. V*, eds. Bouchoule A., Pouvesle J. M., Thomann A.L., Bauchire J.M. and Robert E., (GREMI, CNRS/University of Orleans, Orleans, France 2001), 1937.

124.Harrop R., Harrop P. J., *Thin Solid Films* **3**, (1969) 109

125.Pratt I. H. and Lausmann T. C., *Thin Solid Films* **10** (1972) 151.

126.Tibbitt J.M., Shen M. and Bell ., *Thin Solid Films* **39** (1975) L43.

127.Holland L., Biederman H. and Ojha S.M., *Thin Solid Films* **35** (1976) L19.

128.Biederman H., Ojha S. M. and Holland ., *Thin Solid Films* **41** (1977) 329.

129.Biederman H., *Thin Solid Films* **55** (1978) L11.

130.Lehmann H. W., Frick K., Widmer R., Vossen J. L. and James E., *Thin Solid Films* **52** (1978) 231.

131.Biederman H. *Vacuum* **31** (1981) 285.

132.Maréchal N. and Pauleau Y., *J. Vac. Sci. Technol.* **A10** (1992) 477.

133.Sugimoto I. and Miyake S., *Thin Sol. Films* **158** (1988) 51.

134.Wydeven T., Golub M., and Lerner N.R., *J. Appl. Polym. Sci.* **37** (1989) 3343.

135.Biederman H., Bílková P., Ježek J., Hlídek P., Slavínská D., *Journal of Non-Crystalline Solids* **218**, (1997) 44.

136.Biederman H., Ježek J., Bílková P., Hlídek P., Slavínská D. and Zemek J., in *Proc. of ISPC 13,* Vol. III. ed. Wu C.K., (Peking University Press, Beijing, 1997) 1264.
137.Favia P., d'Agostino R., *Surface and Coatings Technology* **98** (1998) 1102.
138.Biederman H., Ježek J., Bílková P., Hlídek P. and Slavínská D., in *3rd conf. Proceedings Physics and Chemistry of Molecular Systems*, ed. Schauer F., (Faculty of Chemistry, TU Brno, Brno, 1998) 35.
139.Biederman H., Slavínská D., Choukourov A., Boldyreva A., Stelmashuk V., in *15th Int. Symp. On Plasma Chemistry, Symp Proceedings Vol. VI*, eds. Bouchoule A., Pouvesle J. M., Thomann A.L., Bauchire J.M. and Robert E., (GREMI, CNRS/University of Orleans, Orleans, France 2001) 575.
140.Biederman H., Zeuner M., Zalman J., Bílková P., Slavínská D., Stelmashuk V., Boldyreva A, *Thin Solid Films* **392** (2001) 208.
141.Favia P., Cicala G., Milella A., Palumbo F., Rossini P., d'Agostino R., In *15th Int. Symp. On Plasma Chemistry, Symp Proceedings Vol. VI*, eds. Bouchoule A., Pouvesle J. M., Thomann A.L., Bauchire J.M. and Robert E., (GREMI, CNRS/University of Orleans, Orleans, France 2001) 587.
142.Ryan M. E., Fonseca J.L.C., Tasker S. and Badyal J.P.S., *J. Phys. Chem.* **99** (1995) 7060.
143.Golub M.A., Wydeven T., and Johnson A.L., *Langmuir* **14** (1998) 2217.
144.Zeuner M., Hirsch D., Neumann H., Zalman J. and Biederman H., in *Proc. of ISPC 14,* Vol.III, eds. Hrabovský M., Konrád M. and Kopecký V.,(Institute of Plasma Physics AS CR, Prague, Czech Republic 1999) 1527.
145.Mathias E., Miller G.H., *Journal of Physical Chemistry* **71** (1967) 2671.
146.Yamada Y., Tanaka K. and Saito K., *Surface and Coatings Technology* **43/44** (1990) 618.
147.Kitoh M., Honda Y., *Thin Solid Films* **271** (1995) 92.
148.Hismeh G.A., Barr T.L., Sklyarov A. and Hardcastle S., J. *Vac. Sci. Technol.* **A14**, (1996) 1330.
149.Fukushima K., Ikeda Y., Hayashi T., Kikuchi N., Kusano E., KinbaraA., *Thin Sol. Films,***392** (2001) 254.
150.Kholodkov I., Biederman H., Slavinska D., Choukourov A., Trchova M., *Vacuum* **70** (4) (2003) 505.
151.Biederman H., Stelmashuk V., unpublished results
152.Youngblood J.P. and McCarthy J., *Thin Sol. Films* **382** (2001) 95.

CHAPTER 10

BIOMEDICAL APPLICATIONS OF PLASMA-DEPOSITED THIN FILMS

KATHRYN J. KITCHING, VICKI PAN AND BUDDY D. RATNER
University of Washington
Seattle, USA

10.1. Introduction

Over one hundred years ago, Sir William Crooke, the pioneer of "glow tubes", referring to the plasmas in his apparatus, stated *"...the investigation of the fourth state of matter will be one of the most challenging and exciting fields of human endeavor"*.

Indeed, since that time, the technique has developed considerably and plasma deposited thin films have found application in a wide range of technologies. Specifically, they have been explored for biomedical applications since the 1960s. Some common applications for synthetic materials in medicine and biology are highlighted in Table 10.1.

Table 10.1. Medical Devices – Usage in United States.

Device	Annual use
Intraocular lens	2,700,000
Hip and knee prostheses	400,000
Cardiovascular stent	1,500,000
Heart valve	80,000
Pacemaker	100,000
Breast implant	200,000
Contact lens	30,000,000
Dental implant	500,000
Catheters	20,000,000

What unique characteristics make plasma treatments particularly attractive for biomedical applications and what successes have been seen? This review article will address these questions and highlight advances and possibilities in biology and medicine.

The first question, what special properties might these plasma surfaces have, is both easy and hard to answer. The easy answer is that there is a list of characteristics that are clearly desirable for applications in contact with living systems. Plasma films are advantageous for the following reasons:

- dry processing is clean and efficient
- coating times are rapid – often seconds
- films are highly cross-linked, pin-hole free and provide good barrier properties
- coatings are conformal, important for complex geometries
- treatments can be done on a continuous basis
- coatings exhibit tenacious adhesion to the substrate
- many possible chemistries are possible
- many possible substrates can be coated
- monomer costs are negligible
- the coatings are sterile
- the coatings are non-toxic
- there is much precedent due to widespread application in microelectronics

The more complex answer to the first question addresses the meaning of biocompatibility. Based on contemporary regulatory standards and most accepted definitions, plasma deposited coatings are biocompatible. This subject will be addressed in the first section of the review. Other topics to be discussed in this article include applications for cell culture, controlled release of drugs, biosensors, implanted plasma-treated biomaterials, contact lens treatments, non-fouling surfaces, plasma-deposited films for biorecognition and patterned plasma-deposited films. The second question, addressing the successes that have been seen, is effectively answered in the following sections dealing with the application of plasma modified materials.

10.2. Biocompatibility: relationship to plasma deposited films

Plasma-deposited thin films and plasma treatments have frequently been referred to in the literature as "biocompatible". To add perspective to this review, it is worthwhile examining what this commonly used word means, and how it applies to plasma coatings and treatments.

A widely published definition of biocompatibility, developed during a consensus conference on biomaterial definitions, is "*the ability of a material to perform with an appropriate host response in a specific application*" [1]. This definition, though perfectly accurate, offers us little insight into exactly what biocompatibility is or why plasma films might be biocompatible.

Regulatory agencies have evolved definitions based primarily on safety and acceptable familiar reactions. Basically, if a material or device does not kill or alter living cells (i.e., it's nontoxic) and if upon implantation in an animal it heals with what is referred to as the classic foreign body reaction, it might be designated "biocompatible". The characteristics of an acceptable foreign body reaction are that after approximately 1 month of implantation, the device or material is encapsulated in a thin, avascular, collagenous bag with relatively quiescent white cells in the vicinity of the implant. Note that this description of the foreign body reaction does not address differences that might occur due to different chemical compositions. In fact, there is no documented evidence that different surface compositions will induce different host reactions. All implants, hydrophilic, hydrophobic, charged or neutral, are perceived by the organism as a foreign body and encapsulated or walled off. An extended discussion of issues in biocompatibility has been published [2].

For materials contacting the blood stream, a different set of rules apply. The criteria for biocompatibility expressed above still must be met. But in blood, surface properties strongly influence blood compatibility as manifested primarily by the absence of clotting, activated cells and emboli. The problems with defining which surface chemistries give desirable blood compatibility are discussed elsewhere [3].

Table 10.2.

Surface Functionality For Biomaterials
Lubricity
Blood Compatibility
Reactive groups for attaching biomolecules
Compositions promoting cell growth
Temperature sensitive coatings
Non-fouling (protein resistant) coatings
Antibacterial coatings

Barriers to control drug release rates
Wettable or non-wettable (hydrophobic) surfaces
Patterns of chemistry (typically micron scale)
Surfaces to induce biomineralization
Surfaces strongly adsorbing proteins

With this background, we can ask, why do we need plasma deposited films and treatments for biomaterials? The material deposited from a plasma environment generally meets the criteria outlined above for biocompatibility. Such films (or surface treatments), if sufficiently crosslinked, have very low levels of leachable components. They tend to heal as a thin, quiescent capsule. Such coatings deposited on medical devices or other materials become the interface with the body. The coatings can prevent leachables from the substrate being released into the body.

Finally, the plasma deposited coatings and treatments can provide useful functionality. Some of the functional surface properties that can be imparted on biomaterials and medical devices are summarized in Table 10.2.

10.3. Applications for Cell and Tissue Culture

Plasma deposition has been advantageous for creating new surface chemistries via both grafting and polymerization processes. The following section will review in detail how this technology has developed for enhancing the culturing of cells on surfaces and for modifying protein and cell interactions.

In order for a cell culture surface to be viable it must be able to support the initial cell attachment, followed by cell spreading and growth events. It is well recognized that such cell behaviors are mediated by the layer of adsorbed proteins on the material surface. On contact with the biological environment synthetic materials tend to adsorb proteins at the interface via hydrophobic and electrostatic forces. In some surfaces there is a higher affinity for non-adhesive proteins, like albumin, while others favor cell adhesive proteins, such as fibronectin. Therefore, in addition to the species composition, surface chemistry also plays an important role in determining the properties associated with an adsorbed protein, i.e. bond strength, conformation, exchange and restructuring. Of particular importance for cell culture applications is protein composition and protein conformation, which in turn determine the availability of cell-adhesive

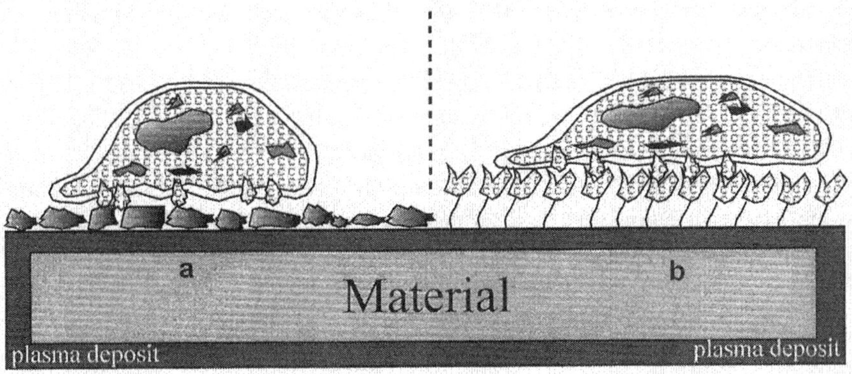

Figure 10.1. Relationship between surface chemistry, protein adsorption and cell processes. In this diagram, a material is uniformly coated with a plasma-deposited thin film. This plasma film surface modification can be used to control the protein adsorption and subsequent cell behavior, (a) proteins non-specifically adsorb to the surface (many proteins, many conformations) – cells can respond to the uncontrolled protein film, (b) the plasma film is used to anchor a tether which binds a protein at a particular site on the protein molecule – all proteins are uniformly immobilized and only the protein of choice is on the surface, thus permitting superior control of cell processes such as spreading.

domains, such as RGD peptide sequences for cell binding events. This relationship between the surface chemistry, protein adsorption and cellular behaviors is illustrated in Figure 10.1.

Glow discharge has been applied commercially since the 1960s for the processing of polystyrene used for tissue culture purposes. Corona discharge (ionized air) or plasma treatment are effective means of achieving this, although the precise parameters depend on the manufacturer and are often proprietary in nature. However, it is accepted that the treatment alters the surface chemistry by an oxidative mechanism and tissue culture polystyrene (TCPS) is a superior surface for primary and secondary cell culture compared with the unmodified material. Early work carried out to enhance polymer surfaces for cell growth by a glow discharge mechanism involved treatment of polystyrene (PS) petri dishes using air plasma [4]. The cell culture capabilities were comparable to that of commercial TC dishes and the effect was not compromised following aging. Another commercially available culture dish under the name of Primaria™ is treated with a process that results in a surface with nitrogen and oxygen-containing moieties. This is consistent with electron spectroscopy for chemical analysis (ESCA) data provided by the supplier (Becton Dickinson) and additional studies performed on the dishes have indicated a plasma treatment, most likely using a nitrogen species [5].

It has been established that the enhanced cell culture capability of substrates, particularly plasma treated surfaces, results from an ability to preferentially adsorb cell-adhesive proteins, such as fibronectin (Fn) [6,7,8] and vitronectin (Vn) [9] from the cell culture serum. In one study it was found that TCPS adsorbed between two and fourfold more Vn than unmodified PS under cell culture conditions – therefore considered to be a major factor contributing to increased cell attachment and growth on TCPS [10]. It was also shown the conformation of adsorbed Vn on TCPS was more favorable for inducing cell attachment than those on unmodified PS. Furthermore, in a study to compare N-containing Primaria™ with regular TCPS it was found that although both surfaces adsorb a similar amount of Vn, Primaria™ adsorbs two to three more times Fn than TCPS. Also, the adsorption of Fn and Vn on Primaria™ was non-competitive and thus it was concluded that different binding sites were involved [11]. The effect of adsorbed proteins on cellular behaviors had previously been investigated and when both Fn and Vn were depleted from the serum used for cell culture, BHK-21 and HT1080 fibroblasts failed to attach to TCPS or Primaria™ surfaces [12]. Interestingly however, cell-line dependence was observed in the case of human dermal fibroblasts. Vn-depletion only affected cell spreading but not attachment, and Fn depletion had no effect on either event. This behavior was consistent with previous studies for human vein endothelial cells and other cell types, including human bone derived cells, such that attachment onto TCPS during the first 90 min of culture is mainly due to adsorption onto the surface of serum Vn [13].

The relationship between cell growth and surface composition is complex and may be a contributing factor to the sub-optimal performance of medical implants. Therefore, glow discharge surface modification has increasingly been applied to improve performance. For example, vascular grafts typically made from poly(ethylene terephthalate) (PET, Dacron®) or expanded polytetrafluroethylene (ePTFE, GoreTex®), frequently suffer from thrombosis and other adverse foreign body reactions associated with incomplete endothelialization on the graft lumen. The surface chemistry of these hydrophobic polymeric grafts has consistently shown low attachment and growth of endothelial cells (EC) [14]. Vascular graft material is thought to show only a moderate enhancement of cell adhesive proteins using pre-adsorption studies since the substrate material possesses only a weak binding affinity toward them. Therefore, one approach that was considered was the use of non-depositing plasma treatments to promote the growth of EC on such biomaterials by altering the surface chemistry. The results showed that following O_2 plasma treatment the surfaces more readily retain Fn and other cell adhesive proteins, and there is enhanced EC attachment [15]. These findings are consistent with other studies in which N_2 plasma treatment was also considered [16]. Clear relationships between these treatments and *in vivo* performance have not as yet been established.

Another non-depositing plasma that has found different levels of success is the ammonia (NH_3) plasma. ESCA analysis revealed the incorporation of C-N, C-O and C=O groups on ePTFE and PTFE after NH_3 plasma treatments [17]. Enhanced attachment of human endothelial cells (HUEC) was found on the plasma treated surfaces and improved performance over human collagen, human fibronectin or collagen-fibronectin coated PTFE was observed. In further studies, plasma treatment, followed by pre-adsorption of Fn or collagen, showed that EC attachment and growth were statistically superior in culture [18]. Enhanced growth of animal and human endothelial cells has been achieved when biodegradable polymers used for tissue engineering scaffolds, such as poly(L-lactic acid) and poly(glycolic acid), were subjected to a NH_3 plasma processing technique [19]. Enhanced EC attachment/growth on vascular grafts *in vivo* was reported using the proprietary Primaria™ surface treatment [20]. A polyurethane graft was modified using this process, pre-seeded with autologous microvessel EC, and implanted for 5 weeks in dogs. A new artery with an endothelial layer, vascular media and adventitia was found encased inside the pre-seeded graft.

Following the improvements achieved using non-depositing plasma treatment, the use of N-containing thin films has also been applied and found to enhance cell culture. Plasma polymers made from amines (heptylamine, butylamine, allylamine) and amide monomers (dimethyl formamide, dimethyl acetamide, dimethyl propionamide) increased EC growth on otherwise poor cell growth fluropolymer surfaces. The main promoter of attachment on these plasma polymers was postulated to be the amide group. A good correlation was found between the EC attachment and the surface N/C ratio as determined by ESCA, while no correlation was observed with the surface O/C ratio [21]. However, the ammonia treatment was found not to be as efficient which may result from surface restructuring over time and loss of treatment effect that occurs in the case of NH_3 treated fluoropolymers. In another example, plasma deposited thin films obtained from pulsed processing of allylamine result in stable surface coatings on polysiloxane substrates and showed improved culture of neuronal cells [22]. For a study using a moving electrode configuration, 50 to 60 cm long ePTFE grafts having ID 4 mm were successfully coated with butylamine [23]. ESCA and Fourier Transform Infrared Spectroscopy (FTIR) analysis confirmed a thin film was deposited onto the graft that contained N and O functionalities. When tested in an artificial circulatory system, using either constant or pulsatile flow conditions, an EC monolayer was observed for the butylamine plasma coated surfaces, but not in the case of untreated graft material.

Cell culture can also be enhanced by the deposition of O-containing plasma polymers. A plasma polymerized 2-hydroxylethyl methacrylate (ppHEMA) thin film significantly enhanced the attachment and growth of rabbit corneal epithelial cells onto silicone rubber surfaces as examined by light microscopy [24]. Cell attachment onto control and argon plasma treated surfaces was negligible. In other studies, different levels of O incorporation in the coating have been achieved by mixing O_2 with organic monomers such as acetone, methanol, and formic acid. Modified PS and poly(ethylene terephthalate) PET surfaces have significantly higher bovine aortic EC growth, being well correlated with an increased oxygen content in the coating as determined by ESCA [25]. Chemical derivatizations and ESCA were used to measure the concentration of hydroxyl, carbonyl and carboxyl groups as compared to EC growth for a range of O-containing plasmas [26]. Correlation was only found to depend on the carbonyl content, which was thought to result mostly from ketones. Interestingly, poly(vinyl methyl ketone) (PVMK) surfaces, rich in carbonyl groups, do not support cell growth. This is evidence that the unique properties of plasma deposited films, particularly the density and highly cross-linked nature, provide suitable cell culture surfaces. In effect, the relationship between cell proliferation and surface chemistry is not a simple one.

Studies have shown acidic plasma polymers can also provide good culture surfaces, although the dependence between surface acid groups and cell culture capability is not linear. In the case of plasma copolymer (PCP) coatings of acrylic acid/octa-1,7-diene (acid-PCP), oxygen functionalized surfaces were obtained. The use of diluent hydrocarbon in the monomer feed allowed for deposition of surfaces with a range of O/C concentrations and correlated using ESCA. PCP surfaces with a low level of carboxylic acid (2.3%) were found to promote human keratinocyte attachment; the performance being similar to that observed for collagen I [27]. In a similar study carried out, osteoblast-like cells achieved a maximum level of attachment for a surface having 3% carboxylic acid [28]. The acid-PCP surface outperforms TCPS and other plasma polymer surfaces such as allyl amine/octa-1,7-diene, where cell attachment was improved with increasing nitrogen content, but failed to reach the levels seen for collagen I. Extensive osteoblast-like cell attachment was found on these optimal acid-PCP coatings and for carboxylic acid-terminated self-assembled monolayers (SAMs); whereas methyl-terminated SAMs and a plasma polymer of octa-1,7-diene (OD-PP) displayed poor adhesion [29]. This behavior was also paralleled using these surfaces to culture human keratinocytes [30]. Interestingly, SAMs and acidic plasma polymers display similar cell culture results although there are significant differences between their chemistries. For example, plasma polymers have only limited carboxyl functionality at the surface, but all the surface groups are carboxylic acids for a well-packed SAM. Nevertheless, the role of carboxyl groups in promoting cell attachment has been observed in studies using SAMs as model systems and an increasing order of EC growth was found on surfaces with these end groups: $-CO_2H >> -CH_3 > -CO_2CH_3 > -CH_2\text{-}OH$ [31,32]. Synergistic effects have been cited as advantageous in cell attachment for SAMs of mixed end groups [33].

The plasma processes so far considered promote cell culture by increasing the adsorption and retention of adhesive proteins. However, adsorbed proteins are known to be eluted by serum proteins and it may be the case that covalently attached proteins and peptides are more effective for cellular processes on the surface. The activation of even the most chemically inert materials can be achieved using the appropriate plasma treatment. In the case of ePTFE, a mixture of Ar/O_2 plasma was explored for the grafting of acrylic acid, followed by sequential attachment of NH_2-terminated poly(ethylene glycols) and RGD peptides to improve endothelial cell affinity [34]. In another study, Ar plasma-induced grafting of poly(acrylic acid) (pAA) and immobilization of type III collagen was carried out on silicone biomaterial surfaces [35]. Cell adhesion on silicone rubber controls was negligible, but showed improvement for pAA-grafted surfaces although cell growth was depressed probably due to the acid environment of the membrane. The collagen-linked surfaces showed significant attachment and growth while cell morphology was normal for primary cells. The collagen did not appear to be denatured and thus may provide an appropriate surface for clinical applications. The introduction of amino (NH_2) functional groups onto materials to provide a site for biomolecule immobilization can be achieved using NH_3 or alkylamine plasmas. The covalent immobilization of insulin and transferrin growth factors onto polyurethane surfaces using NH_3 treatment showed increased growth of bovine vascular EC compared with immobilized collagen and unmodified controls. These surfaces were also more efficient than free or adsorbed growth factors, and the effect was long lasting (ca. 9 months) for controlling the cell morphology and biochemical state when co-immobilized with collagen [36].

In addition to enhanced cell culture capability, surface modification can also impart the device with new functionality. For example, a surface with both a cell-adhesive state and a non-adhesive state that can be controlled by a small change in temperature may be used for the harvest of an intact cell sheet without the need for enzymatic digestions. Such "smart devices" with built-in "release agents" may be advantageous for numerous tissue engineering applications. The surfaces are typically based on grafted poly(*n*-isopropyl acrylamide) (NIPAM) [37] that has a phase transition temperature, or lower critical solution temperature (LCST), of 31 °C. Above 31 °C, the surface is moderately hydrophobic and adsorbs appropriate proteins for cell attachment and growth. Below 31 °C, the surface is a hydrogel that resists protein adsorption and cell attachment. Various methods of chemical grafting, photochemical grafting, and e-beam grafting have been investigated [38]. Recent work involving plasma polymerization of NIPAM showed retention of the monomer structure from ESCA and Fourier Transform Infrared Spectroscopy (FTIR) analysis. The phase transition property, having a value of 29 °, was measured using a novel technique based on shear force modulation atomic force microscopy (AFM) and provided further evidence that plasma polymerization does not affect the NIPAM structure responsible for the thermal behavior [39]. In cell culture experiments, similar bovine aortic EC attachment and growth were observed on NIPAM treated and untreated TCPS. Cultured cell sheets were successfully detached from the dish with plasma deposited NIPAM at room temperature [40].

This new breed of plasma polymerized coatings, together with improved methodology for treating materials of different chemistry and geometry, has numerous advantages and applications for cell and tissue studies. This "clean" one-step process that is largely independent of the substrate material, in addition to exceptional functionality, is desirable for fabricating real-world biomedical materials and devices where cell attachment and growth is important over other surface coating techniques.

10.4. Applications for Controlled Release

Controlled release can be defined as: a method in which one or more active agents or constituents are made available at a desired site and time at a specific rate. Such technology can benefit many fields including environmental, agricultural, consumer and industrial; but it is particularly relevant in the biomedical domain, as will be considered here. Where a bioactive agent must be delivered, for example a drug, the main advantage of such systems include: extending the active lifetime, reducing or eliminating undesirable reactions, maximizing the overall efficiency. There are many ways to control diffusion, reaction rate or physiochemical parameters. Designing the appropriate material, manipulation of the biological barriers, or targeting the fate of the agent are all viable methods that have achieved some success. In each case, the interaction between the delivery system and the barrier must be understood to achieve the desired release of active substance. There are three main mechanisms: diffusion, degradation, and swelling followed by diffusion – any combination may occur in a given release profile that can also be sub-divided into matrix and reservoir-type systems. In a matrix system the active constituent is homogeneously dissolved or dispersed and release is dependent on geometry. A rate-controlling barrier surrounds an active species in the case of reservoir delivery. Diffusion occurs when the agent passes through the barrier that forms the controlled release device and can occur on a macroscopic or molecular scale. The release rate normally decreases with time in a matrix system since the active agent has progressively farther to travel and longer diffusion time is required. For reservoir systems, the coating surrounding the reservoir has a uniform and non-changing thickness and diffusion remains fairly constant. Degradation – when an active agent is released from a system during erosion – occurs via bulk (uniform) and surface (proportional to area) mechanisms. In swelling-controlled systems the active agent is not released until placed in a thermodynamically compatible medium – absorption of fluid causes swelling followed by diffusion. The theory behind controlled release can be found in a number of texts [41,42,43]. Delivery systems may also be classified depending upon whether they involve physical or chemical methods of release of the active agent [44] and for biological systems may be accomplished in a number of ways: subcutaneously, intravenously, transdermally, mucosally or orally [45].

The controlled release of biologically active agents has numerous applications ranging from small molecular weight antibiotics to fight device-related infections, through to high molecular weight peptides and oligonucleotides for novel protein/gene therapy. The manner in which a substance is released is of critical importance with regard to a particular application. Although this is not always the case in the majority of biological applications, a steady and consistent rate over time is preferable. This release profile, as it is known, is most often associated with the diffusion of drugs through a porous or hydrogel based matrix. However, associated with such processes, there is an initial burst effect when a drug is first released [46]. To eliminate this potential for overdosing and to prolong the device lifetime, physical properties of the matrix, like the cross-linking density, porosity, and swelling rate, are important factors to control. By using plasma pretreatment processes, such surface coatings can be incorporated into the design and may act as a diffusion barrier. For example, plasma polymerized 2-hydroxyethyl methacrylate (HEMA) was applied over a porous polyurethane (BioSpan®) matrix containing hirudin, a potent anti-thrombotic protein, to reduce coagulation at the site of cardiovascular implants [47]. BioSpan® matrices that had undergone HEMA plasma deposition resulted in a thin, highly cross-linked film on the surface that functioned as a diffusional barrier whereby the initial burst and subsequent release of hirudin was found to be greatly reduced.

In another study, the effect of coating chemistry and coating thickness for controlled release applications was carried out using plasma polymerized *n*-butyl methacrylate (BMA) [48]. An antibiotic, ciprofloxacin™, was formulated in porous polyurethane matrices, and a strong burst release was observed on uncoated devices [49]. To achieve the desired zero order release kinetics, plasma polymerized BMA deposited under intermediate reactor power (40 W) was shown to be the most effective. Coatings with low cross-linking density deposited at a low power (5 W) did not pose a sufficient diffusion barrier, whereas those achieved at higher powers (50 W, 60 W) were not as efficient – perhaps due to plasma related damages on the drug-loaded device. An increase in deposition time led to thicker films and a linear relationship between these parameters and release rate was shown. The bactericidal properties of plasma polymerized BMA and triglyme coated ciprofloxacin™ loaded matrices were investigated over 24 hours in a flow system with the species *Pseudomonas seruginosa* and found to be much improved [50].

Another area in which plasma polymerized films may find application is for delivery of agents from long-term or permanent implantable materials, where the pharmaceutical is not the main function of the device but may aid in initial wound healing and subsequent tissue integration. The materials most widely used in such implants i.e. metals, polyurethanes and fluoropolymers are not suited for diffusion-controlled delivery of drugs. Conversely those that are, e.g. polymer matrices of lactides and hydrogels, do not possess the necessary mechanical or biodegradation properties for long-term applications. Systemic injection of agents such as liposomes is possible in some cases i.e.

anticoagulation, but only a fraction reach the target site and can cause unwanted effects in other areas of the body. Therefore, localized delivery around the implant is preferable particularly in the case of angiogenesis – some important concepts and improvements for such systems have recently been outlined [2]. However, immobilization onto solid carrier materials is desirable and the fabrication and characterization of surface bound PEGylated liposome layers onto tissue culture polystyrene (TCPS) and Teflon (FEP) as models for expanded polytetrafluoroethylene (ePTFE) has been carried out [51]. Since PEGylated liposomes cannot be directly attached to these materials, RF plasma polymerization of acetaldehyde was performed to obtain a thin (~30 nm) polymeric interfacial bonding layer with surface aldehyde groups. Polyethylenimine was covalently bound onto the aldehyde groups by reductive amination, followed by covalently attaching a biotin layer, upon which proteins were bound and finally the PEGylated liposomes. Release characteristics and angiogenesis assays for the surface-bound liposomes were compared to control samples and displayed diffusion control behaviour and angiogenesis inhibition. There is much work to be carried out to further develop such systems, but there is obvious potential for utilizing plasma deposited films as precursors in control of biological recognition to implants, by eliciting a desired response using controlled release, while minimizing or localizing adverse effects.

In effect, these new tailor made biomaterials having desirable functional groups as a result of using plasma polymerization in some capacity will not only find application for innovative drug delivery systems, but a number of other areas as envisioned early on by pioneers in the field of controlled release [52]. Some of these include: genetic engineering and therapy including drug targeting; potential linings for artificial organs; immunology and many dermatological applications. Engineering implants that bioerode after the desired release period would avoid the need for surgical removal or trauma and be particularly beneficial in the treatment of chronic diseases such as diabetes. In the future such technology will undoubtedly be adapted to include a host of other systems that mimic the natural healing and biological mechanisms of the human body.

10.5. Applications for Biosensors and Electrodes

An evolving field of investigation and technology that falls within the biomaterials arena is that of biosensors. Also, stimulatory and measurement electrodes are finding wide use in biology in medicine [53]. The production of functionally active thin films on metal substrates is a way to "bridge" the gap between electronics and biological systems [54]. This may, for example, be achieved by creating membranes over electrodes that model a natural biological counterpart as closely as possible. Considering this example, the functionality and application of biosensors can be deemed quite complex, but the principles upon which they are based and operate are relatively straightforward.

Biosensors have many advantages over other analysis methods for biomolecules. For example, *in vitro* diagnostic systems require significant amounts of time on sample preparation to eliminate interferants, concentrate target molecules and sometimes label target analytes. Biosensors permit direct measurement in situ of anaytes of interest.

Biosensors, and chemical sensors in general, interpret (bio)chemical data into electrical and/or visual information. A basic system could include a sensitive layer coupled to a transducer and signal processor. The sensitive layer interacts with the analyte, either by surface adsorption or bulk absorption, to produce a physical change in the layer. The transducer converts this chemical effect into an electrical signal, which can be further amplified and processed to generate a display. An overview of such systems and recent advances in the design and functionality has been described in detail in a recent publication [55]. Plasma polymers can be integrated into the design of biosensors to impart a number of desirable characteristics including: (1) enhancement of sensor selectivity by reduction of signals associated with interferants; (2) insulation and packaging of the sensor or sensor components; and (3) participation in the fabrication of the sensitive layer.

The biological environment is demanding and biocompatibility issues must be addressed for sensors in the same way as any implantable material or device. There are biomaterial-host interactions to consider. In the case of glucose sensors, for example, surface modification of the membrane using heparin, phospholipid and diamond-like carbon are possibilities that have been considered to control the biological reaction [56]. Sensor encapsulation that can inhibit transport to the sensor element might be addressed by appropriate biological and chemical manipulation of the membrane [57]. One contribution to biocompatibility that can influence the performance of sensors is the non-specific protein adsorption and subsequent cell attachment, referred to as bio-fouling, often resulting in a strong interference signal. Biofouling has considerably impeded the long-term use and stability of implantable biosensors. There are a number of techniques to characterize biofouling of biosensor membranes [58], in addition a recent article has discussed different methods for reducing biofouling on sensors [59]. One technique reviewed is hydrogel coatings of poly(ethylene glycol) (PEG) and poly(hydroxyethyl methacrylate) (PHEMA). Several problems have been identified in the application of hydrogel coatings. These include poor adhesion, insufficient mechanical strength, deleterious effects of coating chemicals on sensor components, and limited analyte diffusion through the hydrogel matrix. Plasma polymerized hydrogels have potential in overcoming some of these problems.

A study has been carried out in which plasma polymerized triethylene glycol dimethyl ether (triglyme) $CH_3O(CH_2CH_2O)_3CH_3$ was deposited onto gold coated surface plasmon resonance (SPR) sensors and resulted in a protein resistant non-fouling surface [60]. The bare gold sensor signal was saturated in the presence of 1mg/mL of bovine serum albumin (BSA), which resisted elution in buffer. In contrast however, plasma deposited triglyme coated sensors showed a significantly lower signal associated with protein adsorption, and the baseline fell close to the original level upon rinsing in buffer. The excellent coating quality of plasma polymers (i.e. ultra-thin, good adhesion) is particularly applicable to the fabrication of SPR sensors due to the inherent shallow sampling depth of a few hundred nanometers (nm). The ability to discriminate between non-specific protein binding and other interactions should provide a suitable basis for the design of a label-free, real-time biosensor. The benefit of having a reference channel in sensor design was subsequently demonstrated using a dual-channel SPR sensor in which non-fouling plasma polymerized tetraglyme (pp4G) was applied over one of the two SPR channels [61]. This served as a reference channel to correct for optical signal change not associated with the bio-specific recognition events, such as bulk refractive index change, upon switching the solution and flow dynamics. The second SPR channel contained biotinylated moieties to probe streptavidin-biotin binding events. The subtracted spectrum between the reference and sensing channel was shown to be free of environmental factors. In a different biosensor experiment, ultra-thin pp4G ($<$20 nm) was applied over planar microelectrode arrays [62]. The coating provided protection against protein adsorption and cell adhesion for *in vitro* experiments, and did not affect the ionic conductivity of the arrays indicating that pp4G does not impede the diffusion of small ions to the sensor.

In order to function effectively in the challenging biological environment, implantable devices require a stable, biocompatible packaging and insulation. For instance, an important factor for biosensors is the insulation of electrical leads that must not be compromised over time in the bodily environment.

Parylene C® poly(monochloro-*p*-xylyene) is a biocompatible, polymeric coating used in the medical device industry to protect sensitive components from corrosive bodily fluids. However, the coating does not adhere satisfactorily to substrate metals to obtain the required lifetimes necessary for many implants. This is especially true when the coating film is subjected to electrochemical gradients [63]. A number of wet chemistry techniques have been developed using compounds that have an organic and inorganic face to bind to the substrate and polymer respectively [64]. The application of plasma polymerization has been carried out using hydrocarbon, CH_4 and C_2H_6, films as primers to enhance the adhesion of Parylene C® to clean tantalum and silicon surfaces, previously coated with a thin layer of silicon dioxide (SiO_2), via covalent bonds generated in the free radical process [65]. In a similar study, a plasma deposited primer layer significantly improved the adhesion of Parylene C®, and the encapsulated device survived saline soak for over a year, whereas Parylene C® alone failed after 60 days [66]. Plasma polymerization and thermal vapor deposition of methane and xylylene monomers was applied to synthesize a flexible coating – Plasmalene®. The insulating property of the coating was not compromised after repeated flexing in saline [67]. Furthermore, a flexible, conformal polymer coating achieved using pulsed plasma-enhanced chemical vapor deposition (PECVD) of hexamethylcyclotrisiloxane and Ar (1:5 ratio) was considered a potential candidate for insulating materials [68]. One coating in particular (plasma duty cycle of 100 ms on / 600 ms off) appeared promising. It was highly flexible when twisted, resisted flaking or cracking; and survived (n=4) 0.9 year of warm saline soak whilst maintaining its insulating property. In addition, films prepared under pulsed conditions appear to retain more cyclic siloxane structure than those prepared via continuous wave (CW) excitation, which may be a significant factor in improved coating performance.

The majority of work focusing on the use of plasma polymers in biosensors has been in the preparation of the sensitive (sensing) layer. This can be further divided into two subcategories: (1) use of the plasma polymer surface for biomolecule immobilization to achieve bio-specific recognition and sensing events; and (2) using the plasma polymer matrix itself as the sensitive layer. In the first subcategory, biomolecules, such as antibodies, have been immobilized onto plasma polymer coated sensors for the recognition and measurement of other biomolecules, i.e. specific antigens. Both covalent [69,70] and physical [71,72] immobilizations of biomolecules like heparin or enzymes have been suggested and demonstrated in studies involving plasma polymers for biosensors [73] and related applications such as immuno-diagnostic assays and non-fouling applications [74]. However, only biosensor applications of plasma polymers shall be considered in this section and for a more detailed discussion of their use in the fabrication of sensitive layers, the reader is directed to a recent review publication [75]. The potential of RF plasma for biosensor devices was investigated using N_2 and NH_3 to modify propylene (Duragard®), poly(vinylidene fluoride) (Durapore®) and poytetrafluoroethylene (Fluropore®) membrane surfaces [76]. Subsequently this allowed immobilization of glucose oxidase (GOD) using glutaraldehyde as a linking agent for covalent attachment. An arrangement incorporating the immobilized enzyme membrane and a dissolved O_2 electrode to monitor enzyme activity resulted in a glucose sensor since the concentration of dissolved O_2 decreased with increasing glucose concentration. In another study, an immunosensor consisting of Hepatitis-B antigen and a piezoelectric quartz crystal (QCR) was developed [77]. The quartz surfaces were first treated with ethylenediamine plasma to form amino groups on the surface that were later activated with glutaraldehyde to covalently immobilize anti-Hepatitis-B (anti-HBs) antibody. The sensor was calibrated for HBs-Ag detection and a linear response was obtained for an antigen concentration of 0.01 to 0.0554 IU/ml in buffer.

In a conventional sensor arrangement where enzymes are immobilized within a matrix, or between two membranes, "sandwich-type", the limited analyte diffusion can adversely affect the linearity in sensor output. Therefore, the single-type recognition layer method has been developed were the surface of the membrane is modified to have active groups for enzyme immobilization [78]. Amylamine plasma has been utilized in the fabrication of a single sensing layer upon an amperometric probe-type glucose electrode for potential on-line monitoring of food and beverage production [79]. Cellulose acetate (CA) membranes were found to be preferable to other surfaces since they overcome the interference of many electroactive compounds. Plasma polymerization was carried out to optimize the glow-discharge parameters for the monomer (5 W / 20 min). Then the membranes were activated in glutaraldehyde and GOD was covalently immobilized. The diffusion limit issue was successfully addressed and the linear response range was found to be 320 mM glucose with a 500 second response time at pH 4. In further work, CA membranes were modified by amine plasmas, i.e. ethylene diamine (EDA) and n-butylamine (n-BA) to obtain a single-layer enzyme electrode [80]. In another study an amperometric glucose sensor prepared using a dry plasma polymerization process was found to have a much faster response time (12 seconds), but effective only up to concentrations of 30 mM glucose [81].

In the alternative "bulk absorption-type" sensitive layer, molecular adsorption and diffusion of analytes through the plasma polymer itself is measured. Studies of "electronic noses" based on QCR sensors and sensor arrays have successfully employed plasma sputtered thin films as the sensitive element. In one study, QCR sensors were coated with thin fluoropolymer films, prepared by sputtering polychlorotrifluoroethylene (PCTFE) targets in Ar plasmas, and are suitable for the sensing of acetone [82]. A higher-power sputtered film was found to produce a reproducible and fast adsorption-desorption profile that was attributed to the coating chemistry obtained under these conditions, since analysis showed a lower incidence of CF_2 and CF_3 moieties. In addition, high-power films display more unsaturation, cross-links, and stable radical sites that are available to interact with target molecules. The "electronic nose" system has been subsequently expanded to include PCTFE and other thin films including plasma sputtered pyrolytic graphite (PG), polyethylene (PE) and polytetrafluoroethylene (PTFE) as the sensor probe [83]. Statistical data analysis was used to classify performance and it was found that plasma polymers containing high concentrations of stable radicals and π-conjugated moieties are useful probe materials for molecular sensing. This approach, in which an array of QCRs were sputter coated with different RF films of amino acids and polymers, was successfully applied in the discrimination of odorants with a botanical nature [84].

The application of plasma polymerized thin films is particularly relevant to biosensor studies because of the versatility of the method permitting many substrates to be treated with a wide variety of surface chemistries. This will then provide a platform technology for designing and fabricating specific biosensors that utilize nature's own biosensing mechanisms. Such an approach should also have applications in screening pharmaceutically active agents. The versatility of RF plasma technology allows it to be adapted according to biological conditions and requirements – this should be a key technology for the future manufacture of biosensors [85].

10.6. Applications for Implanted and Blood-Contacting Biomaterials

In Section 1 of this article, remarks were made about the nature of biocompatibility. The key conclusions were that the definition of biocompatibility is sufficiently general as to be of only limited value and that essentially all materials that do not leach cytotoxic substances will heal in the body with similar reactions – encapsulation in a thin, collagenous bag. This foreign body reaction is especially evident in soft tissue sites. In bone, other types of reactions (osteointegration, bone bonding) are possible, but the collagenous encapsulation reaction can also be seen. In blood, there can be strong reaction to different surface chemistries, at least over shorter implant times (hours). These reactions are elaborated upon below. In the eye, possibly because of low vascularity, relatively little (or very slow) foreign body reaction is seen. In percutaneous (through the skin) implantations, a unique reaction, often referred to as marsupialization, is observed – epithelial cells from the skin attempt to grow down along the implant and push the implant out of the skin. With this preface, the literature on plasma treatment of biomaterials for implantation can be examined.

The *in vivo* response to plasma coated materials is the most relevant way of assessing the biocompatibility of the coatings. However, these studies are generally only carried out once a thorough surface characterization and number of *in vitro* tests have been performed and the material has shown the potential chemical properties, stability and biocompatibility. Conversely none of these preliminary tests can provide the required information that is necessary to understand exactly what happens to a material once implanted. In addition, a material may possess the necessary properties to be used in a particular fashion at one site in the body but not at a different one [86]. The fact that plasma deposited films prepared using radio frequency glow discharge (RFGD) are formed without catalysts, other harmful additives or leachable material makes their application as implant materials potentially desirable. Still there remains a need to carry out detailed biomedical examination of a plasma treated material since the method of modification or the substrate material may affect the resulting biological reaction. For instance, fragments of degraded molecules can react and cause a toxic reaction once the implant is in contact with living tissue.

An early study of the implantation of plasma polymerized films deposited on poly(dimethyl siloxane) (Silastic®) in rat skeletal muscle for periods up to 84 weeks indicated there was no excessive foreign body reaction [87]. Three gases, chlorotrifluoroethylene (CTFE), styrene (S) and ethylene (E), were used to prepare thin polymer films (PPCTFE, PPS and PPE) (<0.5 μm) onto Silastic® medical grade tubing. The coated material and control samples were subsequently sterilized using ethylene oxide then allowed to degas under vacuum for a period of 96 hours before being implanted. The coated material and control samples were implanted into paravertebral muscle of Wistar rats. For each surface, twenty rats were randomly divided into equal groups for necroscopy at 2, 4, 8, 12 and 24 weeks. Two rats having PPS coatings were sacrificed at 75 weeks, three with PPE at 77 and three PPCTFE samples at a time period of 84 weeks for investigation by scanning electron microscopy (SEM) and histopathology. Examination of the major organs showed no evidence of systemic toxicity for any of the samples. The fibrous capsule surrounding the implant – consisting of mainly mature connective tissue and containing a small number of fibrocytes – gave a peak response dependent on the polymer but was largely indistinguishable between the coated and uncoated materials by 24 weeks. Compared to Silastic®, PPS showed more inflammatory cells 2 and 4 weeks after implantation; PPE was much less at 2 weeks; whilst PPCTFE was identical. There was little difference in the number of inflammatory cells in the implant area between the treated and control samples or indeed between the polymers themselves after a time period of 4 weeks. SEM performed on the control shows the collagenous tissue to be less densely packed than capsule tissue surrounding each of the plasma polymerized implants. Overall these films were well tolerated in rat muscle and the fibrous tissue response after steady state conditions had been reached (12–24 weeks) was not different from Silastic®, a material that is inert in biological tissue. Of particular interest, however, is the fact that SEM images of polymer films implanted for 75 weeks show cracking compared with samples stored under ambient conditions whose uniform surface coating is maintained. The overall differences between the respective polymer films once implanted were negligible which suggests that the bulk material properties and application for a specific biomedical device may be more important considerations. In an extension of these studies; implantation of control Pt wires (250 μm diameter, 2.5 cm length) into the central nervous system of rabbits and those having plasma deposited coatings (0.1–0.3 μm) of methane (M), tetrafluoroethylene (TFE) and tetramethyldisiloxane (TMDSO) was carried out [88]. In addition, Parylene N and Parylene C films – used to protect electrical circuits and electrodes that are implanted – were vacuum-deposited onto the Pt wire and compared with the polymer coatings. Interestingly unlike the skeletal muscle implantations where each of the polymer coatings show similar behavior to the Silastic® control, once steady state conditions have been reached, reactivity of coated wires in the cortex is dependant on the polymer type. TMDSO was more reactive in the cortex than TFE and the methane coatings showed minimal reactivity. Parylene C displays a reactivity about the same magnitude as the TFE coatings, while the Parylene N reaction was severe, showing extensive capsule thickness, infiltration of reactive cells around the implantation site and fibrous tissue –

most likely growing down from the meninges. Parylene C has been shown to be nonreactive [89] and although Parylene N had not previously been implanted, the difference between the two forms is surprisingly larger than the respective polymers having completely different chemistry. This effect may be attributed to chlorine in the ring. However, careful toxicology was not performed to assess for soluble leachants or cytoreactive degradation products – these could also account for differences in the noted reactivity Still, the work demonstrates plasma polymerized coatings respond favorably *in vivo* and suggests that the nature of the particular coating needs to be tailored to the specific physiology and anatomy of the organ in which it is to be implanted.

A study to improve adhesion onto inorganic substrates of a hydrophilic film of plasma polymerized N-vinyl-2-pyrrolidone (NVP) was carried out and the resulting substrate surfaces examined *in vivo* [90]. The rationale for this work is in achieving suitable coatings that can act as a biomaterial interface on a number of non-organic substrates, such as metals, which when in contact with the body can form toxic by-products following corrosion processes [91]. Although hydrophilic NVP surfaces had been achieved using plasma polymerization [92] the adhesion with inorganic substrates is compromised in aqueous solution since water molecules penetrate the films. Therefore, samples were treated in the RF plasma reactor using argon and then a thin film of hexane (110 nm) was deposited followed by a thicker (420 nm) NVP coating. The use of PPHex as an initial interface layer could improve the adhesion of the hydrophilic PPNVP and *in vitro* cell growth was observed directly on the coatings. The samples were implanted in rats subcutaneously using the cage implant system for determining *in vivo* biocompatibility of medical device materials [93]. The inflammatory exudate that accumulated within the implanted cages was investigted at 4, 7 and 21 days by culturing on brain-heart infusion agar plates to check for infection and white cell concentrations and counting performed. There was no evidence for increased leukocyte concentrations on the PPNVP coatings as compared with polyethylene and empty cage controls – these findings suggest the films are noncytotoxic and biocompatible. A subsequent study to investigate the *in vivo* response to plasma polymerized interface materials having both hydrophobic and hydrophilic surfaces indicated none of the samples were cytotoxic, did not affect the acute and chronic inflammatory responses, or inhibit the normal wound healing response [94].

An *in vivo* screening to test the biocompatibility of polyhydroxybutyrate (PHB) and polyethylene (PE) discs that had been treated with argon plasmas containing perfluorohexane (PFH) and PFH+H_2 was performed by implantation in the peritoneum of Wistar rats [95]. After 3 months the rats were sacrificed and observation of the fibrous capsule that developed around the treated implants was compared with untreated control materials using SEM and an anatomo-pathological analysis. Untreated PE material exhibited a thick, dense, fibrous, highly vascularized capsule having a large number of inflammatory cells and fibroblasts. PHB discs that had not been treated were sometimes found far from the area of implantation and *in situ* observation indicated no capsule formation. However, microscopy revealed a very thin poorly adherent capsule that contained no inflammatory cells. Capsules were porous maintaining communication between the polymer and the biological fluids. Each of the plasma treated PHB samples gave similar results to the untreated material except that the fibrous capsule was no longer porous. The treated PE samples resulted in a thick capsule formation with possible contribution from the underlying bulk polymer. ESCA analysis of the materials showed one of the faces of the discs had been covered in a thinner plasma-deposited layer, but overall the capsule was poorly adherent to implant and contained no inflammatory cells.

Subcutaneous and peritoneal implantation of poly(ethylene terephthalate) (PET) discs, that were made more hydrophilic by argon (Ar) plasma etching or more hydrophobic by depositing a perfluorohexane compound, in rats was investigated [96]. In the abdominal integuments, discs remained loose in the pockets, covered by a thin transparent membrane, and tiny blood vessels were observed in the surrounding muscle tissue. Upon opening the peritoneal cavity the samples were most often found near the liver or bowl loop with blood vessels attached to the surface. Histological analysis 3 days following subcutaneous implantation on tissues directly surrounding the discs was found to be independent of film nature – an exudate cellular-unspecific inflammatory response near the implant site made up of mononuclear lymphocyte type cells and homogeneous pink matter was observed. Histology on the material taken from the peritoneal cavity was identical for the Ar treated and control PET with a small amount of lymphocyte-type mononuclear cells and fibrinous-cellular exudates. There was a slightly increased amount of cellular exudate seen in the case of the hydrophobic PET surface. Both subcutaneous and peritoneal specimens examined 14 days post-implantation were similar for all the samples and were surrounded by a thin band of freshly grown connective tissue made up of fibroblasts, some fibrocytes and lymphocytes – collagen fibers were clearly visible following staining. In conclusion, modification of the PET surfaces using

Ar or perfluorohexane did not indicate toxicity *in vivo* and the biocompatibility of the materials was not affected.

Organ culture experiments were carried out to demonstrate the efficacy of argon treatment for epithelial cell migration and proliferation on a poly(vinyl alcohol) (PVA) copolymer hydrogel [97]. These surfaces were implanted into rabbit eyes and no inflammatory response was detected for a period of over 14 weeks *in vivo*. In a follow-up study the biological response of three porous materials was investigated in which fifty per cent of each of the materials were modified using Ar radio frequency plasmas [98]. The materials implanted into the corneas of rabbits as 6 mm discs were poly(butylene polypropylene) (80:20), PET and expanded polytetrafluoroethylene (ePTFE). Immunohistochemical analysis was carried out at 6 and 12 weeks. In general, the Ar surface treatment decreased corneal edema, neovascularization and the recruitment of inflammatory cells – except in the case of ePTFE, which may be attributed to smaller fiber diameter in this material.

A study in which C_2F_4 plasmas formed on the visceral side of hexamethylene diisocyanate-tanned dermal sheep collagen resulted in a degradable biomaterial with a hydrophobic surface [99]. Consequently, abdominal wall defects generated in male Wistar rats that were repaired using the modified and unmodified implants, showed incorporation into surrounding muscle tissue without bowel adhesions after 4 weeks implantation in the case of the former material. These non-adhesive C_2F_4 plasma surfaces also gave rise to fewer herniations than in the unmodified materials, resulted in less degradation over time and thus a potentially improved biomaterial.

A comparison was made *in vivo* for silicone and poly(L-lactic acid) implants with microtexture; random surface roughness or smooth surfaces following RFGD; a fibronectin (Fn) pre-coating or no pre-treatment [100]. The samples were implanted into the subcutaneous pockets on the flanks of goats for 1, 3, or 12 weeks. There was fibrous capsule formation around all the implants and though the Fn treated surfaces showed more early inflammatory cell reaction, there was no significant effect on tissue reaction via any of these treatments or between the surfaces.

There are other notable examples involving implantation studies of plasma treated surfaces, in fact, too many to report all of them in this review. Two unique and interesting studies include a plasma polymerized ethylene film that has improved the surface properties of acrylic teeth such that *in vivo* plaque accumulation was not observed [101] and argon plasma-treated silicone rubber voice prostheses that enhanced *in vivo* biofilm formation – therefore it was concluded that biofilm formation was highly dependent on the hydrophobicity of the biomaterial surface [102].

The majority of the studies on the *in vivo* implantation of plasma-deposited and plasma treated samples support the contention that surface properties modified by plasmas (or any non-biological surface modification, for that matter) have little effect on the foreign body response [103,104,105]. Yet, by the classical definition of biocompatibility [1], plasma films can be defined as biocompatible in that they typically show low leachables that can trigger cytotoxic reaction and heal in a relatively thin avascular, acellular foreign body capsule. In studies that do report effects of plasma treatment on *in vivo* biocompatibility, it is worthwile examining whether certain other key variables were controlled for (cytotoxicity assay, endotoxin assay, biodegradation, implant movement, edge/shape effects).

Blood compatibility is a complex sub-set of biocompatibility, often with unique considerations. Articles reviewing blood compatibility issues are available [3,106,107,108]. There are similarities to interaction with any implanted biomaterial – protein adsorption followed by white cell interrogation followed by cell interaction. Unique and pertinent aspects include reaction occurring in the presence of fluid flows, a high concentration of fibrinogen, blood coagulation, the presence of platelets, other specialized proteins and the specific structure of the blood vessel wall. A few key points are important here. First, macroscopically visible reactions are rapid (i.e. clotting or coagulation) and such coagulation is generally (though not always) disastrous for the performance of medical devices in the blood stream. Second, there is a clear dependence on material surface properties. Third, different coagulation mechanisms are activated in rapidly flowing blood (arterial circulation) and in slower or stagnant blood flows (venous) [3]. Fourth, although some materials do not adhere to blood elements, they can activate them – thus, looking at adhesive events is not always predictive of performance. Fifth, after some fifty years of exploration of blood compatibility, there are still no clear rules as to which surfaces are blood compatible and which are not. There is a long history of plasma surface treatments being used to address blood compatibility [109]. A few select examples will be presented in this review.

Interactions between platelets and glass surfaces coated with various organosiloxanes – including tetramethoxysilane (TMOS, $(CH_3O)Si$) – polymerized by plasma induced reactions have been investigated [110]. The monomers, contained in vials or coated on glass plates, were subjected to glow discharge in the organosiloxane vapor. The samples were washed with methanol to remove unpolymerized monomers and low-molecular-weight oligomers, then dried under vacuum. The adhesion of platelets on the plasma deposited coatings was between 10 and 30 % less than the glass surfaces.

Fluorine-containing inorganic gases and monomers used to incorporate fluorine atoms onto surfaces have found use in blood interaction applications since they produce hydrophobic, Teflon®-like materials. It is also possible to incorporate a wide range of F/C ratios from the plasma polymerization process of various fluorocarbon monomers such that different surface types can be engineered. For the immobilization of various proteins onto poly(ethylene terephthalate) (PET) coverslips, C_2F_4 plasma polymers were used to modify the surfaces resulting in high retention of albumin and consequently low fibrinogen adsorption and platelet adhesion [74]. In another study, C_2F_4 plasma-treated Dacron® vascular grafts were shown to be less thrombogenic and albumin adsorbed on these surfaces was resistant to elution by sodium dodecyl sulfate [111]. The surface free energy of surfaces in air can be related to retention of albumin i.e. low-energy surfaces retain a larger fraction than the higher energy analogues. The low-energy C_2F_4 plasma treated surfaces possess high interfacial energies in water and high driving forces for protein adsorption through hydrophobic interactions.

Treatment of Gore-Tex®, Sauvage™ and DeBakey™ Dacron® vascular prostheses using Ar and C_2F_4 plasmas were studied for thrombotic response in an *ex vivo* femoral shunt baboon model since this has clinical relevance to man [112]. The grafts that had been coated displayed improved blood compatibility as compared with those that were untreated such that the rate of platelet consumption was drastically reduced and they exhibited high blood flow rates during the course of the experiment due to the reduction of thrombus build-up on the surfaces. An electron spectroscopy for chemical analysis (ESCA) study of these materials indicated a high degree of carbon bound to less than two F atoms and therefore, the plasma polymer coating was likely to be highly crosslinked. Evidence of these low-energy groups was further displayed by critical surface tension measurements, showing a value of about 13.0 dyn cm^{-1} compared with that of Teflon at 18.5 dyn cm^{-1}. In contrast, blood compatibility assays revealed adverse effects following Ar-plasma treatment, indicating gas selection is of major importance for improving surface properties. In a similar study, C_2F_4, C_2F_6, C_2F_6/H_2, and CH_4 plasma polymers were deposited onto silicone rubber, Silastic®, (smooth-walled) and expanded Teflon® Gore-Tex® (fabric) vascular grafts [113]. Both acute and chronic arterial thrombosis parameters were evaluated in the baboon model. For steady-state arterial thromboembolism with chronic A-V shunts, platelet destruction induced by the plasma-modified Silastic® grafts ranged from 1.1 to 5.6 × 10^8 platelets cm^{-2} per day. This was considered relatively non-thrombogenic. Gore-Tex® grafts inserted into the shunt system resulted in acute thrombus formation with approximately 1 × 10^{10} platelets deposited during a 1 hour exposure. In the case of Gore-Tex® grafts modified using C_2F_6/H_2 plasma, although graft surface morphology remained unchanged, platelet deposition was reduced by 87 %. This illustrates that both surface structure and surface chemistry are important contributory factors in the process of thrombogenesis. However, overall, this work indicated smooth-walled vascular grafts covered with a fluorocarbon plasma polymer overlayer were relatively nonthrombogenic.

10.7. Applications for Contact Lenses

There are vast numbers of contact lens wearers (>70,000,000 worldwide), in part due to recent advances in manufacture and technology; as a consequence the fact that they are biomaterials with similarities to implantable devices is often overlooked. Contact lenses adsorb protein layers and are interrogated by white cells (in the closed eye) with reactions analogous to those seen on implants. Thus, materials placed in the eye face similar biocompatibility issues as implants plus a host of other challenges to designers. These challenges include gas permeability, manufacturability, refractive index, comfort and cost.

Ocular tissue is particularly delicate and damage or infection can lead to a number of conditions that may ultimately result in blindness. Conversely, artificial lenses – including intraocular lenses (IOLs) – improve the quality of life for many patients who suffer visual conditions ranging from mild myopia (nearsightedness) through to those people facing progressive disorders (e.g. cataracts) and an eventual loss of sight. Therefore, this dichotomy poses a challenge for the production of suitable, cost effective ophthalmologic devices that can be safely used both cosmetically and medically. Many of the same principles can be applied to any device in the eye, however depending on the required function, the material and processing parameters will be different. This section will deal with advances in the chemistry of contact lenses; in particular how plasma treatment has benefited wearers and can continue to improve contact lenses – one of the most widely used biomaterial devices.

The contact lens concept was illustrated by Leonardo da Vinci in 1508 and Rene Descartes suggested the corneal contact lens in 1632. In the 1820s the

English astronomer Sir John Hershel discovered that vision could be corrected by placing a lens directly on the cornea and this led to the manufacture of contact lenses made from glass that covered the entire eye. In the mid nineteenth century, Adolf Fick experimented with contact lenses in animals and fitted them to human eyes. In the 1940s several optometrists and researchers began to work with plastic materials and poly(methyl methacrylate) (PMMA) became the material of choice over glass for lenses that were were designed to cover only the central area of the eye. PMMA has the bulk chemical, mechanical and optical properties required for ocular use i.e. hardness, high refractive index, biocompatibility and as such is still used today. It does, however, have some drawbacks that include hydrophobicity and low oxygen permeability. Since a contact lens is interfaced with the corneal epithelium and tear film, the wettability of the lens surface is an important factor, otherwise discomfort to the wearer results [114]. In order to modify a surface from hydrophobic to hydrophilic, surface coating or grafting techniques are possible. However, the requirements for a satisfactory contact lens coating are resistance to delamination, extremely thin and transparent (so that the optical properties are retained) and reasonably abrasion-resistant. The grafting of hydrophilic polymers onto hydrophobic substrates can result in thick layers whose optical clarity can easily be compromised when exposed to an aqueous or high humidity environment and can also affect the bulk properties of the underlying material. Therefore, obviously these kinds of modifications are not viable for contact lens applications even if a more wettable surface results. This is where plasma processes, including glow-discharge deposition and surface oxidation, have found important applications since it is possible to prepare an ultra-thin layer (<1000 Å) that is tightly bonded to the surface without interfering with the bulk properties.

Plasma deposition has successfully been applied to PMMA surfaces to increase hydophilicity so that a more wettable, comfortable surface results. The deposition of a thin plasma polymerized film onto a PMMA lens was carried out using a mixture of acetylene, N_2 and H_2O [115]. The wettability of the treated surface was found to increase as determined by contact angle goniometry. The surface was more hydrophilic than several hydrogel (soft contact lens) materials, including those whose water content was over 90 %. Tissue cultures did not indicate toxicity for the modified surfaces. Also, there was a decrease in the adhesion of corneal epithelial cells compared with the untreated control material. Furthermore, there was no accumulation of mucous matter on the treated lenses following insertion into rabbit eyes for an extended period and optical clarity was maintained following four months of continuous wear. In contrast, uncoated lenses displayed accumulation after a much shorter time, i.e. 1 week. Studies have also been carried out, in which different water vapor plasma conditions were investigated, to selectively induce hydrophilic character into the hydrophobic PMMA material and a copolymer thereof – poly(methyl methacrylate)/poly(methacrylic acid) (PMMA/PMAA). The use of Ar/H_2O plasma appeared to modify the surface by a form of "degradation" whereby Ar ion bombardment resulted in reduction/decarboxylation followed by adsorption of H_2O at high energy sites on the PMMA surface [116,117]. Despite the fact there was chemisorption of water on the modified material there was lack of evidence to support actual hydrolysis on the polymer surface. Therefore, further work was carried out to specifically introduce OH functionality [118]. PMMA substrates were treated with O_2/H_2O and H_2O plasma glow discharges and the resultant surfaces investigated using a number of surface techniques including electron spectroscopy for chemical analysis (ESCA), contact angles, and attenuated total reflectance Fourier transform infrared spectroscopy (ATR-FTIR). The data suggests incorporation of hydroxyl (OH) and carbonyl (CO) functionalities; particularly in the case of H_2O radio frequency glow discharge (RFGD) modifications where contact angle data shows an increase in polarity and the addition of chemically bonded OH is likely.

The problem of visual acuity – whereby discomfort to the eye results from extended wear of contact lenses – generally occurs from protein deposits (and occasionally calcium deposits) on the lens interface. Such deposits contribute to cell adhesion on the lens, and the mechanism of initial attachment of corneal epithelial cells has been investigated on tissue culture polystyrene (TCPS) surfaces that had a plasma polymerized film of methyl methacrylate monomer as compared with the control material [119]. It was found that attachment of cultured corneal stromal fibroblasts is dependent on the adsorption of both vitronectin and, to a lesser extent, fibronectin. A study was carried out to evaluate the effect of plasma treatment on the protein adsorption of Ar plasma treated PMMA [120]. An "artificial" tear solution containing lysozyme, albumin, and γ-globulin was used to measure the extent of protein adsorption on the treated surface following immersion. Although there was a dependence over time, the total adsorption of protein was suppressed for the Ar treated PMMA due to an increase in the hydrophilic nature of the surface. Despite each of the proteins having a different relative affinity for the surface, particularly with respect to corresponding changes in wettabilty, overall control in protein adsorption was achieved using this method.

The deposition of non-fouling coatings onto the surface of contact lenses using plasma polymerization is a promising method to reduce the adsorption of proteinaceous matter. The dominant constituent in the layer of adsorbed protein is lysozyme, a positively charged protein at physiological pH, and therefore coatings of *n*-heptylamine (*n*-ha) were deposited onto contact lenses from a gas plasma vapor to bring about electrostatic repulsion of the protein [121]. This was not successful. However, immobilizing a thin coating of partially oxidized dextran onto freshly deposited *n*-ha coatings showed a reduction in lysozyme adsorption. A method of surface modification for PMMA lenses using phosphoryl choline by the use of a plasma glow discharge has been described [122]. Phosphoryl choline is a component of a naturally occurring difatty-acid derivative, lecithin, which is immunologically inert and hydrophilic. Cell adhesion on the lens is greatly reduced following treatment, the coating does not significantly affect the power of the hard contact lens even following exposure to UV and saline and ultimately biocompatibility was reported to be enhanced. In addition, the use of non-fouling oligoglyme [CH_3-O-(CH_2-CH_2-O)$_n$-CH_3] plasma-deposited coatings have been proposed for use in optical applications [123].

Recently a number of surface techniques have been utilized to characterize and detect the adsorption of protein binding events on a number of contact lens materials even at very low concentrations. Human serum albumin (HSA) was spin labeled with 4-maleimido-tempo (MSL) and the irreversible adsorption of HSA-MSL to a number of contact lenses was determined by changes in the electron spin resonance spectroscopy (ESR) signal compared with HSA-MSL in solution, such that immobilzation could be identified at the contact lens surfaces [124]. ESCA and matrix-assisted laser desorption ionization mass spectrometry (MALDI-MS) have investigated initial adsorption onto hydroxyethyl methacrylate (HEMA)-based contact lenses and it was found many proteins adsorb within 10 minutes, as detected by the inherent sub-monolayer sensitivity of the techniques [125, 126].

In addition, for optimum health, the corneal epithelial layer must breathe making oxygen permeability necessary. PMMA is a gas impermeable lens material and as such there is poor O_2 exchange so that frequent removal to allow oxygen to reach the cornea has limited the wear of lenses fabricated from this material. Therefore oxygen permeable lenses, "soft" contact lenses, have been fabricated using elastomeric materials including silicone rubber. The mechanical properties of these compounds are excellent, however, they are hydrophobic, "tacky" and permeable to lipids. There is also a great deal of chain mobility in this class of elastomers and even though processes such as oxygen plasma treatment render the surface more hydrophilic, the effect decays over time. This is most likely due to the fact that the hydrophilic groups are subject to surface reorientation to obtain a lower energy conformation following storage in air [127]. Plasma polymerization processes result in a crosslinked overlayer limiting reorganization and more stable surfaces result. The deposition of such films onto silicone rubber have included studies where a thin layer of methane polymer (50 Å) was generated under RF conditions [128,129,130]. Lenses were more hydrophilic in nature, tackiness and lipid penetration were reduced, while at the same time the O_2 permeation rate was maintained (<15% reduction). Under the optimum conditions the coating was also found to withstand severe and repeated flexing of the contact lens. Silicone rubber contact lens surfaces exposed to a plasma polymer of N-vinyl pyrrolidone (NVP) displayed considerable enhancement in wettability as compared with the non-treated material [131]. The optimum plasma conditions were 40 W for 60s yielding a deposited layer that was not affected by drying at room temperature, boiling in water, rubbing from cleaning processes and maintained oxygen exchange during clinical wearing tests [132].

Improved oxygen permeability (Dk) has also been addressed by the introduction of copolymer materials. These so called "gas-permeable" hard contact lenses have been developed around PMMA by covalent addition of organosiloxane components. However, the presence of a silicon-bearing material leads to a reduction in the lens wettability. RF oxygen plasma treatment of a PMMA/polysiloxane co-polymer (BX3) was carried out to obtain a more hydrophilic surface [133,134]. ESCA analysis showed that the surface was substantially depleted in carbon while the oxygen and silicon were enhanced. There is an increase in the C-O and C=O surface species following plasma treatment that appeared to be independent of treatment time and power. The silicone of the polysiloxane component is not surface depleted during the RF treatment, however, there seems to be increased oxygenation of the silicon containing species. The binding energy (BE) shift of the Si 2p peak after treatment strongly suggests the polysiloxane component of the copolymer is converted to an inorganic silica-like phase. These changes were further substantiated by analysis of the plasma modified surfaces using static secondary ion mass spectrometry (SSIMS) [135]. There is relatively little change in surface composition of the PMMA control material under plasma treatment conditions while the depth of the modified region for the PMMA/polysiloxane copolymer is dependent on processing parameters and can be attributed to the polysiloxane component. The increase of wettabilty of treated PMMA compared to the untreated material, as shown by contact angles, was also less than the change in treated copolymer compared to the unmodified copolymer. In another study, a commercially available lens material, Menicon O_2 – an alternative alkyl acrylate/siloxane copolymer to BX3 – was investigated following oxygen plasma treatment [136]. SSIMS analysis on the oxygen plasma treated surface revealed progressive conversion of the surface organosiloxane species to an inorganic silica phase and were consistent with the previous studies for the structurally related polymer. The effect of aging on oxygen plasma treated siloxane-modified PMMA contact lenses and polyetheretherketone (PEEK) was investigated following storage in air and water [137,138]. Contact angle measurements showed an expected increase in wettability following treatment that decayed on storage. However, the effect was dependant on the conditions and changed over time, whereby there was also a recovery of hydrophilicity. ESCA carried out to monitor changes in the surface chemistry suggests reorganization in a 20 Å region within the outermost 50 Å of the sample.

Further improvements to hydrophobic contact lens materials that have high oxygen permeability (Dk>90) but are hydrophobic in nature i.e., siloxanes, perfluoropolyethers and therefore limited for extended wear applications have been addressed using RFGD. In one study, the monomer methyl methacrylate (MMA) was plasma polymerized to obtain PMMA-like coatings and wettability similar to that of hydrogel lenses [139]. The films deposited at higher power underwent much more pronounced oxidation, whereas at lower powers the plasma-MMA is similar to PMMA but with a three-dimensional crosslinked structure rather than linear. Similarly, contact lenses were first modified using ammonia (NH_3) plasma or alternatively a thin plasma polymer coating of *n*-heptylamine was deposited onto the lenses. Following removal from the reactor the respective samples were immersed into solutions of a number of polysaccharides; partially oxidized dextran (OD), hyaluronic acid (HA), chondroitin sulfate (CS) and carboxymethylated dextrans (CMs) [140]. The clinical performance of the resultant polysaccharide-coated lenses displayed equivalent surface wettability/drying compared with hydrogel lenses and the additional advantage of better oxygen permeability. The influence of ultra-thin carbon tetrafluoride (CF_4) plasma polymerized coatings on siloxane-based hydrogel contact lenses has been investigated by electrochemical impedance spectroscopy (EIS) [141]. The coatings were found to affect the extent of hydration and salt intrusion of the lenses in a measurable manner – although there was significant modification in the surface characteristics, the O_2 and ion permeability was maintained and therefore deemed a viable method for modifying the surface characteristics of hydrogels. Ion migration through a CF_4 coating is surprising and may imply defects or cracks in the coating.

Recently, a new generation of contact lens materials has been developed. Hydrogel contact lenses are comfortable in the eye, but have limited oxygen permeability. Silicone-acrylate-type contact lenses have good oxygen permeability but are rigid and can be uncomfortable. By interpenetrating a hydrogel network into a siloxane-acrylate network, materials have been developed that have molecular-scale, continuous water pathways and siloxane pathways through the lens. Such materials are soft, transparent and have high Dk values. Since the surface of these lenses tended to revert to a hydrophobic state, they have been plasma treated to form a silica-like outer layer that resists reversal to the hydrophobic condition. Little has been published on the chemistry of these lenses or their surface modification, though they are successful in clinical practice.

Advances in contact lens fabrication and plasma technology have played an important role in evolving superior contact lenses. Surface modifications have been frequently applied to address biocompatibilty, bioactivity and biomimicry. These surface treatments perhaps have had a more significant role than the bulk materials themselves in advancing contact lenses. This is particularly relevant as the industry strives for extended wear lenses. For example a glow discharge downstream plasma process using Ar as plasma and carrier gas for the "afterglow-feed" of vinyl monomers has been developed [142]. Radical polymerization of the vinyl monomers is the predominant reaction forming surface grafts of true vinyl polymer structure. Under conventional plasma conditions there is considerable fragmentation of the organic species that can be minimized by such a technique. The ability to control the coating composition, thickness, topography and crosslinking has been investigated using atomic force microscopy AFM, ESCA, ATR-FTIR, contact angle and permeability measurements for poly-N-vinyl-2-pyrrolidone (PVP) grafts. These PVP-coated contact lenses had excellent permeability, stability, and biocompatibility and therefore the application of such a process offers exciting opportunities in the field. Also, improved plasma research on traditional contact lenses can form a basis for further development of other ocular implant materials. For instance, high water content, oxygen permeable contact lens materials are good choices for intracorneal implants but have been limited in their use because they inhibit the regrowth of a stable epithelium over the implant. NH_3 plasma modification of a poly(2-hydroxyethyl methacrylate)/methacrylic acid (PHEMA-MAA) copolymer has effectively enhanced attachment and growth of rabbit corneal epithelial cells [143]. In conclusion, plasma processes have made contact lenses a viable alternative to regular eye glasses or refractive surgery options for many patients.

10.8. Non-fouling Coatings

As previously discussed, protein adsorption occurs on biomaterials as soon as they contact biological milieu – these protein layers then solicit cellular responses (Figure 10.1). The process, being non-specific, offers the biomaterials designer little control over the biological outcome of medical implants, biosensors and diagnostic devices. Therefore, to better control device biocompatibility, one approach includes the inhibition of undesirable protein and cell adhesive events – referred to here as non-fouling – enabling efficient delivery of pre-programmed and specific biological signals. Another approach is to introduce passivating molecules, including proteins such as bovine serum albumin (BSA) that potentially can improve the blood compatibility of biomaterial surfaces. BSA has few specific interactions with cells and also, a monolayer of proteins will most often resist further protein adsorption to the surface. Thus, a layer of BSA can be doubly effective. For example, amino groups introduced into surfaces of Celgard® polypropylene membranes using anhydrous NH_3 plasma, were shown to attach BSA covalently at the physiological pH of 7.5 [144,145]. There is a wealth of opportunity provided by plasma treatment for modifying the interfacial chemistry to achieve a non-fouling condition. Studies on thin film deposition from glow discharges of monomers for non-fouling surfaces include, but are not limited to, organosilicon, halocarbon, and ethylene oxide repeat units. Plasma polymers are also used as an intermediate layer or activation step for immobilization of poly(ethylene glycols) [146], polysaccharides [147,148] and grafting of phospholipids [149] to reduce protein adsorption. Some developments in achieving such potentially important surfaces and the necessary requirements are outlined in this section. The focus will be on poly(ethylene glycol)-like surfaces as these dominate the literature.

The use of poly(ethylene glycol) (PEG), often called poly(ethylene oxide)(PEO), as a surface protector to reduce protein adsorption and cell adhesion is well documented [150,151]. PEG copolymers, such as the Pluronic® 120 series of poly(ethylene oxide)/poly(propylene oxide)/poly(ethylene oxide) (PEO/PPO/PEO) tri-block surfactants, physisorb readily to surfaces. However, to achieve a "biocompatible" surface in which there are no chemicals leaching from the surface following introduction to the bodily environment, a permanent bond between the substrate and PEG is required. If PEG or PEG-PPO copolymers are only physically adsorbed, they will be removed over time by biofluids, since they are soluble in water and many organic solvents. Also, the physisorption, affinity is inversely proportional to the number of non-fouling copolymer PEO moeities. Thus, additional surface engineering is required to fixture PEG copolymers to substrates and optimize the reduction in protein adsorption. The method of CASING (cross-linking by activated species of inert gases) removes low molecular weight materials that make up the weak boundary layer, or converts them to a high molecular weight by cross-linking reactions and also crosslinks the molecules that must be surface-stabilized resulting in greater joint strength [152,153]. This kind of approach was applied to immobilize coated Pluronic® surfactant on low-density polyethylene (LDPE) substrate by an Ar-plasma treatment and obtain a wettable and non-fouling (protein resistant) surface [154]. The weak boundary layer of surfactant, first physically coated via solvent swelling and evaporation, was then cross-linked to the surfaces and/or each other using the Ar-plasma. A large increase in hydrophilicity and decrease in surface energy was measured using water contact-angles and ESCA. Modified surfaces having the highest O/C ratio displayed the greatest reduction of baboon fibrinogen adsorption.

Another approach to immobilize PEG is covalently linking the polymer to substrates that have been favorably surface functionalized. This was achieved for PET films produced using an RF glow discharge polymer deposition process and then modified by introducing amino or hydroxyl groups from a plasma coupling process with allylamine and allyl alcohol respectively [155]. These surfaces were subsequently activated with cyanuric chloride for the surface immobilization of bis-amino PEG. Although the method did not provide complete surface coverage, significant protein rejection was observed for the high molecular weight PEG (20,000 mass units). The inert character of PEG and the ability to resist adsorption of proteins has been the subject of much speculation. Steric repulsion, where the polymer prevents the protein from reaching the substrate surface to adsorb [156] is frequently involved to explain PEG non-fouling behavior. Others propose that the ability of PEG to bind water molecules, structure them and thus screen the surface from biomolecules is critically important [157,158].

The methods previously described to coat PEG on surfaces are fairly limited since they are only applicable to specific substrate materials and reaction schemes. To achieve a general surface coating procedure for a wide variety of surfaces, a method of grafting PEG onto activated silica film coatings was developed [159]. It was carried out first by plasma-enhanced chemical vapor deposition of SiH_4 and O_2 gases to deposit an amorphous silica film (SiO_2). This was further activated in H_2O plasma, then vapor phase PEG (400 mw) introduced to react with surface Si-OH groups at a temperature of 100 °C over a period of 43 hours. This process proved successful, however, as a general rule room temperature deposition of PEG-like coatings is much more preferable for treating most polymeric biomaterial substrates. Deposition at lower temperature using a gas phase plasma reaction was first investigated with linear glyme [CH_3-O-$(CH_2CH_2O)_n$-CH_3] monomers. Low-power pp-tetraglyme (n=4) has displayed great promise as a non-fouling coating candidate that inhibits protein adsorption, platelet and endothelial cell attachment. Although tetraethylene glycol dimethyl ether (tetraglyme) is relatively low molecular weight (MW 222), the plasma polymerized monomer exhibits resistance to protein adsorption with more than a 30-fold reduction [160]. This work has established that glow discharge plasma deposition (GDPD) can be a simple one-step coating regime for the manufacture of PEO-attached surfaces. Also, a recent study in which RF-GDPD tetraglyme was used to obtain PEO-like coatings on a fluorinated ethylene-propylene (FEP) surface displayed a 20 fold reduction in fibrinogen adsorption and significantly lower monocyte adhesion as compared to the untreated material and tissue culture polystyrene (TCPS) [161]. Therefore, these tetraglyme coatings are ideal candidates for non-fouling biomaterial applications that can inhibit non-specific material-host interactions and possibly reduce the foreign body response. Furthermore, the surface chemistry of plasma polymers of oligoglymes (n=1-4) and crown ethers (1,4-dioxane, 12-crown-4-ether, and 15-crown-5-ether) deposited onto glass coverslips has also been studied [162]. ESCA, Time of Flight Secondary Ion Mass spectroscopy (TOF-SIMS) and dynamic contact angles (DCA) were used to characterize the surface. The pp-oligoglymes consist of $(CH_2CH_2\text{-}O)_n$ repeat units, and for both the glyme and cyclic films, intact precursor molecules in the outermost layers were observed. Crown ethers produce coatings with EO structures and also crosslinked yet intact crown ether molecules that exhibit evidence of ion binding. In general the work showed these PEG-like coatings to be effective in the prevention of bacterial adhesion, although interestingly high ether content was not always consistent with high bacterial resistance. Tetraglyme and 15-crown-5-ether both display rich ether content but the former is less adherent to bacteria – surface energetics are not responsible since the contact angle measurements are similar for both. SIMS data indicates that the two films differ structurally and therefore the outermost surface is the important factor. By contrast, monoglyme and dioxane, the smallest crown ether, produce coatings low in EO characteristic, with significant hydroxyl group content, and do not show non-fouling surface properties. Bacterial attachment and biofilm formation was significantly reduced in a study in which 12-crown-4-ether and tri (ethylene glycol) dimethyl-ether (triglyme) were plasma deposited onto stainless steel surfaces. ESCA indicated deposition of PEG-like structures that were more hydrophilic as suggested by water contact angle analysis and had less surface roughness compared with the untreated material as shown by atomic force micrsocopy (AFM) [163]. This work indicates that RF plasma treatment results in non-fouling surfaces for metal substrates that can be used for medical implants.

Another type of EO-containing monomer has been employed in pulsed plasma polymerization and compared to depositions carried out under continuous wave (CW) mode. A low duty cycle (10/200 ms) at 50W of triethylene glycol monoallyl ether, CH_2=CH-CH_2-(O-CH_2-CH_2)$_3$-OH, retained much of the EO repeat unit, and resulted in a 20-fold reduction in protein adsorption [164]. A comparable level of protein resistance was observed with coatings prepared on PET substrates from a similar dimer, diethylene glycol vinyl ether (CH_2=CH-(O-CH_2-CH_2)$_2$-OH), by low duty cycle pulsed plasmas [165]. However, when a monomer with one EO unit, monoethylene vinyl ether (CH_2=$CHOCH_2CH_2OH$), was used the coating does not retain enough EO characteristic to be an effective non-fouling coating. This observation seems to be borne out by many studies: small monomers with only one EO unit tend to form coatings with a high hydrocarbon content and do not create non-fouling coatings. However, when the monomer contains at least two EO units, under the proper deposition conditions, protein resistant coatings can be obtained. The number of EO units in the surface film is highest when the power input during deposition is decreased, especially at low plasma duty cycles. In another study, the application of pulsed plasma to obtain films with maximum EO character and optimize non-fouling was compared for a number of cyclic ethers [166]. 12-crown-4- and 15-crown-5-ethers yielded polymeric films rich in EO units that were highly resistant to protein adsorption and displayed good stability in aqueous solution. This was not the case for dioxane and the differences in the polymer films obtained are thought to result from their respective initial low electron impact fragmentation patterns – as sampled by mass spectra – and the important role of ion radicals. This will be an interesting area to be considered in the future for elucidating the main reaction steps involved in polymer formation under discharge conditions, and may further be achieved by interfacing systems to a number of structural analysis tools such as Fourier transform infrared spectrometers (FTIR).

The use of plasma coatings and processes in generating non-fouling surfaces has made significant progress and there is reasonable understanding in what parameters are required for a successful outcome. The simplicity of GDPD and potential to achieve materials that are both non-fouling and plasma-sterilized allow easy scale-up and batch processing.

10.9. Plasma Surface Treatments For Biorecognition

Biological recognition is generally considered to be a process where a specific chemical species interacts with a precise complementary chemical structure leading to a "lock and key" fit, i.e., the recognition event. Both molecular shape and chemical complementarity are important for this process to occur. Examples from nature are widespread: protein-antibody, lectin-carbohydrate, RNA-DNA, integrin-protein, etc. Many non-biological recognition systems have also been synthesized – inorganic catalysts are a prime example.

Biorecognition has wide application in medicine, biotechnology and basic research. Specific applications where biorecognition is needed include biosensors, surface diagnostics (for example, ELISA assays), biomolecule separations, biocompatible implant materials and enzymatic bioreactors. The importance of biorecognition for medical implant materials has been recently emphasized [167,168].

Plasma systems, considering their non-specific chemistries with a broad, stochastic distribution of functional groups, seem unlikely candidates to generate materials showing high levels of chemical recognition. Yet, plasma deposited surfaces have been used to create surfaces with high biorecognizability. One method for doing this has already been covered in a number of other sections of this review. That is, any surface can be functionalized using plasmas to introduce chemical groups that can be used to chemically immobilize recognizable biomolecules. As just one example consider the chemical immobilization of glucose oxidase enzyme onto radio-frequency plasma-modified polyurethanes for glucose measurement [169]. This process in essence takes the non-specific surface chemistry of the polyurethane or the plasma treatment and creates a highly specific, bioactive surface.

An entirely different approach to using plasma deposited layers for creating biorecognition surfaces has recently been demonstrated. The idea uses molecular-scale imprinting to create recognition sites for proteins [171,172]. Imprinting has been found useful for synthesizing recognition sites for low molecular weight compounds [173]. Because of the labile and macromolecular nature of proteins, only limited success has been achieved with these large molecules. In the experiments to be described here, tailored protein-binding cavities have been created by imprinting taking advantage of some key properties of RF-plasma deposition of organic thin films, namely, their ability to deposit on most surfaces, their smooth conformality and their chemical reactivity. The method is illustrated as a cartoon in Figure 10.2. The structure of the nanopit surface created by this process has been confirmed by a range of methods including transmission electron microscopy, AFM, ESCA, time of flight (Tof) SIMS, contact angle methods and light microscopy. Consistent with Figure 10.2, analytical evidence suggests a polysaccharide-like surface with imprinted "nanocavities" complementary to the shape and chemistry of protein

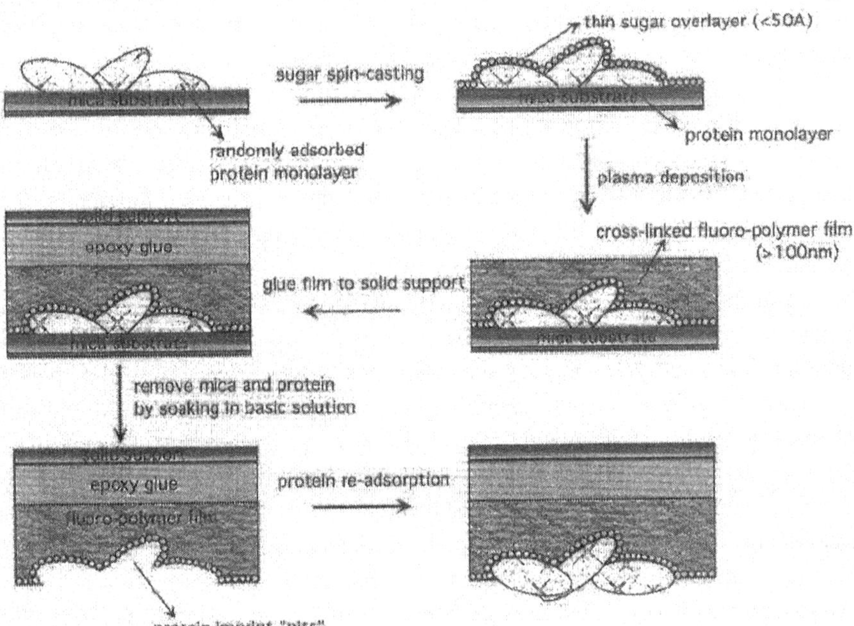

Figure 10.2. Schematic representation of the process used to create protein imprints [170].

molecules. The nanopit is always coated with roughly 1nm of sugar, the chemical recognition template.

Once the nanocavities are prepared and crosslinked in shape (and surface chemistry) by the reactive plasma, only template protein molecules can fill the cavities and bind strongly. Bound protein in its "pit" cannot readily exchange with dissolved protein because of strong non-covalent interactions (H-bonds, hydrophobic bonds) and steric hindrance. Non-template protein molecules that weakly adsorb on the surface are displaceable. Competitive adsorption experiments with binary protein mixtures show selective adsorption of template proteins to their imprints. In one example, the recognition specificity was demonstrated by the fact that lysozyme imprints have a greater than 20 fold increase in adsorption of lysozyme (MW 14,600) versus ribonuclease A (MW 13,700). Recent experiments have shown that recognition surfaces for a very large protein (fibronectin) can also be created by this method [174]. The fibronectin protein, "recognized" by the nanocavities, retains biological activity as suggested by endothelial cell attachment studies.

10.10. Patterned Plasma-Deposited Films

Patterned surfaces that provide regions of differing biological activity are important in biomedical applications. Some of these have already been considered in this article, including uses for cell culture and biosensors. In addition, the field of surface patterning has displayed much potential for state of the art implant biomaterials and may even find a niche in tissue engineering. Such materials can be fabricated using standard photolithography (PL), soft lithography, and other micro stenciling or shadowing techniques. These microfabrication processes allow the biological and topographical parameters of a surface to be controlled on a micrometer-level – in particular, the nature of cells and their interaction with one another, the surrounding medium and the substrate. The patterning of cells can be: via preferential attachment using a template i.e. molecular recognition, or by means of a removable physical barrier that will not damage the cells. Photolithography, on which all such technology is based, involves the patterning of a layer of photosensitive polymer, otherwise known as "photoresist", by means of light and is analogous to developing film in a darkroom [175]. However, complex equipment is needed for micropattering by photolithographic means – particulates, such as dust, interfere with the photoresist layer, the chemicals used are toxic to cells and damaging to biomaterials. For these reasons, "soft lithography" techniques have emerged [176]. In this case, a microstructured surface of poly(dimethylsiloxane) (PDMS) is utilized and also acts as a stamp in "microcontact printing" to transfer organic molecules only to select areas of a surface [177]. A detailed account of micropatterned surfaces for biomedical applications may be found in a recent review article [178]. Plasma polymerization has more recently been employed in the manufacture of patterned surfaces; an important factor in the success as a surface modification technique is the adhesion of the resulting film to the substrate, and in turn how this may be selectively manipulated. However, in most cases plasma polymer films can basically be patterned by combining with photolithographic processes. This section will consider some criteria and examples of plasma systems used in this way.

Fluoropolymers including plasma polymerized tetrafluorethylene (C_2F_4, TFE), have been shown to strongly bind proteins and have non-thrombogenic properties [179,180] – the surface fluorine to carbon ratio, wettability and film composition being contributory factors [181]. In particular, surface coatings of hexafluoropropylene (C_3F_6, HFP) were patterned using photolithography (PL), and resulted in features ranging from 5 to 40 microns when investigated and modified using scanning force microscopy (SFM) [182]. Direct writing of features using atomic force microscopy (AFM) was carried out in an earlier investigation in which fluoropolymer films were deposited using radio-frequency glow discharge (RFGD). The tip acted as a lithographic tool to pattern the surface, such that surface defects were introduced into the film to control interactions with biological adsorbates [183]. In order to achieve a non-adhesive background, non-fouling plasma polymerized tetraglyme (pp4G) was first deposited onto a substrate surface followed by standard photolithography. Subsequently, adhesive domains of plasma polymerized TFE were deposited onto the patterned photoresist and fluoropolymer features ranging from 5 to 100 microns following lift-off were observed [184]. This heterogeneous surface has displayed different binding affinities. Investigation of a model protein (BSA) and peptide conjugated to a PEG-fluorocarbon surfactant showed preferential binding to the flurocarbon pattern, while the tetraglyme areas inhibit background adsorption – resulting in domain structures [185]. This has provided a template to control the shape and size of vascular smooth muscle cells (SMC) and therefore investigate the influence on cell physiology [186].

Exposed, uncoated SiO_2 substrates adsorb proteins readily and therefore pp4G was patterned using PL to achieve non-fouling regions [187]. As a result, attachment and growth of bovine aortic endothelial cells was limited only to adhesive SiO_2 domains. These micro-patterned "cell surfaces" were active for 3 weeks in culture, until overcrowding resulted in cell death. An example of micro-scale cell cultures achieved using patterned pp4G is shown in Figure 10.3. Photolithography provides excellent alignment of surface domains to allow structural moieties to be neatly fabricated within a system – of great advantage for many biosensors and bio-micro-electro-mechanical systems (bioMEMS). In cases where spatial alignment is not required soft lithography techniques, such as microcontact printing (μCP) of adhesive proteins onto plasma deposited films, can provide a fast yet elegant technique to achieve micro-scale patterns of cells and proteins. This has been carried out for the non-fouling pp4G surface [188]. Surprisingly, proteins that have been micro contact printed onto the normally protein resistant pp4G surfaces adhere well for long periods of time, even in cell culture.

Since plasma is a conformal process, placing a removable mask on the substrate can prevent deposition/etching underneath, and control the patterning results. This has been reported for plasma polymerized methanol and *n*-hexane onto a number of substrates; including mica. By using transmission electron microscopy (TEM) grids as masks, hydrophilic and hydrophobic patterned structures were obtained respectively. In addition, patterned plasma etching and surface activation were reported using water plasma [189,190]. The techniques result in surface patterns possessing various functional groups on the

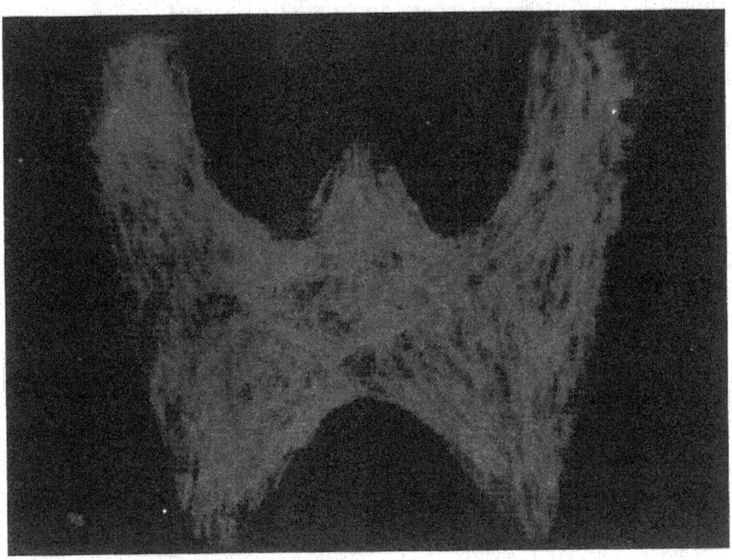

Figure 10.3. Micro-scale cell cultures of bovine aortic endothelial cells obtained using patterned pp4G [187].

micrometer scale, as evidenced by scanning electron microscopy (SEM). The SEM micrographs show different levels of brightness for the different plasma polymers and therefore, have potential for characterizing the distribution of functionality in such deposited thin films. Since it is also possible to achieve micro-patterning using plasma surface activation as well as thin film deposition, such methods are often complementary. The merits of these processes are compared with other techniques for micropatterned cell growth in a brief review article on this topic [191]. However, a greater level of control of surface structure is likely by effectively combining all these processes and should provide endless opportunities in conjunction with other methodologies [192].

10.11. Perspectives and Conclusions

Plasmas have been explored as surface modification tools for biomaterials since the 1960s. Now, they are often used in the materials engineering and fabrication of clinically used medical devices. Unfortunately, many of the applications of plasmas for biomaterials are trade secrets and are not advertised or promoted by companies manufacturing medical devices. However, it is well acknowledged that plasmas are used in medicine and biology for cleaning, sterilization, pyrogen/endotoxin vaporization, adhesive bond promotion, biomolecule immobilization, wettability, electrical insulation, degradation protection, rate limiting barriers, lubricity, cell growth, blood compatibility and biocompatibility. The best known applications are for contact lenses, cell culture dishes, sterilization and for modifying blood interactions.

The potential of plasma treatments for biomedical applications is high. This statement can be supported with the following justification. Medical device and diagnostics manufacturers have heavy investments in fabrication, user (physician, researcher, etc.) training, regulatory approval and marketing. They will not change product lines easily without good justification. Since plasma provides a surface modification that can be implemented in the gas phase on most existing devices and can improve performance without requiring a complete redesign, manufacturers will be less reluctant to incorporate novel product enhancements. Furthermore, the treatment is relatively low cost, uses well-known processing equipment and gives a sterile product. For these reasons, plasmas will most likely see increasing use in biomedicine over the next few years.

The versatility of plasmas for surface treatment will also increase their application in biology and medicine. The versatility comes from the fact that any molecule (or mixture) that can be put into the gas phase can be deposited as a coating, especially where surface chemistry can be precision-controlled [193,194]. For example, recently, poly(lactic acid)-like surfaces that are biodegradable were synthesized in a plasma by using a cyclic lactide molecule as a precursor [195]. The limitation imposed by volatility is also being challenged by recent experiments showing that an electrospray system could be used to introduce non-volatile, high molecular weight biological molecules into the gas phase of a plasma system [192]. Such molecules could be deposited on plasma activated surfaces or introduced directly into the plasma glow for modification and deposition.

In conclusion, it is clear that plasma technology has made significant advances in many fields of surface preparation, particularly biomedical devices. The flexibility that plasmas offer in treating a wide range of different substrate materials make plasmas a desirable processing technique for addressing biocompatibility and biorecognition issues, while at the same time accommodating the more complex materials and geometries that evolve. Furthermore, continued progress affords the technique a uniquely important and versatile role in the production of designer surfaces and the next generation of biomaterials.

Acknowledgements

This work was supported by the University of Washington Engineered Biomaterials (UWEB, an NSF Engineering Research Center, NSF EEC-9529161) and NESAC/BIO (an NIH National Resource for surface analysis, NIH RR-01296).

References

1. Williams D.F. (Ed.), Definitions in Biomaterials. Progress in Biomedical Engineering, 4. Elsevier, Amsterdam. 1987, 72 pages.
2. Ratner B.D., *J. Controlled Rel.*, **78** (2002) 211.
3. Ratner B.D., *J. Biomed. Mater. Res.*, **27** (1993) 283.

4. Amstein C.F, Hartman P.A., *J. Clin. Microbiol*, **2** (1975) 46.
5. Ratner B.D., Chilkoti A., Lopez G.P., Plasma Deposition, Treatment and Etching of Polymers, d'Agostino R. (Ed.), Academic Press, San Diego, 1990, p496.
6. Yamada K.M., Olden K., *Nature* **275** (1978) 179.
7. Klebe R.J., Bently K.L., Schoen R.C., *J. Cell. Physiol*, **109** (1981) 481.
8. Bently K.L., Klebe R.J., *J. Biomed. Mater. Res.* **19** (1985) 757.
9. McKeownlongo P.J., Panetti T.S., *Trends in Glycoscience and Glycotechnology*, **8** (1996) 327.
10. Steele J.G., Dalton B.A., Johnson G., Underwood P.A., *J. Biomed. Mater. Res.*, **27** (1993) 927.
11. Steele J.G., Dalton B.A., Johnson G., Underwood P.A., *Biomaterials*, **16** (1995) 1057.
12. Steele J.G., Johnson G, Underwood P.A., *J. Biomed. Mater. Res.*, **26** (1992) 861.
13. Steele J.G., McFarland C., Dalton B.A., Johnson G., Evans M.D.M., Howlett C.R., Underwood P.A., *J. Biomater. Sci. Polymer Edn.*, **5** (1993) 245.
14. William S.K., Jarrell B.E., Friend L., Radomski J.S., Carabasi R.A., Koolpe E., Mueller S.N., Thornton S.C., Maxinucci T., Levine E., *J. Surg. Res.*, **38** (1985) 618.
15. Mooradian D.L., Trescony P., Keeney K., Furcht L.T., *J. Surg. Res.*, **53** (1992) 74.
16. Dekker A., Reitsma K., Beugeling T., Bantjes A., Feijen J., van Aken W.G., *Biomaterials*, **12** (1991) 130.
17. Sipehia R., Martucci G., Barbarosie M., Wu C., *Biomat. Artif. Cell. Im.*, 21 (1993) 455.
18. Sipehia R., Martucci G., Lipscombe J., *Artif. Cell. Blood Sub.*, 24 (1996) 51.
19. Chu C.F.L., Lu A., Liszkowski M., Sipehia R., *BBA-Gen. Subjects*, **1472** (1999) 479.
20. Williams S.K., Carter T., Park P.K., Rose D.G., Schneider T., Jarrell B.E., *J. Biomed. Mater. Res.*, **26** (1992) 103.
21. Griesser H.J., Chatelier R.C., Gengenbach T.R., Johnson G., Steele J.G., *J. Biomater. Sci. Polymer Edn.*, **5** (1994) 531.

22. Harsch A., Calderon J., Timmons R.B., Gross G.W., *J. Neuroscience Methods*, **151** (2000) 135.
23. Tseng D.Y., Edelman E.R., *J. Biomed. Mater. Res.*, **42** (1998) 188.
24. Hsiue G.H., Lee S.D., Wang C.C., Shiue M.H.I., Chang P.C.T., *Biomaterials*, **14** (1993) 591.
25. Ertel S.I., Ratner B.D., Horbett T.A., *J. Biomed. Mater. Res.*, **24** (1990) 1637.
26. Ertel S.I., Chilkoti A., Ratner B.D., Horbett T.A., *J Biomater. Sci. Polymer Edn.*, **3** (1991) 163.
27. France R.M., Short R.D., Dawson R.A., MacNeil S., *J. Mater. Chem.* **8** (1998) 37.
28. Daw R., Candan S., Beck A.J., Devlin A.J., Brook I.M., MacNeil S., Dawson R.A., Short R.D., *Biomaterials*, **19** (1998) 1717.
29. Daw R., Brook I.M., Devlin A.J., Short R.D., Cooper E., Leggett G.J., *J. Mater. Chem.*, **8** (1998) 2583.
30. Haddow D.B., France R.M., Short R.D., MacNeil S., Dawson R.A., Leggett G.J., Cooper E., *J. Biomed. Mater. Res.*, **47** (1999) 379.
31. Franco M., Nealey P.F., Campbell S., Teixeira A.I., Murphy C.J., *J. Biomed. Mater. Res.*, **52** (2000) 261.
32. Tidwell C.D., Ertel S.I., Rater B.D., Tarasevich B.J., Atre S., Allara D.L., *Langmuir*, **13** (1997) 3404.
33. Tidwell C.D., Endothelial cell interactions with model surfaces effect of surface chemistry, surface mobility, and the adsorbed protein layer, Ph.D. dissertation, 1999, University of Washington, Seattle, WA.
34. Baquey Ch., Palumbo F., Porte-Durrieu M.C., Legeay G., Tressaud A., d'Agostino R., *Nucl. Instr. Meth. in Phys. Res. B*, **151** (1999) 255.
35. Lee S.D., Hsiue G.H., Chang P.C.T. and Kao C.Y., *Biomaterials*, **17** (1996) 1599.
36. Liu S.Q., Ito Y. and Imanishi Y., *J. Biomed. Mater. Res.*, **27** (1993) 909.
37. Schild H.G., *Progress in Polymer Science* **17** (1992) 163.
38. Yamada N., Okano T., Sakai H., Karikusa F., Sawasaki Y. and Sakurai Y., *Makromol. Chem. Rapid Commun*, **11** (1990) 571.
39. Pan Y.V., Wesley R.A., Luginbuhl R., Denton D.D., Ratner B.D., *Biomacromolecules*, **2** (2001) 32.
40. Pan Y.V., Unpublished results.

41. Baker R.W., Controlled Release of Biologically Active Agents, Wiley-Interscience, New York, 1986.
42. Fan L.T. and Singh S.K., Contolled Release A Quantitative Treatment. Springer-Verlag, 1989.
43. Peppas N., Recent Advances in Drug Delivery Systems, Anderson J.M., Kim S.W. (Eds.), Plenum Press, 1984, 279.
44. Zeoli L.T., Kydonieus A.F., in Controlled Release Technology Das K.G. (Ed.), John Wiley & Sons, 1983, 61.
45. Langer R., *Science*, 249 (1990) 1527.
46. Huang X., Brazel C.S., *J. Control. Rel.*, **73** (2001) 121.
47. Kim D.D., Takeno M.M., Ratner B.D., Horbett T.A., *Pharmaceut. Res.*, **15** (1998) 783.
48. Kwok C.S., Horbett T.A., Ratner B.D., *J. Control. Rel.*, **62** (1999) 301.
49. Kwok C.S., Wan C., Hendricks S., Bryers J.D., Horbett T.A., Ratner B.D., *J. Control. Rel.*, **62** (1999) 289.
50. Hendricks S.K., Kowk C., Shen M., Horbett T.A., Ratner B.D., Bryers J.D., *J. Biomed. Mater. Res.*, **50** (2000) 160.
51. Vermette P., Meagher L., Gagnon E., Grieseer H.J., Doillon C.J., *J. Control. Rel.*, **80** (2002) 179.
52. Langer R.S., Peppas N.A., *Biomaterials*, **2** (1981) 201.
53. Heiduschka P., Thanos S., *Progress in Neurobiology*, **55** (1998) 433.
54. Sackmann E., *Science*, **271** (1996) 43.
55. Dinh T.V., Cullum B., *Fresenius J. Anal. Chem. B.*, **366** (2000) 540.
56. Schlosser M., Ziegler M., Biosensors in the body. Fraser D. (Ed.) John Wiley & Sons, New York, 1992, 138.
57. Reichert W., Saavedra S., Medical and dental materials. Williams D. (Ed.) VCH publishers Inc. New York, 1992, 303.
58. Wisniewski N., Moussy F., Reichert W.M., *Fresenius J. Anal. Chem. B.*, **366** (2000) 611.
59. Wisniewski N., Reichert M., *Coll. Surf. B.*, **18** (2000) 197.
60. Mar M.N., Ratner B.D., Yee S.S., *Sensors Actuators B*, **54** (1999) 125.
61. Lu H.B., Homola J., Pan Y.V., Ratner B.D., Campbell C.T., Yee SS., *Sixth World Biomaterials Congress Transactions*, 2000, 151.
62. Pan Y.V., Hanein Y., Leach-Scampavia D., Bohringer K.F., Ratner B.D., Denton D.D., *The 14th IEEE International Conference on MEMS Technical Digest*, 2001 435.

63. Nichols M.F., Hahn A.W., James W.J., Sharma A.K., Yasuda H.K., *J. Appl. Polym. Sci. Appl. Polym. Symp.*, **38** (1984) 21.
64. Paralyne conformal coatings Specifications and properties, Indianapolis Speciality Coatings, 1994, 45.
65. Yamagishi F.G., *Thin Solid Films*, **202** (1991) 39.
66. Nichols M.F., Hahn A.W., *ISA Trans.* **27** (1991) 331.
67. Nichols M.F., *ISA Trans.*, **29** (1993) 77.
68. Lewis, H.G.P., Edell D.J., Gleason K.K., *Chem. Mater.*, **12** (2000) 3488.
69. Sperling C., Konig U., Hermel G., Werner C., Muller M., Simon F., Grundke K., Jacobasch H.J., Vasilets V.N., Ikada Y., *J. Mater. Sci. Mater. Med.*, **8** (1997) 789.
70. Favia P., d'Agostino R., Palumbo F., *J. Physique IV*, **7** (1997) 199.
71. Yuan S., Szakalasgratzl G., Ziats N.P., Jacobsen D.W., Kottkemarchant K., Marchant R.E., *J. Biomed. Mater. Res.*, **27**(1993) 811.
72. Chen J.P., Kiaei D., Hoffman A.S., *J. Biomat. Sci. Polym. Ed.*, **5** (1993) 167.
73. Kampfrath G., Hintsche R., *Anal. Lett.*, **22** (1989) 2423.
74. Kiaei D., Safranj A., Chen J.P., Johnston A.B., Zavala F., Deelder A., Castelino J.B., Markovic V., Hoffman A.S., *Rad. Phys. Chem.*, **39** (1992) 463.
75. Muguruma H., Karube I., *Trends Anal. Chem.*, **18** (1999) 62.
76. Kawakami M., Koya H. and Gondo S., *Biotechnology and Bioengineering*, **32** (1988) 369.
77. Mutlu S., Saber R., Kocum C., Piskin E., *Anal. Lett.*, **32** (1999) 317.
78. Mutlu M., Mutlu S., Alp B., Boyaci I.H., Piskin E., Plasma Processing of Polymers, d'Agostino R., Favia P., Fracassi F. (Eds.), Kluwer Academic, Dordrecht, 1997, 477.
79. Alp B., Mutlu S., Mutlu M., *Food Res. Intl.*, **33** (2000) 107.
80. Biederman H., Boyaci I.H., Bilkova P., Slavinska D., Mutlu S., Zemek J., Trchova M., Klimovic J., Mutlu M., *J. Appl. Polm. Sci.*, **81** (2001) 1341.
81. Hiratsuka A., Muguruma H., Sasaki S., Ikebukuro K., Karube I., *Electroanalysis*, **11** (1999) 1098.
82. Sugimoto I., *J. Appl. Phys.*, **70** (1991) 2887.
83. Sugimoto I., Nakamura M., Kuwano H., *Sensors Actuators B*, **10** (1993) 117.
84. Kasai N., Sugimoto I., Nakamura M., *Sensors Actuators B*, **65** (2000) 114.

85. Hiratsuka A., Karube I., *Electroanalysis*, **12** (2000) 695.
86. Lyman D.J., Roland S.H., Biomaterials in Polymers Biomaterials and Medical Applications, Encyclopedia Reprint Series, Kvoschwitz J.I. (Ed.) John Wiley & Sons, New York, 1981, 52.
87. Nichols M.F., Hahn A.W., Easley J.R., Mayhan K.G., *J. Biomed. Mater. Res.*, **13** (1979) 299.
88. Hahn A.W., York D.H., Nichols M.F., Amromin G.C., Yasuda H.K., *J. Appl. Polym. Sci Applied Polymer Symposium,* **38** (1984) 55.
89. Loeb G.E., Bak M.J., Salcman M., Schmidt E.M., *IEEE Trans. Biomed. Eng.*, **24** (1977) 121.
90. Marchant R.E., Johnson S.D., Schneider B.H., Agger M.P., Anderson J.M., *J. Biomed. Mater. Res.*, **24** (1990) 1521.
91. Merrit K., Brown S.A., Biological effects of corrosion products and metals, in Corrosion and Degradation of Implant Materials, ASTM STP 859, Fraker A.C., Griffin C.D. (Eds.), Philadelphia, 1985, 195.
92. Marchant R.E., Yu D., Khoo C., *J. Appl. Polym. Sci A Polym. Chem.*, **27** (1989) 881.
93. Marchant R.E., *Fundam. Appl. Toxicol.*, **13** (1989) 217.
94. Johnson S.D., Anderson J.M., Marchant R.E., *J. Biomed. Mater. Res.*, **26** (1992) 915.
95. Clarotti G., Schue F., Sledz J., Aoumar A.A.B., Geckeler K.E., Orsetti A., *Biomaterials*, **13** (1992) 832.
96. Piglowski J., Gancarz I., Staniszewska-Kus J., Paluch D., Szymonowicz M., Konieczny A., *Biomaterials*, **15** (1994) 919.
97. Latkany R., Tsuk A., Shue M-S., Loh I-H., Trinkaus-Randall V., *J. Biomed. Mater. Res.*, **36** (1997) 29.
98. Wu X.Y., Tsuk A., Leibowitz H.M., Trinkaus-Randall V., *Br. J. Opthalmol.*, **82** (1998) 569.
99. Vanderlaan J.S., Lopez G.P., Wachem P.B., Nieuwenhuis P., Ratner B.D., Bleichrodt R.P., Schakenraad J.M., *Int. J. Artificial Organs*, **14** (1991) 661.
100. Parker J.A.T.C., Walboomers X.F., Von Den Hoff J.W., Maltha J.C., Jansen J.A., *Biomaterials*, **23** (2002) 3545.
101. Ballanda M., Cassinelli C., Morra M., *J. Biomed. Mater. Res.*, **36** (1997) 216.

102. Everaert E.P.J.M., van De Belt-Gritter B., van Der Mei H.C., Busscher H.J., Verkerke G.J., Dijk F., Mahieu H.F., Reitsma A., *J. Mater. Sci. Mater In Medicine*, **9** (1998) 147.
103. Ratner B.D., *J. Biomed. Mater. Res.*, **27** (1993) 837.
104. Martinson L., Goessl A., Ciridon W., Pan Y.V., Sigle R., Savinit G., Leach-Scampavia D., Leach K., Kyriakides T., Shen L., Branca A., Carter W., Bornstein P., Sage H., Giachelli C., Horbett T., Ratner B.D., Proceedings of the World Biomaterials Congress 2000, 1289.
105. Ratner B.D., *J. Controlled Release*, **78** (2002) 211.
106. Ratner B.D., *J. Biomater. Sci. Polymer. Edn.*, **11** (2000) 1105.
107. Hanson S., Ratner B.D., Testing of blood-materials interactions, in Biomaterials science: An introduction to materials in medicine. Ratner B.D., Hoffman A.S., Schoen F.J., Lemons J.E. (Eds.), Academic Press, San Diego, 1996, 228.
108. Llanos G.R., Sefton M.V., *J. Biomater. Sci. Polymer. Edn.*, **4** (1993) 381.
109. Hollahan J.R., Stafford B.B., Falb R.D., Payne S.T., *J. Appl. Polym. Sci.*, **13** (1969) 807.
110. Ishikawa Y., Sasakawa S., Takase M., Iriyama Y., Osada Y., *Makromol. Chem Rapid Commun.*, **6** (1985) 495.
111. Kiaei D., Hoffman A.S., Horbett T.A., *J. Bio. Sci. Polym. Ed.*, **4** (1992) 35.
112. Garfinkle A.M., Hoffman A.S., Ratner B.D., Reynolds L.O., Hanson S.R., *Trans. Am. Soc. Artificial Internal Organs*, **30** (1984) 432.
113. Yeh Y.S., Iriyama Y., Matsuzawa Y., Hanson S.R., Yasuda H., *J. Biomed. Mater. Res.*, **22** (1988) 795.
114. Royce F.H. Jr, Ratner B.D., Horbett T.A., in Biomaterials Interfacial Phenomena and Applications, ACS Advances in Chemistry Series, Vol 199. Cooper S.L., Peppas N.A. (Eds.), American Chemical Society, 1982, 453.
115. Yasuda H., Bumgarner M.O., Marsh H.C., Yamanashi B.S., Devito D.P., Wolbarsht M.L., Reed J.W., Bessler M., Landers M.B., Hercules D.M., Carver J., *J. Biomed. Mater. Res.*, **9** (1975) 629.
116. Hook T.J., Gardella J.A. Jr, Salvati L. Jr., *J. Mater. Res.*, **2** (1987) 117.
117. Hook T.J., Gardella J.A. Jr, Salvati L. Jr., *J. Mater. Res.*, **2** (1987) 132.
118. Hook T.J., Gardella J.A. Jr, Salvati L. Jr., *J. Polymer. Sci.*, **A27** (1989) 1267.
119. Steele J.G., Johnson G., Griesser H.J., Underwwod P.A., *Biomaterials*, **18** (1997) 1541.

120.Kamiya K., Yanagihara Y., Takai T., Kuzuya M., *Chem. Pharm. Bull.*, **42** (1994) 1896.
121.Kingshott P., Chatelier R.C., St John H.A.W., Griesser H.J., *ACS. Polym. Mater. Sci. Eng.*, **76** (1997) 77.
122.Sunny M.C., Sharma Chandra P., *Biomater. Artif. Cells Immobilization Biotechnol.*, **19** (1991) 599.
123.Johnston E.E., Ratner B.D., Bryers J.D., Absracts, 214th ACS National Meeting, Las Vegas, NV, September 7-11 1997.
124.Garrett Q., Griesser H.J., Milthorpe B.K., Garrett R.W., *Biomaterials*, **20** (1999) 1345.
125.Kingshott P., St John H.A., Chatelier R.C., Griesser H.J., *J. Biomed. Mater. Res.*, **49** (2000) 36.
126.McArthur S.L., McLean K.M., St John H.A., Griesser H.J., *Biomaterials*, **22** (2001) 3295.
127.Valiant P.L., Grobe G.L., McGee J.A., Ammon D.A. Jr, Rabke C.E., in Polymer Surfaces and Interfaces Characterization, Modification and application. Mittal K.L., Lee K.W. (Eds.), VSP, Utrecht, Netherlands, 1997, 71.
128.Koziol J.E., Peyman G.A., Yasuda H., *Arch. Opthal.*, **101** (1983) 1779.
129.Ho C.P., Yasuda H., *ACS. Polym. Mater. Sci. Eng.*, **56** (1987) 705.
130.Ho C.P., Yasuda H., *J. Biomed. Mater. Res.*, **22** (1988) 919.
131.Masamitsu K., *J. Photopolym. Sci. Technol.*, **6** (1993) 337.
132.Masamitsu K., Hasebe M., Miwa Y., Mizutani Y., *Kobunshi Ronbunshu*, **42** (1985) 841.
133.Fakes D.W., Newton M., Watts J.F., Edgell M.J., *Surf. Interf. Anal.*, **9** (1986) 416.
134.Fakes D.W., Newton J.M., Watts J.F., Edgell M.J., *Surf. Interf. Anal.*, **10** (1987) 416.
135.Davies M.C., Fakes D., Brown A., Newton J.M., *ACS. Polym. Mater. Sci. Eng.*, **56** (1987) 704.
136.Davies M.C., Fakes D.W., Brown A., Newton J.M., *Surf. Interf. Anal.*, **13** (1988) 233.
137.Munro H.S., McBriar D.I., *ACS. Polym. Mater. Sci. Eng.*, **56** (1987) 337.
138.Munro H.S., McBriar D.I., *J. Coat. Technol.*, **60** (1988) 41.
139.Gengenbach T.R., Pasic P., Zientek P., Griesser H.J., Davies T., Papas E., Austen R., *ACS. Polym. Mater. Sci. Eng.*, **76** (1997) 75.

140.Griesser H.J., Chatelier R.C., Dai L., St John H.A.W., Davies T., Austen R., *ACS. Polym. Mater. Sci. Eng.*, **76** (1997) 79.
141.Weikart C.M., Matsuzawa Y., Winterton L., Yasuda H.K., *J. Biomed. Mater. Res.*, **54** (2001) 597.
142.Chabrecek P., Lohmann D. Surface modification of extended wear contact lenses by plasma–induced polymerization of vinyl monomers. Special publication of the Royal Society of Chemistry 235 (Fundamental and Applied Aspects of Chemically Modified Surfaces), 1999, 223.
143.Sipehia R., Garfinkle A., Jackson W.B., Chang T.M., *Biomater. Artif. Cells Artif. Organs*, **18** (1990) 643.
144.Sipehia R., Chawala A.S., *Biomat. Med. Dev. Art. Org.*, **10** (1982) 229.
145.Sipehia R., Chawala A.S., Chang T.M.S., *Biomaterials*, **7** (1986) 471.
146.Lopez G.P., Ratner B.D., *Plasmas and Polymers*, **1** (1996) 127.
147.Dai L., St John H.A.W., Bi J., Zientek P., Chatelier R.C., Griesser H.J., *Surf. Interf. Anal.*, **29** (2000) 46.
148.McLean K.M., Johnson G., Chatelier R.C., Beumer G.J., Steele J.G., Griesser H.J., *Coll. Surf. B*, **18** (2000) 221.
149.Hsiue G-H., Lee S-D., Chang PC-T., Kao C.Y., *J. Biomed. Mater. Res.*, **42** (1998) 134.
150.Lee J.H., Lee H.B., Andrade J.D., *Prog. Polym. Sci.*, **20** (1995) 1043.
151.Andrade J.D., Hlady V., Jeon S.I., *Adv. Chem. Ser.*, **248** (1996) 51.
152.Hansen R.H., Schonhorn H., *J. Polym. Sci. B*, **4** (1966) 203.
153.Schonhorn H., Hansen R.H., *J. Appl. Polym. Sci.*, **11** (1967) 1461.
154.Sheu M.S., Hoffman A.S., Feijen J., *J. Adh. Sci. Tech.*, **6** (1992) 995.
155.Gombotz W.R., Guaghui W., Horbett T.A., Hoffman A.S., *J. Biomed. Mater. Res.*, **25** (1991) 1547.
156.Gölander C.G., Herron J.N., Lim K., Claesson P., Stenius P., Andrade J.D., Properties of immobilizated PEG films and interaction with proteins, in Poly(ethylene glycol) chemistry, biotechnical and biomedical applications, Harris J.M., (Ed.), Plenum Press, New York, 1992, 221.
157.Wang R.L.C., Kreuzer H.J., Grunze M., *J. Phys. Chem. B*, **101** (1997) 9767.
158.Harder P., Grunze M., Dahint R., Whitesides G.M., Laibinis P.E., *J. Phys. Chem. B*, **102** (1998) 426.
159.Alcantar N.A., Aydil E.S., Israelachvili J.N., *J. Biomed. Mater. Res.*, **51** (2000) 343.

160. Lopez G.P., Ratner B.D., Tidwell C.D., Haycox C.L., Rapoza R.J., Horbett T.A., *J. Biomed. Mater. Res.*, **26** (1992) 415.
161. Shen M., Pan Y.V., Wagner M.S., Hauch K.D., Castner D.G., Ratner B.D., Horbett T.A., *J. Biomater. Sci. Poly. Ed.*, **12** (2001) 961.
162. Johnston E.E., Ratner B.D., Bryers J.D., RF plasma deposited PEO-like films surface characterization and inhibition of *Pseudomonas Aeruginosa* accumulation, in Plasma Processing of Polymers, d'Agostino R., Favia P., Francessco F., (Eds.), Proceedings of the NATO Advanced Study Institute Program, Vol. 346, 1996, 465.
163. Denes A.R., Somers E.B., Wong A.C.L., Denes F., *J. Appl. Polym. Sci.*, **81** (2001) 3425.
164. Beyer D., Knoll W., Ringsdorf H., Wang J-H., Timmons R.B., Sluka P., *J. Biomed. Mater. Res.*, **36** (1997) 181.
165. Wu Y.J., Timmons R.B., Jen J.S., Molock F.E., *Coll. Surf. B*, **18** (2000) 235.
166. Wu Y.J., Griggs A.J., Jen J.S., Manolache S., Denes F.S., Timmons R.B., *Plasmas and Polymers*, **6** (2001) 123.
167. Ratner B.D., *J. Biomed. Mater. Res.*, **27** (1993) 837.
168. Ratner B.D., *J. Mol. Rec.*, **9** (1997) 617.
169. Danilich M.J., Kottke-Marchant K., Anderson J.M., Marchant R.E., *J. Biomater. Sci. Polymer. Edn.*, **3** (1992) 195.
170. Ratner B.D., Shi H., Wang J., Pan S., Recognition Templates for Biomaterials with Engineered Bioreactivity in Frontiers of Multifunctional Nanosystems, Buzaneva E., Scharff P. (Eds.), Kluwer Academic Publishers, Dordrecht, The Netherlands, 2002, 77.
171. Shi H., Tsai W-B., Ferrari S., Ratner B.D., *Nature*, **398** (1999) 593.
172. Shi H., Ratner B.D., *J. Biomed. Mater. Res.*, **50** (1999) 1.
173. Shi H., Ratner B.D., *Current Opinions in Solid State & Materials Science*, **4** (1999) 395.
174. Wang J., Protein template-imprinting to enhance specific protein adsorption and cell adhesion, Master's Thesis, University of Washington, Seattle, Washington, 2001.
175. Ristic L., (Ed.), Sensor Technology and Devices. Boston, MA Artech House, 1994, 524.
176. Xia Y.N., Whitesides G.M., *Annu. Rev. Mater. Sci.*, **28** (1998) 153.
177. Kumar A., Whitesides G.M., *Appl. Phys. Lett.*, **63** (1993) 2002.

178.Folch A., Toner M., *Annu. Rev. Biomed. Eng.*, **2** (2000) 227.
179.Kiaei D., Hoffman A.S., Hanson S.R., *J. Biomed. Mater. Res.*, **26** (1992) 357.
180.Kiaei D., Hoffman A.S., Horbett T.A., Lew K.R., *J. Biomed. Mater. Res.*, **29** (1995) 729.
181.Favia P., Perez-Luna V.H., Boland T., Castner D.G. and Ratner B.D., *Plasmas and Polymers*, **1** (1996) 299.
182.Garrison M.D., Luginbuhl R., Overney R.M., Ratner B.D., *Thin Solid Films*, **352** (1999) 13.
183.Garrison M.D., Ratner B.D., *ACS Polymer Preprints*, **38** (1997) 989.
184.Goessl A., Garrison M.D., Lhoest J-B., Hoffman A.S., *J. Biomat. Sci. Polym. Ed.*, **12** (2001) 721.
185.Goessl A., Golledge S.L., Hoffman A.S., *J. Biomat. Sci. Polym. Ed.*, **12** (2001) 739.
186.Goessl A., Bowen-Pope D.F., Hoffman A.S., *J. Biomed. Mater. Res.*, **57** (2001) 15.
187.Pan Y.V., Hanein Y., Leach-Scampavia D., Bohringer K.F., Ratner B.D., The 14th IEEE International Conference on MEMS, Technical Digest, 2001, 435.
188.McDevitt T., Pan Y.V., unpublished data.
189.Dai L., Mau A.W.H., Gong X., Griesser H.J., *Synthetic Metals*, **85** (1997) 1379.
190.Dai L., Griesser H.J., Mau A.W.H., *J. Phys. Chem. B*, **101** (1997) 9548.
191.Ohl A., Schroder K., *Surf. Coatings Tech.*, **116-119** (1999) 820.
192.Kitching K.J., Lee H-N., Elam W.T., Turecek F., Ratner B.D., Johnston E.E., Miller R.J., *Rev. Sci. Instr.*, (to be submitted).
193.Lopez G.P., Ratner B.D., *J. Appl. Polym. Sci. Appl. Polym. Symp.*, **46** (1990) 493.
194.Savage C.R., Timmons R.B., Lin J.W., *Chem. Mater.*, **3** (1991) 575.
195.Favez D., RF-Plasma Deposition and Surface Characterization of a Biodegradable Thin-Film Coating, Diploma Thesis, École Polytechnique Fédérale De Lausanne, Lausanne, Switzerland, 2002.

INDEX

www.ingramcontent.com/pod-product-compliance
Ingram Content Group UK Ltd.
Pitfield, Milton Keynes, MK11 3LW, UK
UKHW050429080125
453372UK00006B/33

9 781860 944673